SERIES EDITOR
Bob Digby

Cameron Dunn

David Holmes

Dan Cowling

Endorsement statement

In order to ensure that this resource offers high-quality support for the associated Pearson qualification, it has been through a review process by the awarding body. This process confirms that this resource fully covers the teaching and learning content of the specification or part of a specification at which it is aimed. It also confirms that it demonstrates an appropriate balance between the development of subject skills, knowledge and understanding, in addition to preparation for assessment.

Endorsement does not cover any guidance on assessment activities or processes (e.g. practice questions or advice on how to answer assessment questions), included in the resource nor does it prescribe any particular approach to the teaching or delivery of a related course.

While the publishers have made every attempt to ensure that advice on the qualification and its assessment is accurate, the official specification and associated assessment guidance materials are the only authoritative source of information and should always be referred to for definitive guidance.

Pearson examiners have not contributed to any sections in this resource relevant to examination papers for which they have responsibility.

Examiners will not use endorsed resources as a source of material for any assessment set by Pearson.

Endorsement of a resource does not mean that the resource is required to achieve this Pearson qualification, nor does it mean that it is the only suitable material available to support the qualification, and any resource lists produced by the awarding body shall include this and other appropriate resources.

OXFORD
UNIVERSITY PRESS

Great Clarendon Street, Oxford, OX2 6DP, United Kingdom

Oxford University Press is a department of the University of Oxford. It furthers the University's objective of excellence in research, scholarship, and education by publishing worldwide. Oxford is a registered trade mark of Oxford University Press in the UK and in certain other countries

Series editor: Bob Digby

Authors: Cameron Dunn, David Holmes, Dan Cowling

The moral rights of the authors have been asserted

Database right of Oxford University Press (maker) 2023

First published in 2016
Second edition 2023

British Library Cataloguing in Publication Data
Data available

978 138 202919 3

978 138 202920 9 (ebook)

10 9 8 7 6 5 4 3 2 1

The manufacturing process conforms to the environmental regulations of the country of origin.

Printed in Great Britain by Bell and Bain Ltd., Glasgow

Acknowledgements

The publisher and authors would like to thank the following for permission to use photographs and other copyright material:

Cover: Rost9/Shutterstock. Photos: **pp6-7:** Mave / Alamy Stock Photo; **p9:** Andrew Roland/Shutterstock; **p10:** Stocktrek Images/Getty Images; **p14:** ©EUMETSAT 2021; **p16:** GSFC/SVS/NASA; **p18:** © Bryan & Cherry Alexander Photography/Arctic Photo; **p22:** © Tom Bradwell, 2021; **p23:** Xinhua/Alamy Stock Photo; **p25:** Seti Afoa/AP Images; **p27:** Stefano Garau/Shutterstock; **p29(tr):** Xinhua/Alamy Stock Photo; **p29(m):** Dinodia Photos / Alamy Stock Photo; **p30(br):** Claudia Weinmann / Alamy Stock Photo; **p30(tr):** MUNIR UZ ZAMAN/AFP/Getty Images; **p31:** Neil Cooper / Alamy Stock Photo; **p33(t):** RGB Ventures / SuperStock / Alamy Stock Photo; **p33(b):** UPI / Alamy Stock Photo; **p34 (l):** Natural History Museum, London/Science Photo Library; **p34(r):** Trevor Clifford Photography/Science Photo Library; **p35:** The Asahi Shimbun/Getty Images; **p37:** NG Images / Alamy Stock Photo; **p38:** ASP GeoImaging/NASA / Alamy Stock Photo; **p43:** Keith Tsuji/Getty Images; **p44:** Sayyid Azim/AP Images; **p49:** Julian Bound/Shutterstock; **p56:** AMOS GUMULIRA/AFP/Getty Images; **p57:** Joseph Mizere/Xinhua/Alamy Stock Photo; **p59:** John Warburton-Lee Photography / Alamy Stock Photo; **p60:** robertharding / Alamy Stock Photo; **p64:** Namas Bhojani/Bloomberg/Getty Images; **p65:** Steve Bloom Images / Alamy Stock Photo; **p67:** dpa picture alliance archive / Alamy Stock Photo; **p68:** david pearson / Alamy Stock Photo; **p69:** Joerg Boethling / Alamy Stock Photo; **p70:** PRAKASH SINGH/AFP/Getty Images; **p71:** Manjunath Kiran/AFP/Getty Images; **p75:** Amit Dave/Reuters; **p76:** Joerg Boethling / Alamy Stock Photo; **p77:** Robert Nickelsberg/Getty Images; **p79(t):** Vivek Renukaprasad / Alamy Stock Photo; **p79(b):** Westend61 GmbH / Alamy Stock Photo; **p80:** Janusz Gniadek / Alamy Stock Photo; **p82:** Planetobserver / Science Photo Library; **p85:** Gina Rodgers / Alamy Stock Photo; **p86:** Stuart Boulton / Alamy Stock Photo; **p87(t):** Thomas Koehler/Photothek/Getty Images; **p87(b):** Bob Digby; **p88:** Bob Digby; **p89:** Bob Digby; **p91:** PLEIADES © CNES 2016, Distribution Airbus DS; **p92:** Dinodia Photos / Alamy Stock Photo; **p94(l):** UtCon Collection / Alamy Stock Photo; **p94(r):** Rapp Halour / Alamy Stock Photo; **p95(t):** Universal Images Group North America LLC / Alamy Stock Photo; **p95(b):** SNEHIT PHOTO/Shutterstock; **p96:** Frank Bienewald/Alamy Stock Photo; **p98(t):** TRDEL/AFP/Getty Images; **p98(b):** Elena Odareeva/Shuterstock; **p100:** Peter Horree / Alamy Stock Photo; **p101(t):** SEBASTIAN D'SOUZA/AFP/Getty Images; **p101(b):** INDRANIL MUKHERJEE/AFP/Getty Images; **p103:** Jeffrey Isaac Greenberg 3+ / Alamy Stock Photo; **p104(t):** Medicshots / Alamy Stock Photo; **p104(b):** Gary Yim's photography/Moment Open/Getty Images; **p105:** Lok Seva Sangam; **pp106-107:** Martin Gillespie/Shutterstock; **p108:** George Hopkins/Alamy Stock Photo; **p109:** Stephen Warren; **p110(a):** Kevin Wells Photography/Shutterstock; **p110(b):** Matt Gibson/Shutterstock; **p110(c):** geogphotos / Alamy Stock Photo; **p110(d):** Kevin Eaves/Shutterstock; **p110(e):** geogphotos / Alamy Stock Photo; **p111(b):** Sinclair Stammers/Science Photo Library; **p111(c):** Ric Ergenbright / Alamy Stock Photo; **p112(tr):** Bob Digby; **p112(mr):** pkphotography/iStockphoto; **p112(br):** Philip Bird LRPS CPAGB/ Shutterstock; **p113(t):** Sciencephotos/Alamy Stock Photo; **p113(b):** Education Images/ Universal Images Group/Getty Images; **p114(t):** Craig Joiner Photography / Alamy Stock Photo; **p114(b):** clayborough photography/Moment Open/Getty Images; **p115:** geogphotos / Alamy Stock Photo; **p116(t):** Bill Copland, CC BY-SA 2.0; **p116(b):** Kevin Eaves/Shutterstock; **p117(t):** Stephen Dorey / Alamy Stock Photo; **p117(b):** Ernie Janes / Alamy Stock Photo; **p118:** Bob Digby; **p119:** Bob Digby; **p123(t):** Dan Burton Photo / Alamy Stock Photo; **p123(b):** cgwp.co.uk / Alamy Stock Photo; **p124(b):** Peter Smith/ Alamy Stock Photo; **p124(t):** Tony Watson / Alamy Stock Photo; **p125(t):** Cambridge Aerial Photography / Alamy Stock Photo; **p125(b):** Bob Digby; **p127:** JUSTIN TALLIS/AFP/Getty Images; **p129:** Peter D Noyce / Alamy Stock Photo; **p130:** incamerastock / Alamy Stock Photo; **p131:** Bob Digby; **p133:** Mike McEnnerney / Alamy Stock Photo; **p135:** Ordinance Survey; **p136:** Bob Digby; **p138(t):** Gary Clarke / Alamy Stock Photo; **p138(b):** Bob Digby; **p140:** Bob Digby; **p142:** Ashley Cooper pics / Alamy Stock Photo; **p145:** Ordinance Survey; **p146:** Bob Digby; **p148:** Matt Cardy/Getty Images; **p149(t):** Remy Boprey / Alamy Stock Photo; **p149(b):** JMF News / Alamy Stock Photo; **p150:** Rotherham investment and development office (RiDO); **p151:** David Dixon / Alamy Stock Photo; **p158(a):** Bob Digby; **p158(b):** Bob Digby; **p159:** Tyne & Wear Archives & Museums / Bridgeman Images; **p160:** mikecphoto/Shutterstock; **p162(t):** Paul Thompson Images/Alamy Stock Photo; **p162(b):** Alistair Laming / Alamy Stock Photo; **p164:** copied from Ordnance Survey map 1801; **p165:** Media24/Gallo Images/Getty Images; **p166:** A.P.S. (UK) / Alamy Stock Photo; **p167(t):** Bob Digby; **p167(b):** High Level Photography Ltd; **p172(t):** MS Bretherton / Alamy Stock Photo; **p172(b):** Bob Digby; **p173:** Bob Digby; **p175:** Bob Digby; **p176:** Bob Digby; **p178:** Bob Digby; **p179:** Ordinance Survey; **p180:** Transport for London / Alamy Stock Photo; **p182:** Angela Hampton Picture Library / Alamy Stock Photo; **p183:** High Level Photography Ltd; **p184:** Andrew Lloyd / Alamy Stock Photo; **p185(t):** Bikeworldtravel/ Shutterstock; **p185(b):** incamerastock / Alamy Stock Photo; **p186:** Bob Digby; **p187:** Bob Digby; **p188:** Bob Digby; **p189(t):** Bob Digby; **p189(b):** Barney & Camilla Fawcett; **p190:** Bob Digby; **p191:** Ordinance Survey; **p192:** David Holmes; **p194:** David Holmes; **p195:** David Holmes; **p197:** Content is the intellectual property of Esri and is used herein with permission. Copyright © 2021 Esri and its licensors. All rights reserved.; **p199:** David Holmes; **p200:** Andrew Fox / Alamy Stock Photo; **p201:** David Holmes; **p202:** David Holmes; **p204:** David Holmes; **p206:** David Holmes; **p207:** Copyright © 2021 Esri and its licensors. All rights reserved.; **p209:** David Holmes; **p210:** Andrew Fox / Alamy Stock Photo; **p211:** David Holmes; **p212:** David Holmes; **p214:** David Holmes; **p215:** David Holmes; **p216:** Copyright © 2021 Esri and its licensors. All rights reserved.; **p219:** David Holmes; **p220:** keith morris / Alamy Stock Photo; **p221:** Bob Digby; **p222:** David Holmes; **p224:** David Holmes; **p225:** David Holmes; **p229:** David Holmes; **p230:** Indigo Images / Alamy Stock Photo; **p231:** Bob Digby; **pp232-233:** Andrea Marzorati / Alamy Stock Photo; **p234:** NOAA-NASA GOES Project; **p239(t):** NorthScape / Alamy Stock Photo; **p239(m):** Oleg Znamenskiy/Shutterstock; **p239(b):** francesco de marco/Shutterstock; **p240:** Wendy Stone/The Image Bank Unreleased/Getty Images; **p244:** Anton Deev / Alamy Stock Photo; **p248:** Minden Pictures / Alamy Stock Photo; **p252:** John E Marriott / Alamy Stock Photo; **p255:** NASA; **p259(l):** The Earth Observatory/NASA; **p259(r):** Copyright 2021 Google Earth; **p260:** WorldFoto / Alamy Stock Photo; **p261(a):** D. Kucharski K. Kucharska/Shutterstock; **p261(b):** Cristi Croitoru/Shuterstock; **p261(c):** USDA/Forestry Service; **p261(d):** Avalon.red / Alamy Stock Photo; **p262:** loflo69/Shutterstock; **p265:** Universal Images Group/Getty Images; **p266:** Prisma by Dukas Presseagentur GmbH / Alamy Stock Photo; **p268:** Visuals Unlimited/Nature Picture Library; **p270(t):** Ulga/Shutterstock; **p270(b):** Neil lee Sharp / Alamy Stock Photo; **p271(l):** Justin Sullivan/Getty Images; **p271(r):** bodom/Shutterstock; **p272(t):** incamerastock/ Alamy Stock Photo; **p272(b):** © Lu Guang / Greenpeace; **p273(l):** Abaca Press/Alamy Stock Photo; **p273(r):** Phil Clarke Hill/In Pictures/Corbis/Getty Images; **p274:** Matthew Taylor / Alamy Stock Photo; **p279:** Danita Delimont / Alamy Stock Photo; **p280:** Antoine Gyori/AGP/Corbis/Getty Images; **p282:** Julian Nieman / Alamy Stock Photo; **p283:** REUTERS / Alamy Stock Photo; **p284(t):** ITAR-TASS News Agency / Alamy Stock Photo; **p284(b):** © Denis Sinyakov / Greenpeace; **p286:** Eamon Mac Mahon/AP Images; **p287(t):** Athatbasca Oil Sands / Alamy Stock Photo; **p287(b):** Photo: Rob Tucker; **p289(t):** KeyWorded/Alamy Stock Photo; **p289(b):** Justin Kase zsixz / Alamy Stock Photo; **p290(l):** Lee Foster / Alamy Stock Photo; **p290(r):** Bob Digby; **p291(l):** George Gutenberg / Alamy Stock Photo; **p291(r):** ZUMA Press, Inc. / Alamy Stock Photo; **p292(a):** FooTToo/ Shutterstock; **p292(b):** Jeff Kowalsky/Bloomberg/Getty Images; **p292(c):** dbabbage/ iStockphoto; **p292(d):** Denton Rumsey/Shutterstock; **p295:** Robert Kneschke / Alamy Stock Photo; **p297:** Fernando Bizerra Jr/EPA/Shutterstock; **p299:** Reuters/Alamy Stock Photo; **p303:** Design Pics Inc / Alamy Stock Photo.

Artwork by Q2A Media, Mike Connor, Barking Dog Art, Simon Tegg, Angeles Peinador, and Oxford University Press.

The Ordnance Survey map extracts on pages 135, 145, 179, 191, and 327 are reproduced with the permission of the Controller of Her Majesty's Stationery Office
© Crown Copyright.

Index compiled by James Helling.

Contents

How to use this book

This book is written for the Edexcel GCSE (9-1) Geography B Specification. Every chapter is written to cover the content required for this specification. It is divided by Components and subdivided into Topics. Each Topic (Chapter) is then covered by a sequence of double-page spreads designed to cover the Detailed Content in the specification.

The following features will prove useful:

- Each double page has an **objective** outlining which part of the specification it covers. Each objective is targeted at a part of the specification.
- **Your questions** are designed to help students make sense of the text and resources, and also to take them to a level required for the examination, e.g. by analysing text and data. They are there to help prepare students to answer exam questions on that part of the specification.
- **Exam-style questions** are designed to draw the contents of each spread together and provide suitably targeted questions of the standard that are required to meet the GCSE examination.
- **Skills spreads** (shown with the ✦ icon in the contents) and **Geography skills** questions (also shown with the ✦ icon in Your questions) help to meet the requirements for skills in the specification.

COMPONENT ONE
Global Geographical Issues

Lava in the crater of Mount Nyiragongo, which erupted in 2021

What is Component One?

- Pearson Edexcel's GCSE Geography specification B consists of three Components.
- Each Component consists of three Topics, making nine in all.
- Each Component is assessed by its own exam paper – so Component One is assessed by Paper 1.

What Topics will I study in Component One?

- **Topic 1 Hazardous Earth** is about the global climate system, climate change, extreme weather (e.g. tropical cyclones) and tectonic hazards.
- **Topic 2 Development dynamics** is about global development, and includes a detailed case study of one of the world's emerging countries – India is included in this book.
- **Topic 3 Challenges of an urbanising world** is about the world's rapid urbanisation and urban trends, and includes a detailed case study of one of the world's many growing megacities in developing and emerging countries – Mumbai is included in this book.

You'll also learn several **geographical skills** such as interpreting maps, satellite images, diagrams, statistics, and photos.

What is Paper 1 like?

- **Time:** 1 hour 30 minutes
- **It has three Sections: A, B, and C:** one for each of the three Topics above.
- **It's worth 94 marks:** 90 marks are split equally between the three Topics, with another 4 for Spelling, Punctuation, Grammar and use of specialist geographical terminology (SPaG) which is assessed on an 8-mark question in Section B.
- **It counts for:** 37.5% of your final grade.

Where can I get help in preparing for Paper 1?

Chapter 11 gives you all the guidance that you need on all three exam papers. It includes advice relevant to Paper 1 about:

- the exam format, handling different sections, and how exam papers will be marked (sections 11.1 and 11.2)
- how to answer shorter questions worth 1–4 marks (sections 11.3 and 11.4)
- how to answer longer 8-mark questions in Paper 1 (section 11.5)

1.1 Global temperatures – it's a breeze

In this section, you'll understand how winds and ocean currents affect global temperatures.

Keeping the Earth habitable!

Imagine the Earth without wind! Figure 1 shows how the amount of sun's energy (called **solar insolation**) varies over different parts of the Earth. The angle of the sun's rays makes solar insolation very intense at the Equator, but dispersed over a wider area at the Poles. Winds play an important part in making the Earth more habitable by redistributing heat. Without them, the Equator would get unbearably hot, while the Poles would become even colder!

Heat is redistributed globally in two ways: by air movements caused by **pressure differences**, and by **ocean currents**.

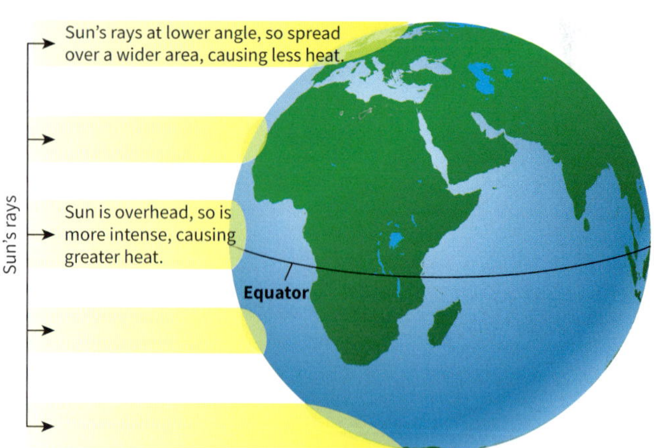

▲ ***Figure 1*** *How the sun's energy (solar insolation) varies around the Earth*

1 Pressure differences

Land and sea heat up differently. On land, dark surfaces (e.g. soil, forest) absorb sunlight, which is converted to heat. Land heats quickly in summer, but only at the surface, so it cools quickly in winter. When it heats, it also heats the air above, so air expands, becomes lighter, and rises. This forms areas of **low pressure** over large landmasses in summer.

The sea behaves differently. Some sunlight is reflected from the surface, while some is absorbed to 30 m depth. It therefore takes the sea longer to heat, but also longer to cool. In summer, air over the sea remains cooler and denser, forming areas of **high pressure**.

Differences in pressure causes air to move – from high to low pressure, creating wind. It moves in a circular way because of the Earth's rotation (see section 1.10).

Figure 2 shows how pressure varies around the world. In January (summer in the southern hemisphere), low pressure areas form over continents A, B and C, which are warmer than the oceans. Oceans X, Y and Z remain cooler, so form areas of high pressure. Notice how wind blows from high pressure to low. Notice too that it's the reverse in the northern hemisphere winter, with high pressure over continents D, E and F, and low pressure over warmer oceans P and Q. In July, it's exactly the other way round.

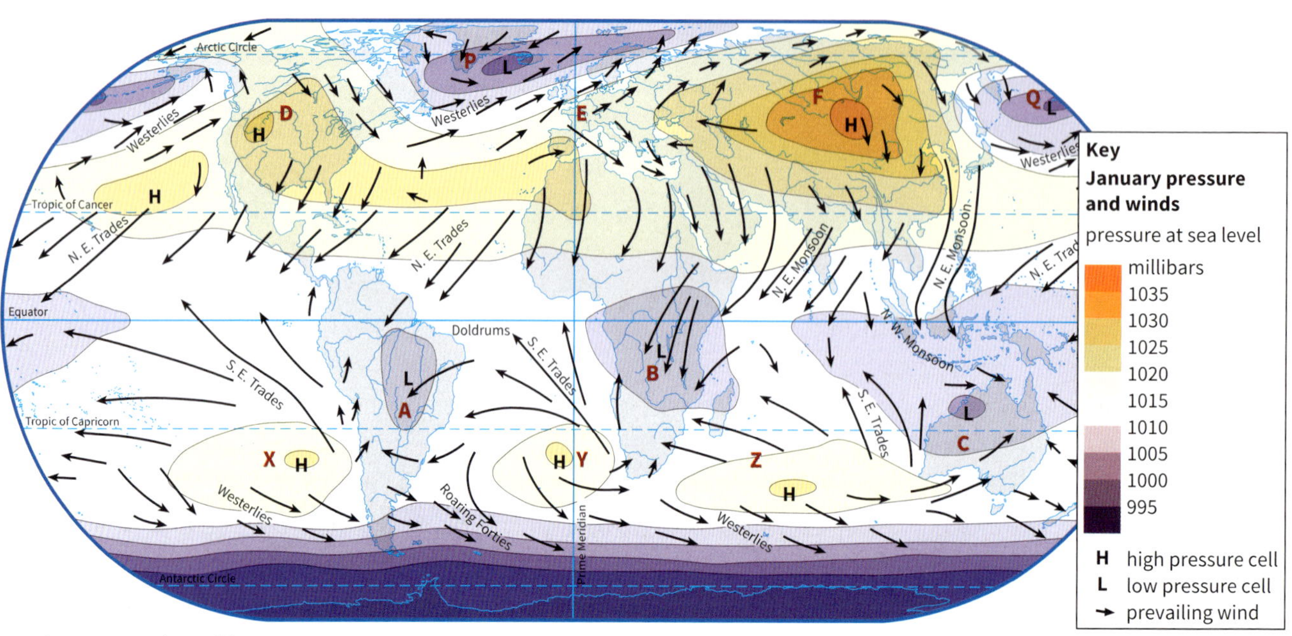

▲ ***Figure 2*** *High and low pressure areas in January*

2 Ocean currents

▲ ***Figure 3*** *Subtropical plants on Tresco, the Scilly Isles*

Welcome to the Isles of Scilly. 70 km west of Cornwall, they are Britain's most southerly location, at 49.5° N. Winters are much warmer than the rest of the UK. Daffodils and new potatoes are picked three months earlier than elsewhere in the UK, and the island of Tresco is warm enough for subtropical plants (see Figure 3).

St John's, Newfoundland, is Canada's most easterly point. At 47.6° N it's south of the Scillies. But it averages 18 days of snow in January, and night temperatures there average -9°C, 15 degrees colder than the Scillies!

The differences are caused by the Gulf Stream, a warm **ocean current**. Driven by westerly winds, the Gulf Stream begins in the Gulf of Mexico, blowing northeast where a branch of it heads towards Europe to become the North Atlantic Drift. So although the Scillies are 49.5° north of the Equator, the North Atlantic Drift keeps January sea temperatures at a warm 11°C. Around Greenland and northern Canada, the current cools, turns south, and forms the cold Labrador Current, which chills Newfoundland.

What are ocean currents?

The Gulf Stream is one of several ocean currents (see Figure 4). In the north Atlantic, cold, salty water is heavy and sinks. This sets up a convection current, which drags surface water down. The current draws warmer salty water over the ocean surface from areas near the Equator such as the Gulf of Mexico. This cools and sinks in the Labrador and Greenland Seas, and flows south toward the Equator where it is warmed again.

Labrador Current
Glasgow
Isles of Scilly
Newfoundland
N. Atlantic Drift
Gulf of Mexico
Gulf Stream
Tokyo
60°
30°
0°
30°
60°

Key
warm current
cold current

▲ ***Figure 4*** *The world's ocean currents*

Your questions

1 Explain what is meant by **a** high pressure air, **b** low pressure air.

2 Explain why high pressure areas form over land in winter and over sea in summer.

3 Use a blank world map and an atlas to show how air pressure systems would look in July (the northern summer). Name continents and oceans, and label 'High' or 'Low' pressure areas.

4 Explain the following using Figure 4:
- The Titanic met icebergs near Newfoundland in April 1912, but not in the mid-Atlantic.
- The eastern part of the UK is cooler in winter than the western part.
- Winters are colder in Tokyo (35°N) than in Glasgow (56°N).

Exam-style questions

5 Define the term 'solar insolation'. (1 mark)

6 Explain **one** role of ocean currents. (2 marks)

7 Explain **two** reasons for the pattern of air pressure in Figure 2. (4 marks)

1.2 The global circulation

In this section, you'll understand how rainfall in Africa depends on the global circulation of the atmosphere.

Waiting for rain!

It's May 2021 in Timbouctou, Mali (Figure 1), on the edge of Africa's Sahara desert. The heat is intense and the ground dry. Rain last fell on 11 October. As yet, water pumps serve people and cattle well. Who knows when rains will arrive? But there's good news. South-east of Mali, people in Nigeria's Meteorological Agency are studying satellite photos like Figure 1. A band of heavy rain is heading north. It has now reached Kano in northern Nigeria and is heading towards Mali.

Mali's rainy season normally arrives in May (shown in Figure 2). The clouds heading across West Africa are part of the Inter-Tropical Convergence Zone (ITCZ), an area of low pressure bringing rain. It stretches around the Earth (Figure 1), moving north in June when the sun is overhead at the Tropic of Cancer. By December the overhead sun moves south to the Tropic of Capricorn, taking the ITCZ with it. Therefore:

- Mali has a short rainy season.
- It rains in Kano (Figure 2) a few weeks earlier, and rains last longer.
- In Lagos, the rains stay longer still and it rains in every month of the year.

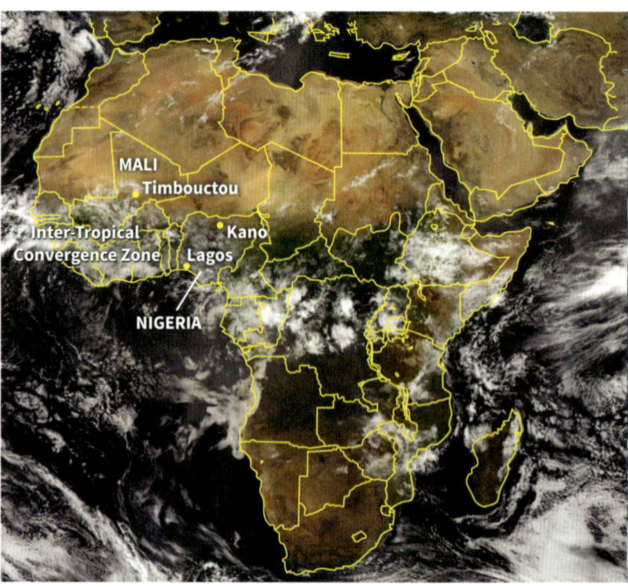

▲ ***Figure 1*** *This satellite photo shows the ITCZ*

▼ ***Figure 2*** *Rainfall in West Africa*

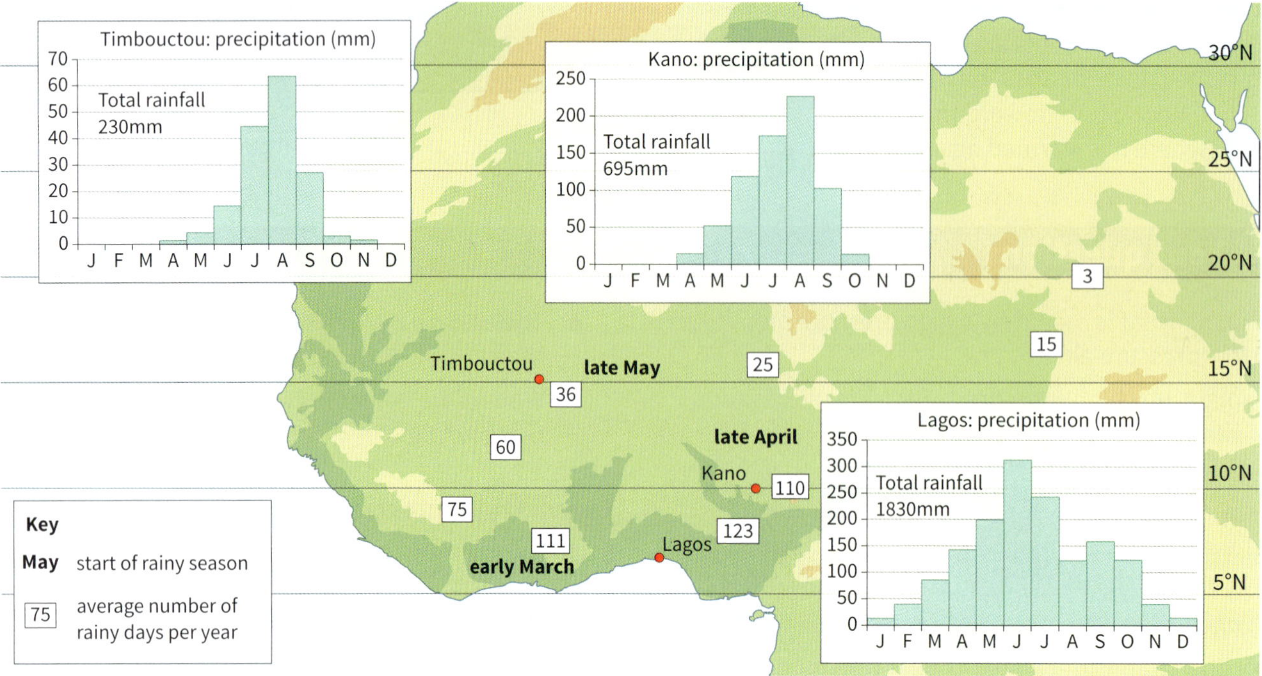

What causes the ITCZ?

The ITCZ forms within the **tropics**, when two masses of air meet – or **converge** (hence its name). It is part of a movement of the atmosphere known as the **global circulation model,** described in section 1.3. This consists of three huge 'cells' of air, the largest of which is the **Hadley Cell.** The cells are caused by heating and cooling, and between them create the world's high and low pressure systems (section 1.1).

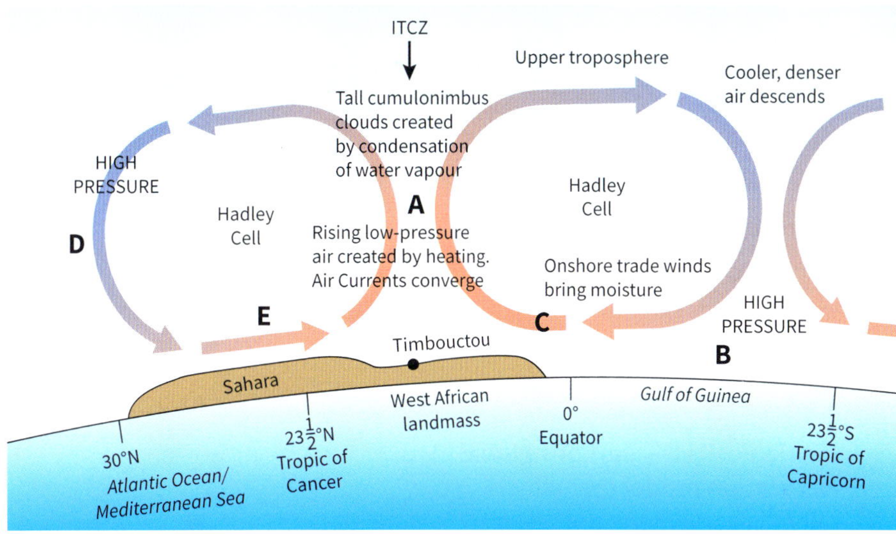

▲ ***Figure 3*** *How the Hadley Cell brings rain to West Africa in July*

In July, the sun is overhead at the Tropic of Cancer, creating summer in the northern hemisphere. At this time, the Hadley Cell works as follows – letters refer to Figure 3:

- In North and West Africa, land is warmer than the surrounding seas because it is summer. Here, the air warms (shown by the red arrows), expands, becomes lighter and rises. This forms a low-pressure area over Mali and the southern Sahara (A).
- Further south, over the Gulf of Guinea (B), an area of cooler, denser, high pressure air forms (shown by blue arrows), because the sea is cooler than the land.
- Winds blow from high to low pressure, towards West Africa and the southern Sahara away from the Gulf of Guinea. These are known as Trade Winds, and carry moisture inland (C). These winds, and the moisture that they contain, are the cause of Mali's rainy season.

However, the Hadley Cell actually consists of two parts, one either side of the Equator. Both of these move together as the sun moves overhead with the seasons. The second lies further north over the Sahara, the Atlantic Ocean and the Mediterranean Sea.

- At the same time as the Cell brings rains to Mali, high pressure also forms over the Mediterranean Sea and Atlantic Ocean (D). This is because the sea is cooler than surrounding lands in the northern hemisphere summer.
- Drawn by the low pressure area at the ITCZ, Trade Winds blow south towards West Africa (E). These Trade Winds converge with those from the Gulf of Guinea at the ITCZ. Forced to rise as they meet, they cool, and water vapour condenses to cause heavy rain.

Your questions

1 Study Figure 1. Describe the pattern of cloud shown on this satellite photo.

2 Complete the following table using Figure 2. 

City	Total rainfall per year	No. of rainy months	Rainiest month	No. of dry months
Lagos				
Kano				
Timbouctou				

3 Explain how the ITCZ affects **a** total rainfall, **b** when the rains arrive, **c** the number of rainy months, and **d** the dry months for cities named.

4 Make a simple copy of Figure 3 and annotate it to show how, and why, Mali and areas to the south have their rainy season in July (the northern summer).

Exam-style questions

5 Study Figure 2. Compare the rainfall distribution throughout the year in Lagos with that in Timbouctou. (3 marks)

6 Explain how global atmospheric circulation determines the location of high and low pressure areas. (4 marks)

1.3

The world's arid regions

In this section, you'll understand about the reasons for the distribution of the world's regions of high and low rainfall.

Completing the cycle

The Hadley Cell is the driving force behind Mali's rainy season in the northern hemisphere's summer (section 1.2). It is also responsible for the dry season in winter. In fact, the Hadley Cell is part of a **global circulation model** that affects rainfall everywhere, not just Mali. How does this happen?

Having brought rain to Mali in June, the Hadley Cell moves south with the changing position of the overhead sun. By **January**, the pattern has reversed completely (see Figure 1).

▲ ***Figure 1*** *The Hadley Cell in January*

- The northern hemisphere winter creates a cool, dense high pressure area over North Africa (A in Figure 1).
- The sun is now overhead at the Tropic of Capricorn, warming the southern hemisphere, and forming a new low pressure area (B).
- Trade Winds blow from high to low pressure, drawing dry air from the Sahara (C), across Mali and causing its dry season.
- These Trade Winds meet others from the southern hemisphere (D) to create the ITCZ, which has shifted southwards. It brings rain to southern Africa at this time.
- To complete the Hadley Cell, rising air from the ITCZ cools high in the atmosphere and becomes denser. Fed by air from below, it spreads out (E) until it subsides as high pressure air (back to A).
- The Hadley Cell is just one part of the global atmosphere. In fact, there are two Hadley Cells (shown in Figure 2). The ITCZ forms when the Trade Winds from each one meet.

The global circulation model is a theory that explains how the atmosphere operates in a series of three cells each side of the Equator.

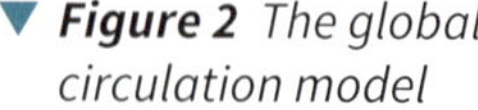

▼ ***Figure 2*** *The global circulation model*

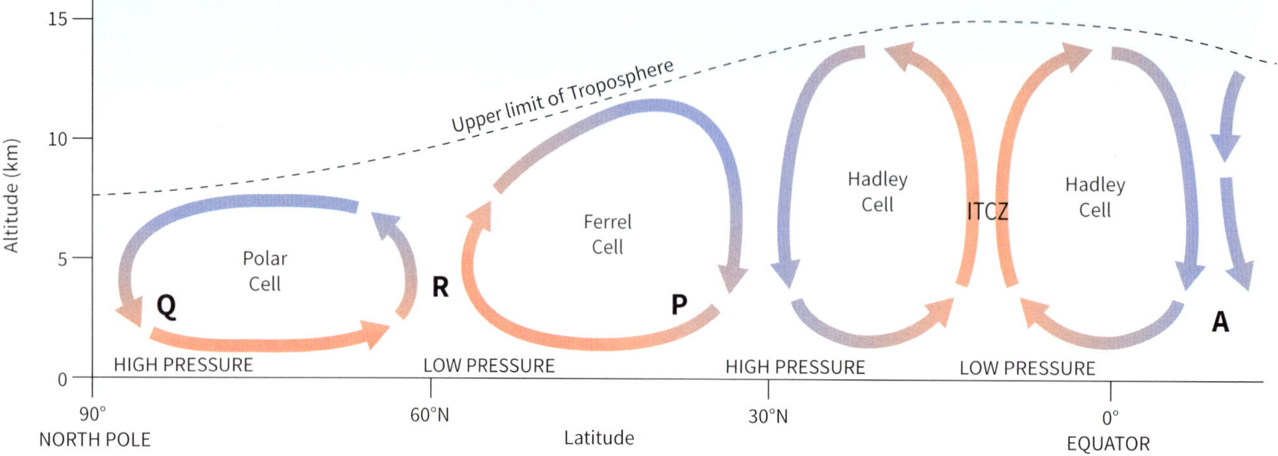

The global circulation model

Like the Hadley Cell, the other cells – the **Ferrel Cell** (30°–60° N and S) and **Polar Cell** (60°–90° N and S) – can be found in each hemisphere (see Figure 3).

The Ferrel Cell is also caused by high pressure over North Africa. As well as blowing south as part of the Hadley Cell, some air also blows towards the poles (Figure 2 – letter P). It's these winds that drive the Gulf Stream (see section 1.1).

These winds collect moisture over the oceans and meet cold, dense air from the **Polar Cell** (Q in Figure 2) between 50°–60° N and S. Just like at the ITCZ, this creates a front bringing high rainfall (R). It's this front that brings much of the UK's rainfall.

North Pole
Descending cool, dry air
Polar Cell
Rising warm, moist air
Ferrel Cell
Descending cool, dry air
Hadley Cell
Rising warm, moist air
Hadley Cell
Descending cool, dry air
Ferrel Cell
Rising warm, moist air
Polar Cell
South Pole
Descending cold, dry air
High
Low
High
High
High
High
Low
High

▲ ***Figure 3*** *The full global circulation model*

Explaining global rainfall

The global circulation model affects climates around the world. One impact is that it leaves some tropical areas with little rain, as Figure 4 shows.

- In June, the rains brought by the ITCZ never reach the Sahara, so it remains dry. In January, a high pressure area brings cooler clear, dry air which means that the Sahara is dry then too.
- Other tropical deserts north of the Equator (e.g. Arabian and Thar deserts) are caused by the same processes.
- South of the Equator, the same is true of the Kalahari Desert (Botswana), the Atacama Desert (Peru and Chile) and Great Sandy Desert (Australia).

The world's tropical deserts all originate in this way.

Similarly, areas of highest rainfall (Figure 4) also lie between 20° N and 20° S and are caused by the ITCZ. Meanwhile the cooler Ferrel Cell (which contains less moisture because it is cooler) takes mild subtropical winds as far as 50–60° N and S, where they meet the colder winds from the Polar Cell. This forms a front, bringing rain. The dry dense air at the Poles creates the **polar desert** of Antarctica and the Arctic ice sheet.

▼ ***Figure 4*** *The world's deserts and areas of high rainfall*

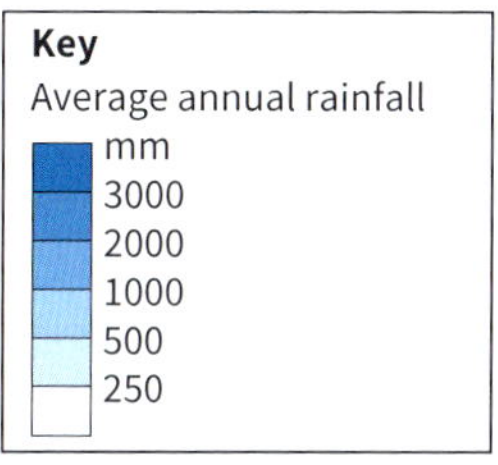

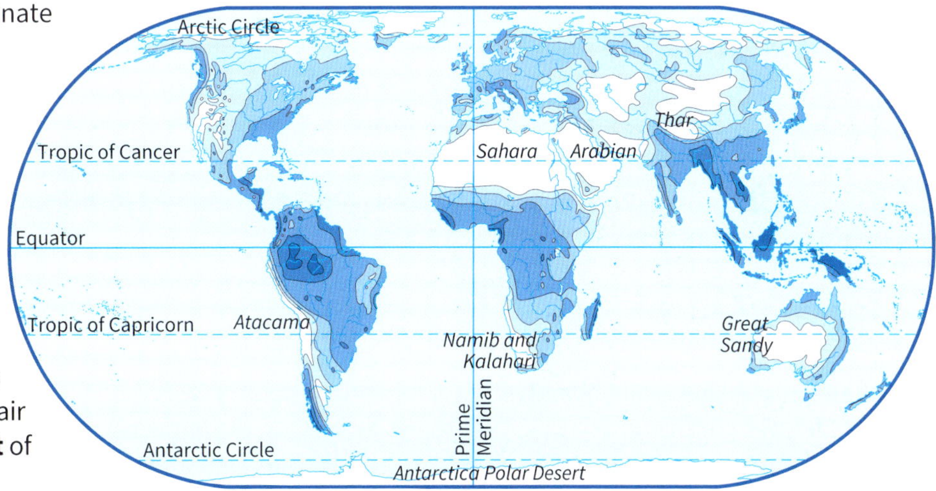

Your questions

1. Copy Figure 1 and annotate it to show how the Hadley Cell moves with the sun's passage in January.
2. On a blank world map, use an atlas to locate, shade, and name the world's **a** hot, and **b** polar deserts.
3. Annotate your map to show how **a** tropical deserts, **b** areas of high tropical rainfall have been formed.

Exam-style question

4. Explain how global circulation influences the location of the world's deserts. (4 marks)

1.4 Geographical skills: learning about climate

In this section, you'll develop some of the skills you need to interpret and understand global climates.

1 Using satellite images

Satellite images are photos taken from space satellites. They've made a huge difference to weather forecasting, helping to make forecasts more accurate.

Satellite images vary in colour depending on the filters used in the camera. Treat them like maps: you need an outline, to know which way is north, and a key. Figure 1 is a black and white image; look carefully and you'll see:

- a large, paler-coloured land area. This is Africa; national borders have been added. You'll see Europe further north
- the sea, which looks darker
- clouds, which appear light grey or white.

The cloud in Figure 1 is approaching West Africa and marks out the Inter-Tropical Convergence Zone – Section 1.2 will help you understand the ITCZ. The importance for West Africa is that it's this cloud that brings the rainy season.

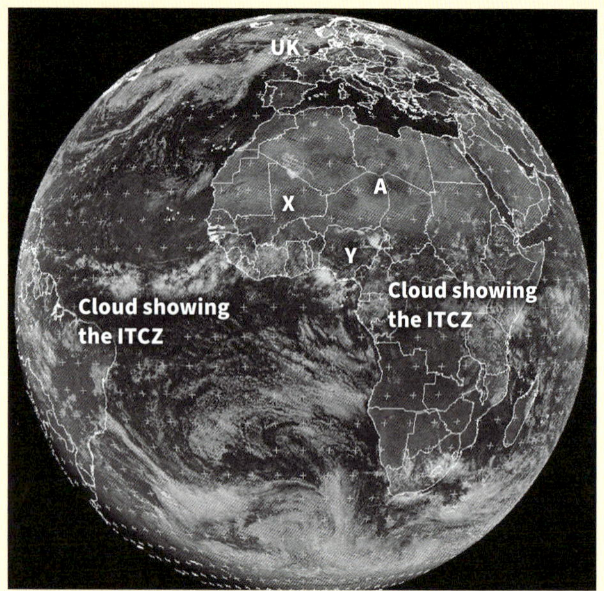

▲ ***Figure 1*** *A satellite image of West Africa, June 2012*

Your questions

1. Use an atlas to identify countries X and Y, and desert A. 
2. Using sections 1.2 and 1.3, explain why **a** the amount of cloud is greater at Country Y than Country X, **b** why there is a desert at A.
3. Describe the likely weather in the UK, and explain your reasons using sections 1.2 and 1.3.

Exam-style questions

4. Study Figure 1. Name the continent in which Desert A is located. (1 mark) 
5. Using Figure 1, describe the distribution of the areas of cloud shown. (2 marks) 
6. Explain the importance of the ITCZ for countries X and Y in Figure 1. (4 marks)  

2 Drawing and interpreting climate graphs

Climate graphs show rainfall and temperature on the same graph. Temperatures are shown as a line, with the scale in °C on the left vertical axis, and rainfall as bars, with the scale in mm on the right. Months are shown on the horizontal x axis. To draw a climate graph:

a) Prepare the axes by looking at how high the rainfall scale should be and round it up. The bottom of the scale is normally 0.

b) Next, prepare the temperature scale. See how high temperatures reach and round it up. The bottom of the scale is normally the lowest temperature, rounded down.

c) The y axis should be 12 cm wide (1 cm for each month), starting with January. Plot rainfall as bars. Each rainfall bar should be 1 cm wide.

d) Plot temperature data as points plotted directly above the centre of each rainfall bar. Join the points.

▼ ***Figure 2*** *A sample climate graph*

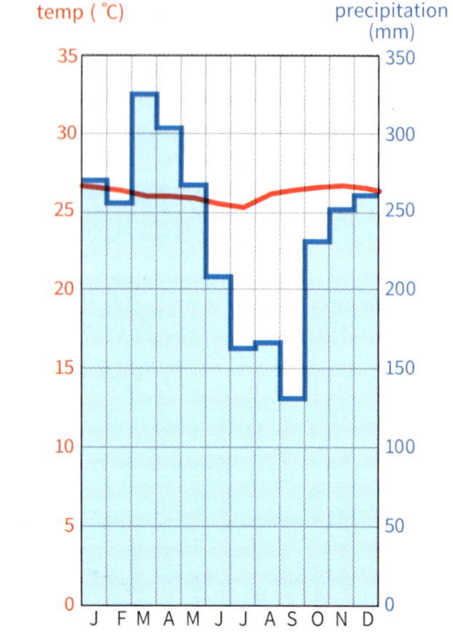

Interpreting temperature graphs is straightforward:

a) Calculate the **mean** (average) for the year, by adding 12 temperatures and dividing by 12.
b) Identify the warmest and coolest months, and calculate the difference. This is the temperature **range**.
c) See how steeply or gently the temperatures rise or fall away from the maximum. These are the **trends**.

For rainfall:

a) Calculate the annual **total**.
b) Identify the **wettest** and **driest** months, and whether the wettest months form a rainy season.
c) See how sharply the amounts vary each month or whether they are similar.

▼ ***Figure 3*** *Climate data for central Mali*

	Jan	Feb	Mar	Apr	May	June	July	Aug	Sept	Oct	Nov	Dec
Temp °C	21.2	24	27.4	30.7	33	33.2	31.8	30.3	30.3	29.4	25.8	21.9
Rain mm	0.1	0.6	1.6	6.3	16.5	40.8	83.5	114.2	64	16.3	1.3	0.2

Your questions

1 Draw a climate graph for the data in Figure 3.

Exam-style questions

2 Study Figure 3.

a Calculate the total rainfall for the year. (1 mark) 

b Describe the rainfall distribution through the year. (3 marks)

c Calculate the temperature range for the year. (1 mark) 

d Calculate the mean monthly temperature for the year. Show your workings and give the answer to one decimal place. (2 marks) 

3 Tracking cyclones

Tracking cyclones is vital for hazard prediction and warnings. A cyclone track shows a series of points joined together to provide a map. Based on the map, other data are then used to help predict which way a cyclone will move next. These data include:

- latest satellite images – where forecasters look for circular cloud patterns (see sections 1.9 and 1.10)
- data on sea surface temperatures, so calculations can be made about any extra heat energy the cyclone may pick up
- a history of all previous cyclones, to understand how any new ones might behave.

The data in Figure 4 gave the Solomon Islands about 2–3 days in which to prepare for Tropical Cyclone Pam in 2015.

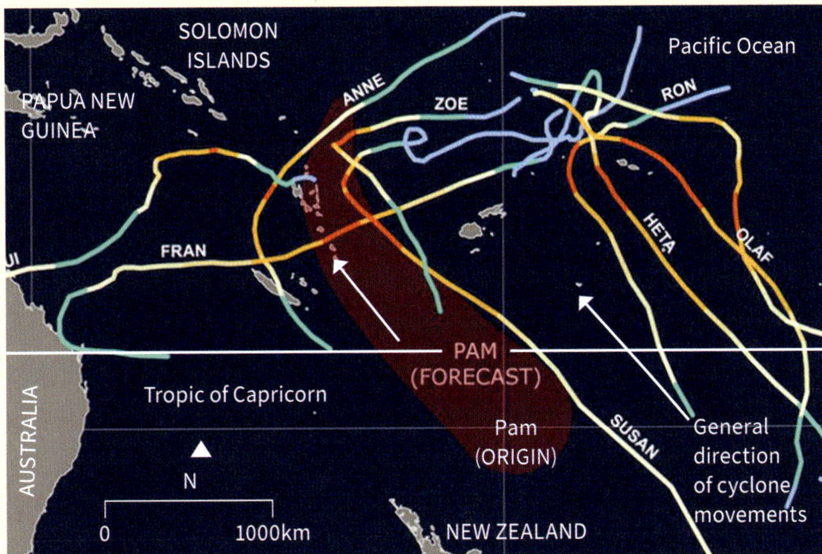

▲ ***Figure 4*** *Category 5 tropical cyclones in the south-west Pacific since 1970. The shaded area shows Tropical Cyclone Pam which devastated Vanuatu and the Solomon Islands in 2015.*

Your questions

Exam-style questions

1 State **two** features of Cyclone Pam in Figure 4. (2 marks) 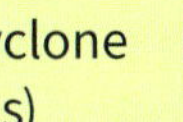

2 State **two** ways in which Cyclone Pam is typical of most tropical cyclones. (2 marks)

3 Explain **one** reason why weather forecasters might want to know about the tracks of previous cyclones when tracking Cyclone Pam. (3 marks)

4 Study Figure 4. Suggest **two** reasons why it would have been difficult to evacuate the Solomon Islands. (4 marks) 

1.5 The causes of climate change in the past

In this section, you'll learn about the four main reasons that climatologists use to explain why climate has changed in the past. These changes are all natural.

Volcanic activity

Big volcanic eruptions can change the Earth's climate. Small eruptions have no effect – the eruption needs to be very large and explosive. Volcanic eruptions produce:

- ash
- sulphur dioxide gas.

If the ash and gas rise high enough, they will be spread around the Earth in the **stratosphere** by high-level winds. The blanket of ash and gas will stop some sunlight reaching the Earth's surface. Instead, the sunlight is reflected off the ash and gas, back into space. This cools the planet and lowers the average temperature.

In 1991, Mount Pinatubo in the Philippines erupted, releasing 17 million tonnes of sulphur dioxide. This was enough to reduce global sunlight by 10%, cooling the planet by 0.5°C for about a year.

Mount Pinatubo was very small-scale compared to the 1815 eruption of Tambora in Indonesia. This was the biggest eruption in human history. In 1816, temperatures around the world were so cold that it was called 'the year without a summer', and up to 200 000 people died in Europe as harvests failed. The effects lasted for four to five years. In general, volcanoes only affect climate for a few years.

Asteroid collisions

Asteroid impacts can alter Earth's climate, but they need to be big. In 1908 an asteroid with a diameter of 100 m exploded in the air 5 km above Tunguska in Russia. The blast flattened 80 million trees but was not large enough to alter the climate. 1 km sized asteroids strike Earth every 500 000 years. An impact of this size would blast millions of tonnes of ash and dust into the atmosphere. This would cool the climate as the dust and ash block incoming sunlight. It would be similar in impact to a large volcanic eruption and its effects could last 5–10 years.

The **stratosphere** is the layer of air 10–50 km above the Earth's surface. It is above the cloudy layer we live in, the troposphere.

A **climatologist** is a scientist who is an expert in climate and climate change.

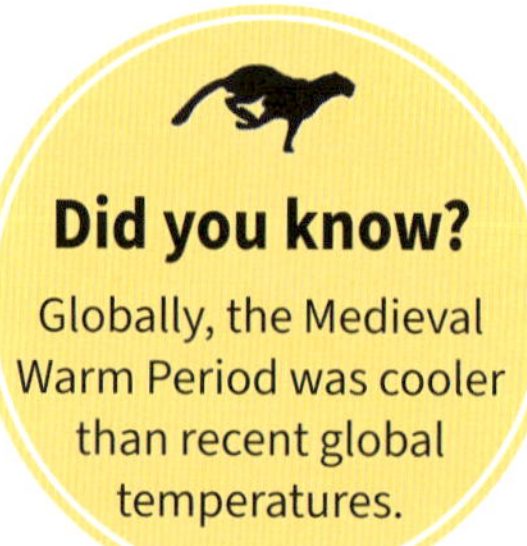

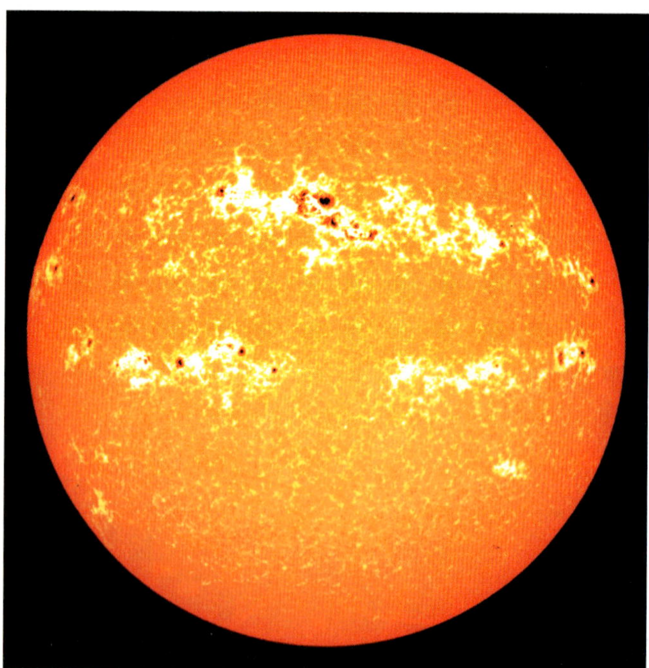

▲ ***Figure 1*** *Sunspots on the surface of the sun in 2003*

Variations in solar output

Over 2000 years ago Chinese astronomers started to record sunspots. These are black areas on the sun's surface (see Figure 1). Sometimes the sun has many spots, at other times they disappear. They tell us that the sun is more active than usual. Lots of spots mean more solar energy being fired out from the sun towards Earth.

Cooler periods, such as the Little Ice Age, and warmer periods, such as the Medieval Warm Period, may have been caused by changes in sunspot activity. Some people think that, on average, there were more volcanic eruptions during the Little Ice Age, and that this added to the cooling. However, climate change on timescales of a few hundred years, and 1–2°C, cannot be explained by volcanoes – but it might be explained by sunspot cycles (see Figure 2).

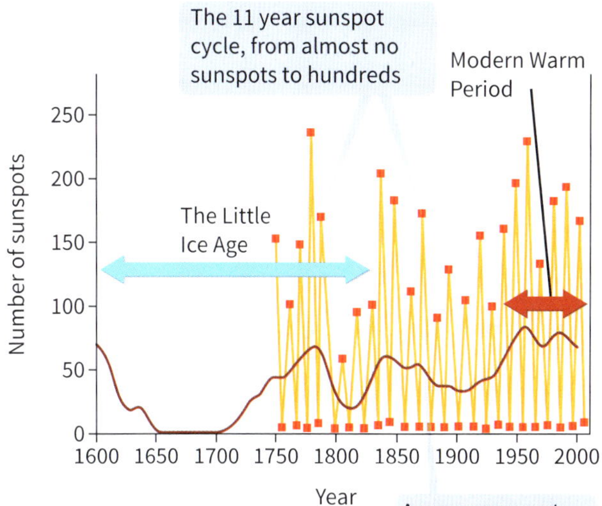

▲ ***Figure 2*** *Sunspot cycles*

The orbital theory

Over very long timescales, there have been big changes in climate. Cold glacial periods and ice ages were 5–6°C colder than today. Some interglacials were 2–3°C warmer than today. Such big changes need a big cause. Scientists think they know what this is – changes in the way the Earth orbits the sun.

You might think that the Earth's orbit does not change, but over very long periods it does, as Figure 3 shows.

- The Earth's orbit is sometimes circular, and sometimes more of an ellipse (oval).
- The Earth's axis tilts. Sometimes it is more upright, and sometimes more on its side.
- The Earth's axis wobbles, like a spinning top.

These three changes alter the amount of sunlight the Earth receives. They also affect where sunlight falls on the Earth's surface. On timescales of thousands of years, the changes would be enough to start an ice age, or end one. These changes are called **Milankovitch Cycles**.

▼ ***Figure 3*** *Orbital changes*

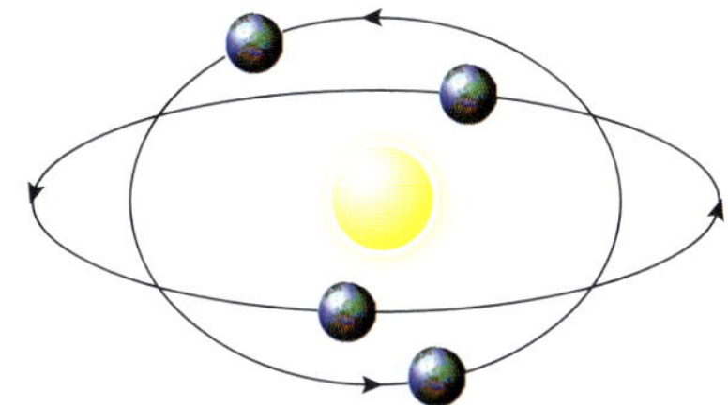

It takes 100 000 years for the Earth's orbit to change from being more circular, to an ellipse, and back again.

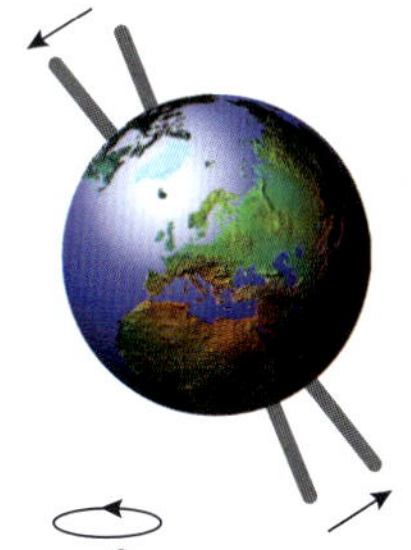

Over a cycle of 41 000 years, continuous change in the Earth's rotational axis occurs, like a spinning top.

Over a cycle 26 000 years, a gradual shift occurs in the orientation of the Earth's axis of rotation, producing a 'wobble'.

Your questions

1 Explain how **a** big volcanic eruptions, **b** asteroid collisions might change our climate.

2 Draw and complete a table to compare the four theories of climate change. Use these headings: **a** What happens? **b** How can this affect climate? **c** Could it cause our current climate change?

Exam-style questions

3 Explain how a volcanic eruption might cause changes to global climate. (4 marks)

4 Explain **two** other natural theories of climate change. (4 marks)

5 Assess the role of natural causes of climate change in explaining past climate change events. (8 marks)

1.6 Past climates – how do we know?

In this section, you'll understand how climate was very different from that of today in both the recent, and distant, past.

What is climate?

The world's winds, ocean currents, and pressure systems change with the seasons. Each day, **weather** changes – temperatures, sunshine, the amount of rainfall, wind direction, etc. But together these changes add up to a seasonal pattern – colder in winter in the UK, or rainier in Mali in June. That's **climate** – the changing patterns which can be predicted. But climate too is changing now, and has also changed in the past. So, how do we know about the past?

▲ ***Figure 1*** *Greenland's ice sheets hold clues to past climates*

The distant past

120 000 years ago, rhinoceroses and elephants roamed around what is now London. At other times in the past, huge ice sheets stretched from the North Pole as far south as London. Scientists know that climate was different in the past. They use physical evidence such as:

- fossilised animals, plants and pollen that no longer live in the UK
- landforms, like the U-shaped valleys left by retreating glaciers
- samples from **ice cores** taken from Greenland (see Figure 1) and Antarctica.

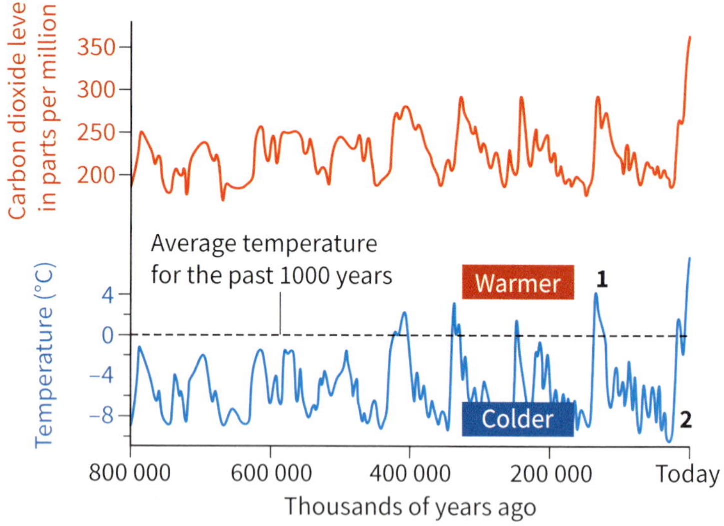

Key

1 Warm periods, called **interglacials**.

2 Cold periods, called **glacials**. Some glacial periods became Ice Ages.

▲ ***Figure 2*** *Past temperatures and carbon dioxide levels*

Ice cores

Ice sheets are like a time capsule. They contain layers of ice, oldest at the bottom, youngest at the top. Each layer is one year of snowfall. Trapped in the ice layers are air bubbles. These preserve air from the time the snow fell. Locked in the air bubble is carbon dioxide. Climatologists can reconstruct past temperatures (shown in Figure 2) by drilling a core through the ice and measuring the amount of trapped carbon dioxide in ice layers.

We know about the period called the Quaternary (the last 2.6 million years) from ice cores. There have been warm periods (interglacials) lasting for between 10 000 and 15 000 years. Cold periods (glacials) vary, but can last as long as 100 000 years. During some glacial periods, it became so cold that the Earth plunged into an ice age. Huge ice sheets extended over the continents in the northern hemisphere. The last time this happened was between 30 000 and 10 000 years ago, in the last ice age (see Figure 3).

Tree rings

In temperate climates such as in Western Europe, trees grow every summer. Periods of growth can be seen from the number of rings in a tree – each ring is a year's growth. Some years are warmer and wetter, when growth is greater; others are cooler or drier. Scientists therefore examine tree rings to learn about past climate conditions, before there were accurate temperature and rainfall data. We can then learn about climatic conditions that the tree experienced during its lifetime. Trees rarely survive more than a few hundred years, but fossils of trees in peat bogs go back thousands of years.

Historical sources

There is evidence for climate change more recently, provided by:

- old photos, drawings and paintings of the landscape
- written records, such as diaries, books and newspapers
- the recorded dates of regular events, such as harvests, the arrival of migrating birds and tree blossom.

These sources are not very reliable, as they were not intended to record climate. However, they do indicate that climate changes every few hundred years. Average temperatures over 2000 years have varied by 1–1.5°C compared to now (see Figure 4).

▼ ***Figure 3*** *Ice cover during the last ice age*

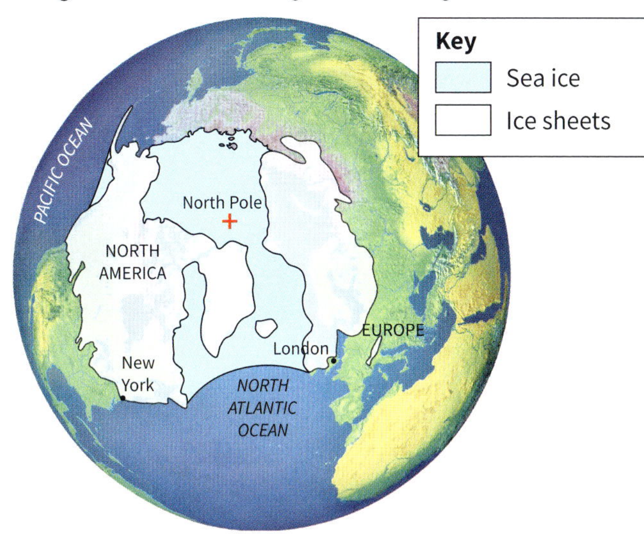

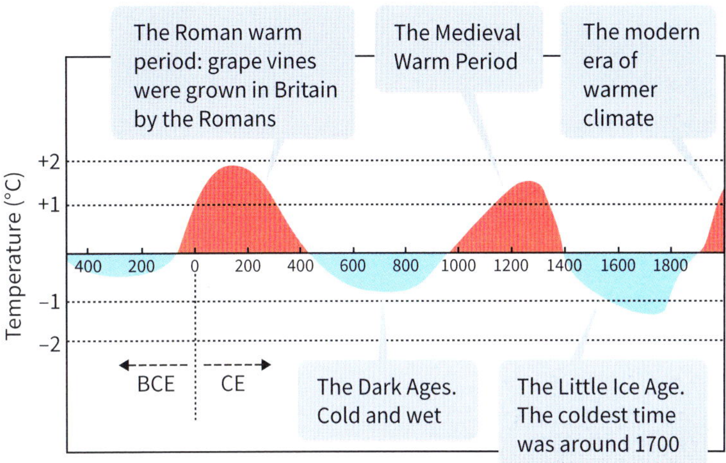

▲ ***Figure 4*** *Temperature change over the last 2000 years*

Your questions

1 Look at the graph of carbon dioxide levels and temperature over the last 800 000 years (Figure 2).

  a Describe the variations shown in the temperature graph.

  b Does the Earth warm up faster, or cool down faster?

  c How closely are temperature and carbon dioxide levels linked?

2 Look at the map of ice sheets (Figure 3). Which parts of the land were covered by ice 20 000 years ago, and are now ice-free?

3 Draw a table of the advantages and disadvantages of tree rings and historical records in tracing past climates.

Exam-style questions

4 Compare the meaning of the terms 'weather' and 'climate'. (3 marks)

5 Study Figure 4. Describe the pattern of temperature change since the year 1000. (3 marks)

6 Explain how **one** piece of evidence can be used to reconstruct past climates. (3 marks)

7 Assess the evidence for natural climate change in reconstructing the UK climate between Roman times and the present day. (8 marks)

1.7 Changing the atmosphere

In this section, you'll understand how our atmosphere is being changed by human activity.

The greenhouse effect

Earth's **atmosphere** is vital to life. The gases which make up the atmosphere are important:

- Nitrogen (78.1%) is an important nutrient for plant growth.
- Carbon dioxide, or CO_2 (0.04%), is taken in by plants, which breathe out oxygen.
- Oxygen (20.9%) is breathed in by animals, which breathe out carbon dioxide.
- Water vapour (1%) forms clouds, essential to the water cycle.

Carbon dioxide is vital, even though it makes up only a tiny fraction of the atmosphere. This is because it helps to regulate temperatures on Earth – it is one of the **greenhouse gases**.

The **atmosphere** is a layer of gases above the Earth's surface.

The **greenhouse effect** is the way that gases in the atmosphere trap heat from the sun. The gases act like the glass in a greenhouse. They let heat in, but prevent most of it from getting out.

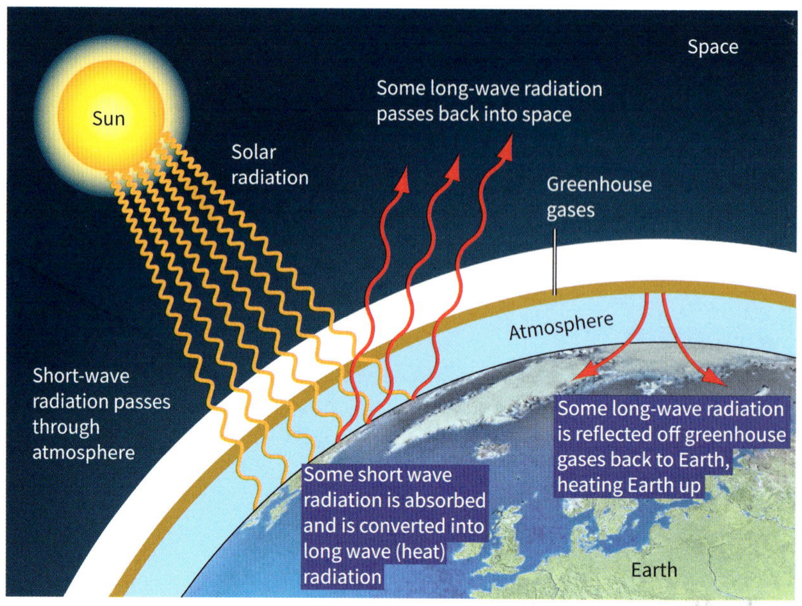

▲ **Figure 1** *The greenhouse effect*

Greenhouse gases

The **greenhouse effect** (see Figure 1) is completely natural. Greenhouse gases help to retain heat in the atmosphere. They make the planet warmer by 16°C and keep the Earth comfortable. Without them, Earth would be a frozen wasteland. CO_2 is the most common greenhouse gas, but there are others shown in Figure 2. The extra greenhouse gases which pollute the atmosphere are produced by humans. In the UK we use many fossil fuels. Burning these produces CO_2, which ends up in the atmosphere as pollution. The main source of this pollution is power stations that produce our electricity (see Figure 3).

▼ **Figure 2** *Greenhouse gases*

Greenhouse gas	% of greenhouse gases produced	Sources	Warming power compared to CO_2	% increase since 1850
Carbon dioxide	89%	Burning fossil fuels (coal, oil and gas), deforestation which releases carbon dioxide.	1	+30%
Methane	7%	Gas pipeline leaks, farming rice in paddy fields, cattle farming and melting permafrost.	21 times more powerful	+250%
Nitrous oxide	3%	Jet aircraft engines, cars and lorries, fertilisers and sewage farms.	250 times more powerful	+16%
Halocarbons	1%	Used in industry, solvents and cooling equipment.	3000 times more powerful	Not natural

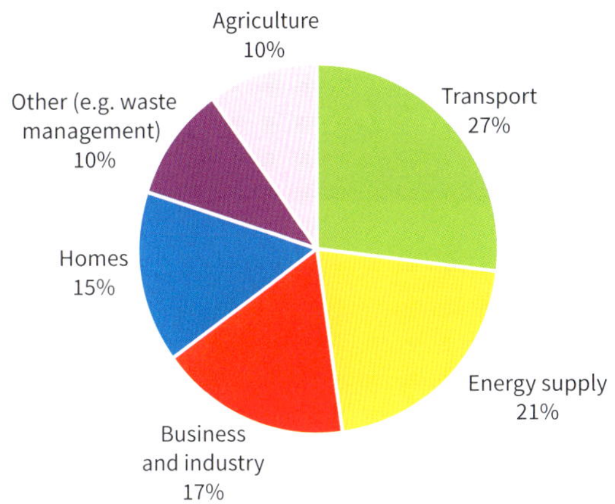

▲ ***Figure 3*** *Percentage of carbon dioxide emissions from different sources in the UK in 2019*

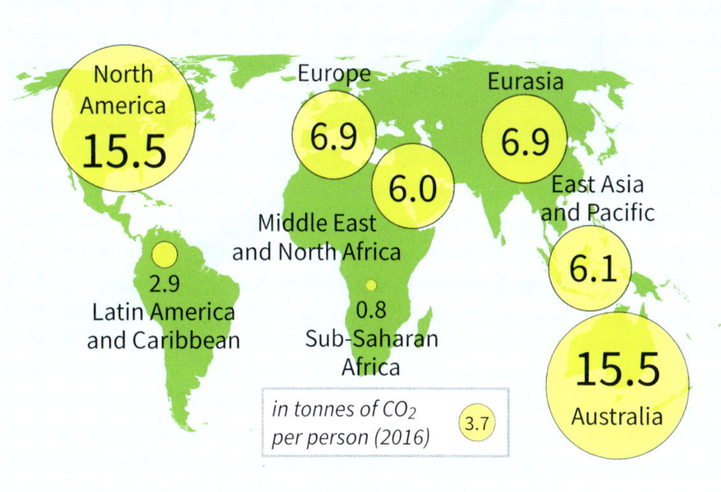

▲ ***Figure 4*** *Global carbon dioxide emissions*

Globally, there are big differences in CO_2 production, with most produced by wealthy consumers, the majority (though not all) of whom live in the industrialised world.

- The EU, USA and Japan emit 27% of all CO_2 emissions.
- China alone emits 29% (and rising) with Russia and India adding another 12% between them.

Consumers in the developing world generally produce 1–3 tonnes of CO_2 per person per year, compared to 7–15 tonnes in the developed world. Figure 4 shows the differences.

Many scientists have concerns about emissions and their effect on climate, such as:

- reducing emissions in the developed world, where lots of fossil fuel is used
- persuading China and India to reduce growth in carbon dioxide emissions (see Figure 5)
- protecting the vulnerable from the impacts of climate change.

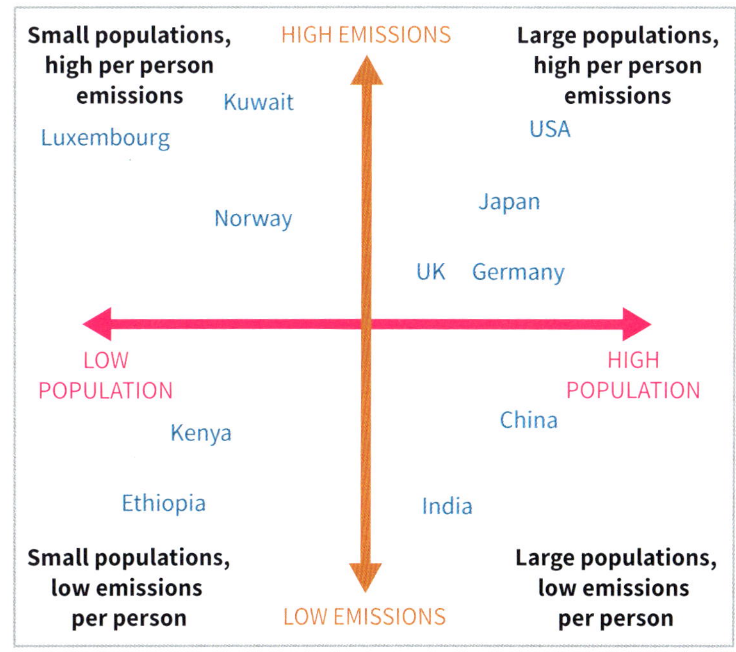

▲ ***Figure 5*** *Carbon dioxide emissions and population size*

Your questions

1 Describe the greenhouse gases, and say why each one is important.

2 **a** Draw your own version of the diagram showing the greenhouse effect (Figure 1).

b Label it to explain how the greenhouse effect works.

3 **a** Make a list of human activities which add extra greenhouse gases into the atmosphere.

b Explain why people in the developing world produce only small amounts of greenhouse gases. Think about their lifestyles and activities compared to the developed world.

Exam-style questions

4 Name **two** greenhouse gases. (2 marks)

5 Study Figure 4. Describe the distribution of carbon dioxide emissions per capita. (3 marks)

6 Study Figure 3. Suggest **two** reasons for the shares of carbon dioxide emissions. (4 marks)

7 Evaluate the view that, 'Most global warming is caused by carbon dioxide emissions from a few rich, developed countries.' (8 marks)

1.8 Changing climate

In this section, you'll understand how pollution of the atmosphere with greenhouse gases has led to the enhanced greenhouse effect, also known as global warming.

A warming planet

All evidence suggests that global climate is changing. This is known as global warming. Global warming means a warming of the Earth's temperatures, and is caused by the **enhanced greenhouse effect**. 'Enhanced' simply means 'working more strongly'. Pollution of the atmosphere with carbon dioxide and other greenhouse gases has given the natural greenhouse effect a boost. Figure 1 shows the increase in carbon dioxide in the atmosphere in Hawaii. More heat is trapped in the atmosphere by the greenhouse gases, and temperatures are rising.

Global warming has been measured:

- Average global temperatures rose by 1°C from 1880 to 2020.
- Sea levels rose by 210–240 mm from 1870 to 2020. As the sea warms, it expands, called **thermal expansion**. Already, melting glaciers and ice sheets have caused rising sea levels; continued melting could cause further significant increases.

Since 1980, global warming seems to have been happening more quickly:

- The seven hottest years on record occurred between 2014 and 2020, with 2020 and 2016 being the hottest.
- The ten warmest years on record have occurred in the second decade of the twenty-first century. Only one year during the twentieth century – 1998 – comes close to twenty-first century temperature levels.
- By 2015, floating sea ice in the Arctic had shrunk by almost 30% compared to average levels during 1981–2010.
- Over 90% of the world's valley glaciers are shrinking. Figure 2 shows photos of the Virkisjökull glacier in Iceland in 1996 and 2009.

Did you know?

Worldwide, 2020 and 2016 were the joint warmest years on record. Climate scientists are in no doubt – rising global temperatures are the result of 250 years of carbon dioxide emissions.

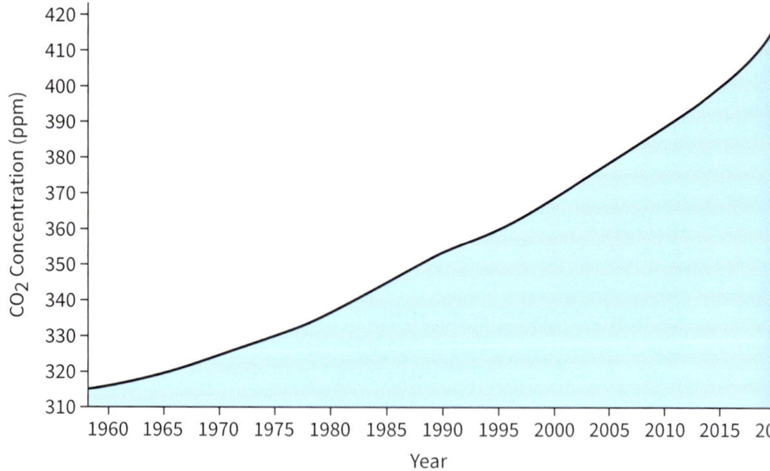

▲ ***Figure 1*** *Carbon dioxide concentrations in the atmosphere as recorded in Hawaii 1959–2020*

▲ ***Figure 2*** *The retreating Virkisjökull glacier in Iceland, which flows from Iceland's highest mountain – the Öraefajökull ice cap*

What do climate experts think?

Many scientists and politicians are worried about global warming. In 2021, over 830 scientists and climatologists from 80 countries wrote a report called *Climate Change 2021: Impacts, Adaptation, and Vulnerability*. They work for the Intergovernmental Panel on Climate Change, part of the United Nations. Their report confirmed earlier research which stated that most of the increase in global average temperatures since the mid twentieth century was very likely due to the increase in greenhouse gases produced by humans. However, there is a tiny minority of scientists who believe that the main causes of global warming are natural.

▲ ***Figure 3*** *Typhoon Vamco which hit the Philippines in November 2020 – will climate change mean more events like this?*

What of the future?

Climate scientists estimate that by 2100 future climate changes will occur (see Figure 4):

- temperatures will rise between 1.1°C and 6.4°C
- sea levels will rise by between 30 cm and 1 metre.

A 'best guess' might be a warming of 3.5°C and a sea level rise of 40 cm by 2100.

Predicting these changes is difficult because we don't know:

- how big the population will be
- whether fossil fuels will continue, or be replaced by renewables
- if people will change lifestyles, using less energy.

Climate change could bring any of the following:

- more storms, cyclones, floods, and droughts
- stronger storms and hurricanes (see Figure 3)
- 'climate refugees', who evacuate vulnerable places.

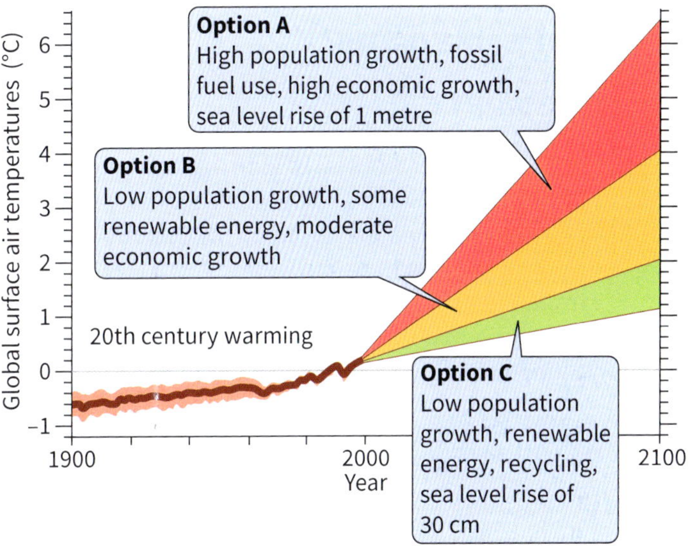

▲ ***Figure 4*** *Possible projected increase in global temperatures by 2100 based on three different situations*

Your questions

1 Explain the difference between the 'greenhouse effect' and the 'enhanced greenhouse effect'.

2 Using Figure 1, describe how carbon dioxide levels have changed in Hawaii since the 1960s.

3 List the evidence for global warming.

4 Use Figure 4 to describe predicted changes in global surface air temperatures from 2000–2100 if **a** Option A, **b** Option B, and **c** Option C take place.

5 Explain why options A, B and C in Figure 4 result in such different predictions.

6 In pairs, draw a spider diagram to show the possible impacts of climate change. Use an atlas to identify areas at risk.

Exam-style questions

7 Define the term 'enhanced greenhouse effect'. (1 mark)

8 State **two** pieces of evidence that the Earth's climate is getting warmer. (2 marks)

9 Explain **two** reasons for rising carbon dioxide levels in the atmosphere. (2 marks)

10 Explain why climate change is leading to rising sea levels. (4 marks)

11 Explain **two** reasons why the predictions of future global temperatures in Figure 4 are uncertain. (4 marks)

1.9 Batten down the hatches!

In this section, you'll understand about the hazards that tropical cyclones bring.

Cyclone alert!

Wednesday 5 February 2020. Residents along the northern coast of Western Australia are warned to prepare for tropical cyclone Damien. A large area at risk has been identified, extending from Dampier on the coast and inland to Wittenoom. Evacuation centres are set up. The storm is what the Australian Bureau of Meteorology calls a Category 3, with winds of 140 km/h. Originating over the Indian Ocean, it makes landfall in Australia on Saturday 8 February, strengthening as it arrives into a Category 4, with gusts reaching 205 km/h. But it is short-lived – by the following day it has dissipated (died out).

What's in a name?

'Tropical cyclone' is a general term used to describe a rotating system of clouds and storms that form, and develop, over tropical or subtropical waters. Once its winds exceed 118 km/h, a tropical cyclone is known as a hurricane, typhoon, or cyclone depending upon where it originates.

- 'Hurricane' is used in the North Atlantic and on the Pacific coast of the USA (Area 1 on Figure 1).
- 'Cyclone' is used in the Indian and South Pacific Oceans (Areas 2).
- 'Typhoon' is used in the western North Pacific (Area 3).

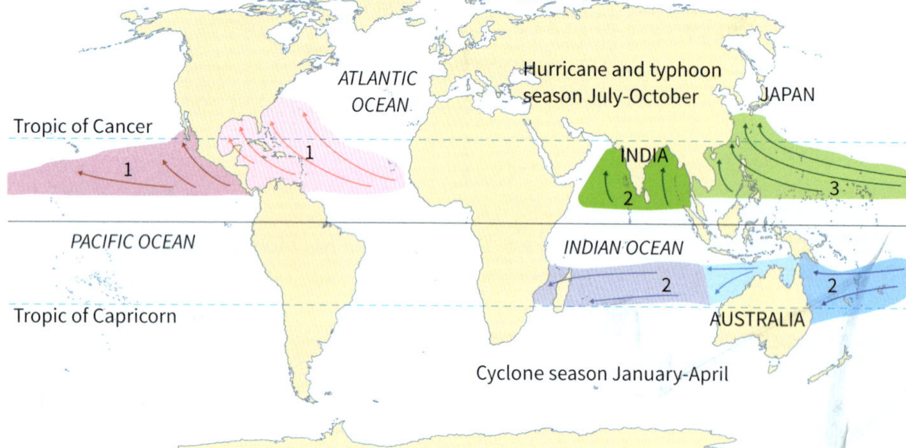

▲ **Figure 1** *The location and pathways of tropical cyclones*

Category	Wind speeds
1	74–95 mph, 119–153 km/h
2	96–110 mph, 154–177 km/h
3	111–129 mph, 178–208 km/h
4	130–156 mph, 209–251 km/h
5	≥157 mph, ≥252 km/h

▲ **Figure 2** *The Saffir-Simpson Hurricane Scale*

As well as varying names, tropical cyclones are also measured and classified differently according to where they occur.

- In the North Atlantic and North-eastern Pacific Oceans, tropical cyclones are classified using the **Saffir-Simpson Hurricane Scale** (Figure 2). It uses five different categories to measure wind strength.
- In the Western Pacific, Japan uses its own **Meteorological Agency's Scale**.
- In Australia, the Australian **Tropical Cyclone Intensity Scale** is used.
- India has its own scale, as does the South West Indian Ocean with a scale devised by the Météo-France Forecast Center on the island of La Reunion.

What hazards do tropical cyclones bring?

They bring a range of hazards, each storm varying depending on location.

- **Strong winds**, which whip up garden furniture, lift roofs, vehicles or caravans, bring down trees, power lines and even destroy whole buildings.
- **Storm surges,** which bring flooding caused by unusually high tides. High tides are even higher than normal during a cyclone because air pressure is so low. Sea level is raised because air pressure is lower and winds drive the waves onshore. High tides extend inland, causing **coastal flooding** – see Figure 3.
- **Intense rainfall**. With thick, dense clouds, it is not unusual for 1000 mm of rain to fall in a single storm! China holds the mainland record – in 1967 Typhoon Carla brought 2700 mm in a single storm. That's over four times London's annual rainfall! The Pacific island of La Reunion dwarfed even that when over 6400 mm fell in 1980 during Tropical Cyclone Hyacinthe, the world's wettest!
- **Landslides.** In 2014, 53 people died in landslides in the Philippines caused by tropical storm Jangmi which saturated, heavy ground, causing it to slump.

Did you know?

The largest tropical cyclone on record is Typhoon Tip in 1979 in the Pacific. At 2220 km across, it was half the size of the United States!

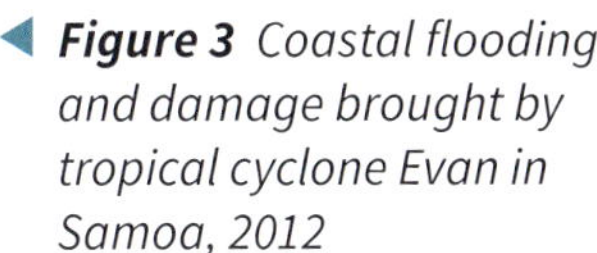

Figure 3 *Coastal flooding and damage brought by tropical cyclone Evan in Samoa, 2012*

Your questions

1. On a world map, shade and label where and when tropical cyclones occur, using the correct names for different parts of the world.
2. Make a sketch of the photograph in Figure 3, and annotate it with the impacts of tropical cyclone Evan. Use the text from 'What hazards do tropical cyclones bring?' to identify hazards in the photo.
3. Write a 200-word script for local radio in Western Australia to warn people what they can expect when tropical cyclone Damien arrives.

Exam-style questions

4. Define the term 'tropical cyclone'. (2 marks)
5. Suggest **two** reasons for the distribution of tropical cyclones shown on Figure 1. (4 marks)
6. Explain **two** economic hazards brought by tropical cyclones. (4 marks)
7. Explain why some impacts of hazards brought by tropical cyclones are longer-term than others. (4 marks)

1.10 How do tropical cyclones form?

In this section, you'll understand the formation and development of tropical cyclones.

Keeping track

Tropical cyclones form in the world's warmest areas between the tropics (see Section 1.9). Their movement – called a **track** – follows pathways driven by global wind circulation (Figure 1). Satellite photography is used to spot where they originate – this helps forecasters to plot where they will go next. 30% of the world's population lives in the area shown in Figure 1 so it's important to track them accurately.

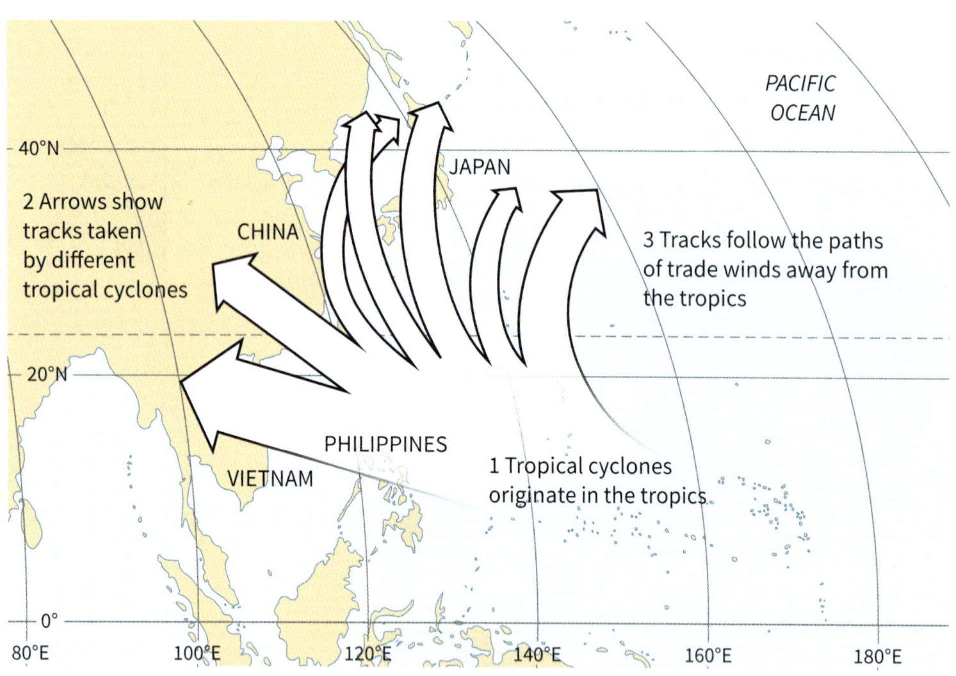

◀ ***Figure 1*** *Tropical cyclone (or 'typhoon') tracks in the north-west Pacific. The thickness of the arrows reflects the average number of tropical cyclones over many years.*

Where do tropical cyclones develop?

Compared to ocean storms, tropical cyclones are actually small, averaging 650 km across. The areas where they form are known as **source regions**, and their formation depends on three conditions occurring at the same time.

1 A large, still, **warm ocean** area whose surface temperatures exceed 26.5°C over long periods. The deeper and broader the area, the more energy is available. Tropical cyclones form in late summer when oceans are warmest – mid-July to September in the northern hemisphere, mid-January to March in the southern.

2 Strong winds high in the troposphere, 10–12 km above the Earth's surface, are needed to draw warm air up rapidly from the ocean surface.

3 A strong force created by the Earth's rotation, called **Coriolis force**. Tropical cyclones do not form near the Equator, where Coriolis force is minimal, but instead where the rotation is stronger, between 5° and 30° latitude.

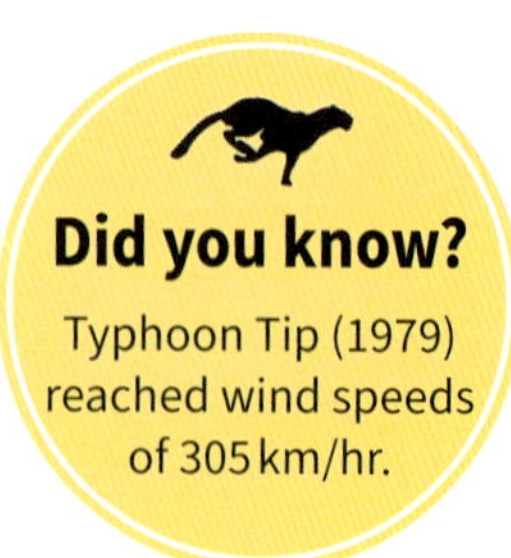

Did you know?

Typhoon Tip (1979) reached wind speeds of 305 km/hr.

Tropical cyclones depend upon a supply of warm moist air. Warm air is lighter than cold. It expands and rises, creating an updraft (Stage 1 in Figure 2). As a result, air pressure falls. The lowest air pressure of a tropical cyclone is always found at its centre. Typically, pressure is as low as 950 mb or even less.

Air pressure in tropical cyclones

Air pressure is the weight of air bearing down on the Earth's surface, and is measured in millibars. 1013 mb is the average air pressure over the Earth's surface at sea level. Tropical cyclones have much lower air pressure than the air surrounding them. The bigger the difference in pressure, the stronger the winds. One of the lowest air pressures ever recorded was 877 mb in Typhoon Ida in the Philippines in 1958, where winds reached 300 km/hr!

But they have an exception! The 'eye' of the storm, in its centre, consists of calm, clear, descending dense air (Stage 3), while just tens of kilometres away, a cyclone rages!

Did you know?

The average hurricane eye stretches 30–60 km across, with some growing as large as 200 km. That's the distance from London to Birmingham!

▼ ***Figure 2*** *Stages in tropical cyclone formation*

Stage 1 Warm air currents rise (number 1 in Figure 3) from the ocean. As the warm air rises, more air rushes in to replace it; then it too rises, drawn by the draught above.

Stage 2 Updraughts of air contain huge volumes of water vapour from the oceans, which condense to produce **cumulonimbus clouds** (number 2 in Figure 3). Condensation releases heat energy stored in water vapour, which powers the cyclone further.

Stage 3 Coriolis force (number 3 in Figure 3), caused by the Earth's rotation, leads to spiralling currents of air around the centre of the cyclone, resembling a whirling cylinder. As it rises, it cools, and some descends to form a clear, cloudless, still, **eye** of the storm.

Stage 4 As the tropical cyclone tracks away from its source, it is fed new heat and moisture from the oceans, enlarging as it does so.

Stage 5 Once it reaches a landmass, it loses its energy source from the ocean. Air pressure rises as temperature falls, winds drop, rainfall decreases, and it **decays** to become a mere storm.

▼ ***Figure 3*** *Huge cumulonimbus clouds associated with tropical cyclones can grow to 12–15 km in height*

Your questions

1 Explain each of the following statements about tropical cyclones:
- They do not form at the Equator.
- They are areas of low air pressure.
- They form over oceans and decay over land.
- Their calmest part is the centre.
- Aircraft avoid flying through them.

Exam-style questions

2 State **two** features of a tropical cyclone. (2 marks)

3 Explain **two** causes of tropical cyclones. (4 marks)

4 Explain **one** reason why tropical cyclones decay over land. (2 marks)

1.11 The impacts of tropical cyclone Amphan

In this section, you'll understand the impacts of Cyclone Amphan on Bangladesh in May 2020. This is the first of two case studies you need for the exam, in this case for a developing or emerging country.

A country under threat

With an area of 144 000 sq km, Bangladesh is about 60% of the size of the UK. But it has a population of 163 million people – over twice that of the UK. In spite of recent economic growth, Bangladesh remains among the world's poorest countries. What makes life challenging is that 80% of Bangladesh is less than 10 m above sea level, see Figure 1, and 10% is below 1 m.

Three of the world's largest rivers, the Ganges, Brahmaputra and Meghna, join within Bangladesh and have created one of the world's largest floodplains. During the annual monsoon, if one river floods, they all flood, inundating much of the coastal area and threatening large numbers of people. Tropical cyclones add to the vulnerability of Bangladesh.

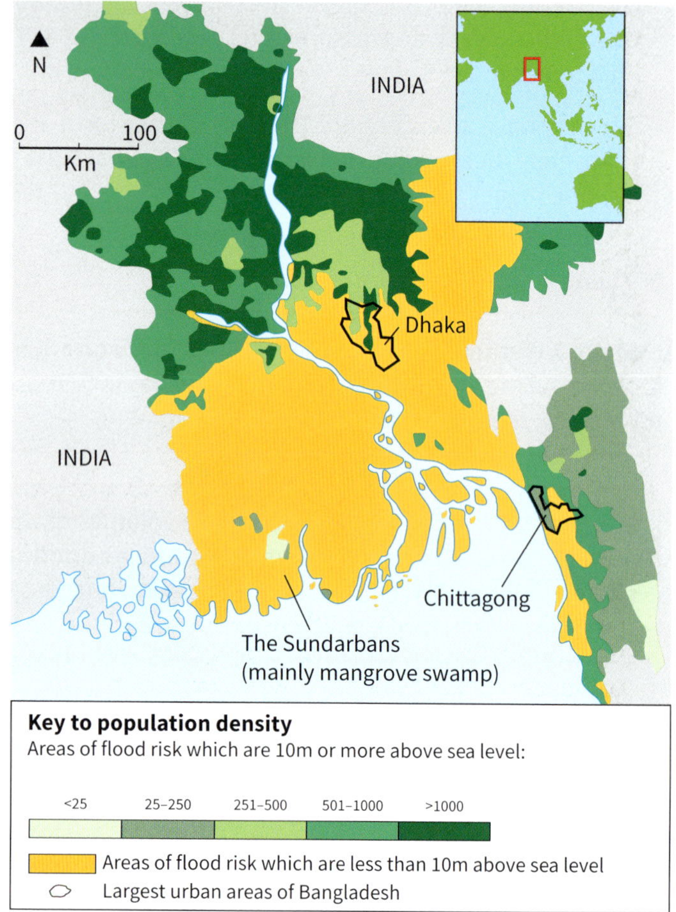

▲ ***Figure 1*** *Areas of high population density in Bangladesh most vulnerable to flooding*

Cyclone Amphan, May 2020

Cyclone Amphan began as a tropical storm in the Bay of Bengal, reaching south-western Bangladesh on 20 May 2020. Its track is shown in Figure 2. It wasn't Bangladesh's wettest cyclone — although 236 mm of rain fell in the Indian city of Kolkata, shown in Figure 2. But three things made this cyclone severe:

- **Intensity of the rain**. On 20 May, the estuary region (shown in Figure 1) received 220 mm of rain in a few hours – that's 40% of London's *annual* total!
- **Wind strength**, which at its peak was 260 km/hr.
- Most seriously, low air pressure (920 mb) out at sea in the Bay of Bengal caused sea level to rise, creating a huge **storm surge**.

Social and economic impacts of the cyclone

Amphan left a trail of destruction. The storm surge raised sea level at high tide by five metres, which submerged and destroyed several villages – see Figure 3. Many of the earth embankments in southern Bangladesh, built to hold back floods and protect people, were washed away.

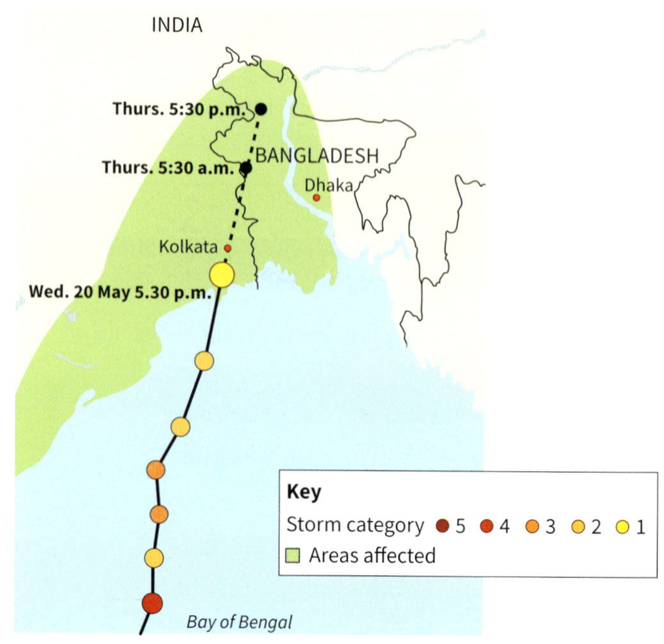

▶ ***Figure 2*** *The track of Cyclone Amphan in May 2020, and areas most affected*

- 20 people were killed.
- 500 000 were made homeless, their homes destroyed either by winds or floods. 90% of those displaced were from Bangladesh's lowest income groups.
- The worst-affected land was on the delta, with 176 000 hectares of farmland flooded, crops killed by inundations of salt water, and damage to over 3000 shrimp and crab farms.
- Over 1 million people were affected, losing either their homes or livelihoods.

Other impacts were longer-term. By mid-2021, many thousands of people had still not returned home and were living in temporary shacks. Many had migrated to cities such as Dhaka, in search of work. Poverty forced them to live in areas of poor housing. Some found that their lack of skills meant they had to take labouring jobs, or pull rickshaws. Homelessness also made it difficult for children to resume normal schooling.

▲ ***Figure 3*** *The impacts of Amphan in Bangladesh*

Environmental impacts of the cyclone

These were severe too.

- Loss of animals meant that animal dung – a source of cooking fuel – was lost, which placed further pressure on firewood sources. (Rising population in Bangladesh meant woodland was already in decline.)
- Sickness and typhoid were problems – flooding meant freshwater was contaminated by sewage, and moist air brought mosquitoes, and therefore malaria.
- The Sundarbans, an area of mangrove forest (see Figures 1 and 4), were badly affected. This area is home to the highly-endangered Royal Bengal tiger, and is a protected reserve. 150 km of mud embankments burst, drowning several tigers.

▼ ***Figure 4*** *A Bengal tiger, native to the Sundarbans*

Your questions

1 Copy the table below. Classify the impacts of Amphan on Bangladesh.

Impact	Short-term impacts (the first weeks)	Longer-term impacts (up to a year)
Economic		
Social		
Environmental		

2 In pairs, decide whether its greatest impacts were **a** economic, social or environmental, and **b** short-term or long-term.

3 Why is it true that the poor were most likely to be killed by the cyclone?

Exam-style questions

4 Study Figure 2. Describe the track of Cyclone Amphan. (2 marks) 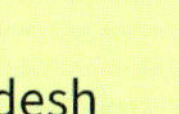

5 Explain **one** reason why people in Bangladesh are especially vulnerable to flooding. (2 marks)

6 Explain **two** economic impacts of Cyclone Amphan. (4 marks)

7 Explain **two** environmental impacts of a named tropical cyclone. (4 marks)

8 Assess the impacts of tropical cyclones on **one** developing or emerging country. (8 marks)

1.12 Planning and preparing for cyclones 1 Bangladesh

In this section, you'll evaluate how well Bangladesh plans and prepares for tropical cyclones.

Poor – and vulnerable

Bangladesh gets just 5% of the world's cyclones each year, but it suffers 85% of the world's deaths and damage caused by them. Its people are poor; GDP per capita (PPP) was only $1900 in 2019, and 22% of its population lives below the poverty line. The poorest of all are rural landless labourers, particularly women. Forced to live on less desirable flood-prone land in poorly-built housing, they suffer most from cyclones. Often their only escape is to the roads built on the embankments designed to protect the country from floods, see Figure 1.

▲ ***Figure 1*** *Temporary homes built by the roadside in southern Bangladesh*

However, Bangladesh *has* developed ways of protecting its population by:

- predicting cyclones (using **forecasting** and **satellite technology**)
- developing **warning systems**
- introducing **evacuation strategies** and building **storm surge defences**.

1 Weather forecasting

Bangladesh's Meteorological Department issues weather forecasts and warnings on TV and radio. The problem is that outside Dhaka, the capital city, few people have access to TV and radio. Households with radios have lower death rates during cyclones than those without. Although digital systems are developing, they are mostly slow outside the cities, though the use of mobile phones to warn people looks promising. In 2019, mobile ownership in Bangladesh reached 54 phones per 100 people.

2 Satellite technology

Weather forecasting is expensive, because it uses digital images from space bought from US, Chinese and Japanese satellites, costing US$12 million a year. Three radar stations transmit live weather updates. This enables cyclone formation and tracking to be carried out quickly and accurately – see Figure 2. In 2018 Bangladesh launched its first space satellite, costing US$248 million.

▲ ***Figure 2*** *Satellite photo showing Cyclone Amphan (see section 1.11) heading for Bangladesh in 2020*

3 Warning systems

Since Cyclone Bhola killed 300 000 people in Bangladesh in 1970 (one of the world's most devastating cyclones), the government has developed an early warning system, enabling coastal communities to be evacuated. It runs awareness campaigns, using a mix of village meetings, posters and leaflets, film shows and demonstrations to spread information about cyclone warning signals and helping people to prepare. There are now 50 000 cyclone warning volunteers, who live and work in threatened areas.

4 Evacuation strategies and storm surge defences

As well as early warning systems, Bangladesh has invested heavily in **evacuation** procedures and safe refuges. The construction of cyclone shelters and coastal embankments by the government has reduced deaths from cyclones.

For those who are evacuated, there are 3500 cyclone shelters, like that shown in Figure 3, in coastal districts, some taking up to 5000 people. All are built from concrete or brick, and people who use them usually survive; death rates are double where there are no shelters. But more are needed.

Embankments are built to protect against storm surges. Bangladesh has 400 km of coastline and thousands of km of low-lying rivers, any of which are liable to flooding by storm surges. Main roads are built on embankments but it is not possible to protect the entire country.

Measuring success

Warning systems and protection are expensive, but Bangladesh has successfully reduced the death toll and damage caused by tropical cyclones through its warning systems, evacuation plans and shelters. But challenges remain – illiteracy means that some do not understand or follow warnings. Instead of moving to cyclone shelters, others believe in 'wait-and-see'. Fear of losing property, and previous false warnings also limit those willing to evacuate.

▼ ***Figure 3*** *A cyclone shelter in Bangladesh*

Your questions

1 Make a large copy of the table and complete it to compare the successes and challenges of cyclone warning and protection systems in Bangladesh.

	What it involves	Its successes	Its future challenges
Weather forecasting			
Satellite technology			
Warning systems			
Evacuation strategies			
Storm surge defences			

2 In pairs, discuss which of the five you think Bangladesh should invest more in to protect itself best in future. Feed your ideas back.

Exam-style questions

3 Describe **one** method used to predict tropical cyclones. (2 marks)

4 Explain **two** ways in which countries affected by tropical cyclones can prepare for them. (4 marks)

5 Evaluate the effectiveness of different methods of cyclone preparation and response in **one** named developing or emerging country. (8 marks)

1.13 Planning and preparing for cyclones 2 The USA

In this section, you'll evaluate how well the USA has been able to plan and prepare for cyclones. This is the second of two case studies you need for the exam, in this case for a developed country.

Managing hurricane risks in the USA

Life along the Gulf of Mexico and the Atlantic coast of Florida, USA, changes every summer between July and October. It's the season for tropical cyclones, called **hurricanes** in the USA. Along these coasts lie cities such as New Orleans (Gulf of Mexico), and Miami (Florida). Florida has a 22% chance that a hurricane will make landfall each year. Occasionally, hurricanes track north along the east coast and can even affect as far north as New York before decaying.

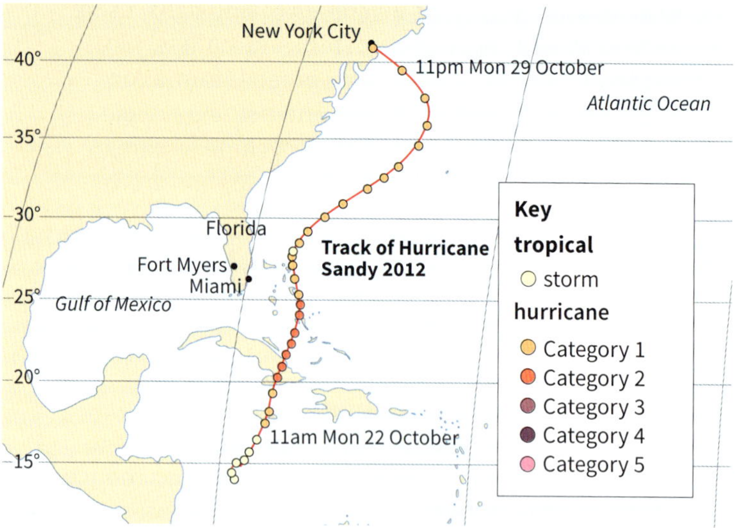

▲ ***Figure 1*** *The Gulf Region and US east coast, where hurricanes are common. The track of Hurricane Sandy, 2012, is shown.*

1 Forecasting, warning and satellite technology

Forecasting is just as essential for the USA as for Bangladesh. The USA has no shortage of **satellite technology**; over 20 weather satellites operate every day. Weather forecasts are frequent, and any warnings issued on TV and radio. Almost everyone has access to media, and mobile phone ownership is high (134 phones per 100 people). There is a National Hurricane Center in Miami, whose job is to:

- issue forecasts and warnings of hazardous weather
- educate people about tropical cyclones.

But the system is not perfect. Satellites are ageing, and in October 2012 one failed to work when Hurricane Sandy (see Figure 1) was first developing. A back-up satellite took over, but its software was out of date, so it gave only broad predictions about Sandy's likely track. A correct prediction of its track was actually made by a weather forecasting centre in Reading, UK! It was these warnings that were used to prepare people in New York, where Sandy proved to be the second-costliest storm in US history because of damage it caused.

▼ ***Figure 2*** *Storm surge risk zones in Fort Myers, Florida*

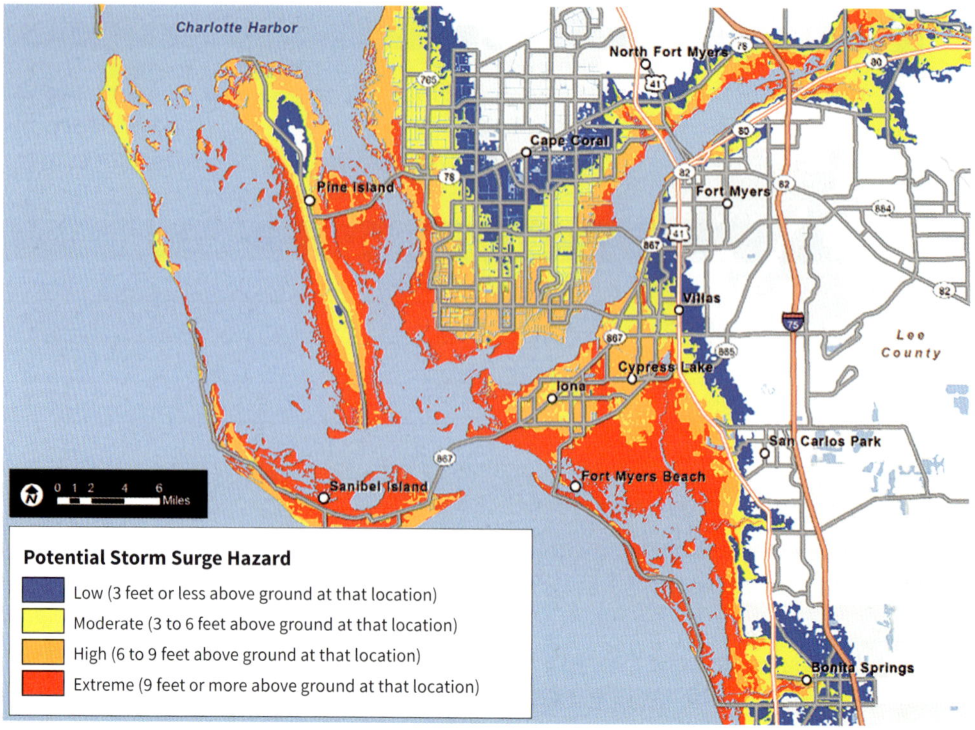

2 Risk and evacuation

The USA has cyclone **warning and evacuation systems** in place to help plan evacuations. In Florida, towns and cities such as Fort Myers (Figure 2) are classified into **risk zones**.

Areas are assessed for risks from high winds or storm surges. Only people who need to leave are evacuated and emergency services concentrate on getting people out without being overwhelmed.

3 Storm surges and defences

In 2005, Hurricane Katrina became the costliest and one of the deadliest hurricanes in US history, killing 1833 people in New Orleans and along the Louisiana coast, and causing US$108 billion in damage.

- The deaths were caused when a 4m-high storm surge flooded 80% of New Orleans. Flooding was caused by the collapse of artificial river embankments (called levées); 700 people drowned in one suburb.
- Government spending cuts had left the levées poorly maintained.
- A million people were evacuated. But those left behind were the poor, the elderly in care homes, the homeless and prisoners abandoned by evacuating guards.

In 2021, Hurricane Ida became the USA's second strongest and sixth costliest hurricane, creating a storm surge and causing over US$50 billion in damage. There were 95 deaths, and over half of these occurred in floods in New York and New Jersey caused by torrential rains as the storm moved north. Embankments held after repairs following Katrina, and evacuation plans moved most people from areas at risk. The successful repairs helped to limit the number of casualties.

▲ ***Figure 3*** *Flooding in New Orleans caused by Hurricane Katrina, 2005*

Your questions

1 Explain the value of a storm surge risk map like the one in Figure 2.

2 Outline the economic, social and environmental impacts of Hurricane Katrina on the USA.

3 Make a large copy of the table and compare the successes and challenges of cyclone warning and protection systems in the USA.

	Its successes	Its future challenges
Weather forecasting		
Satellite technology		
Warning systems		
Evacuation strategies		
Storm surge defences		

4 Take each warning and protection system in turn. In pairs, discuss whether you think the USA or Bangladesh protects people better against tropical cyclones. Feed your ideas back.

Exam-style question

5 Evaluate the effectiveness of different methods of cyclone preparation and response in **one** named developed country. (8 marks)

▲ ***Figure 4*** *Crews restoring power to parts of New Orleans after Hurricane Ida*

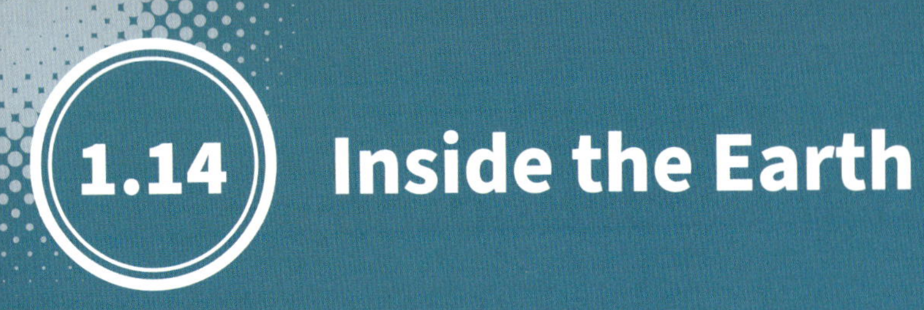

1.14 Inside the Earth

In this section, you'll understand the Earth's structure, its different compositions and physical properties.

Journey to the centre

No one has ever seen the inside of the Earth. The deepest we have managed to get is 3.5 km in the Witwatersrand gold mine in South Africa. Miners would have to drill another 6365 km to reach the Earth's centre!

What we know about the Earth's interior comes from direct and indirect evidence. We can get direct evidence from the Earth's surface. Indirect evidence, like earthquakes and material from space, also helps us to understand the Earth.

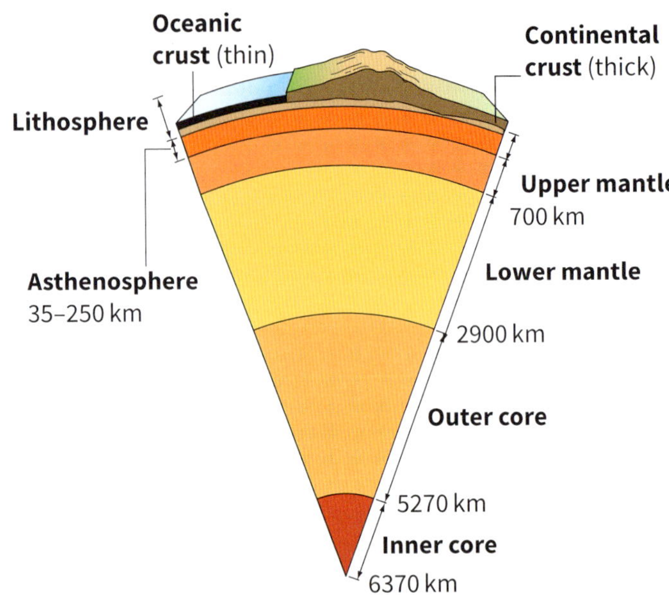

▲ *Figure 1 The structure of the Earth. Geologists (scientists who study the Earth and its structure) believe the Earth has a layered structure, like an onion*

Examining the crust

Figure 1 shows the structure of the Earth. The crust forms the Earth's surface. It is a rock layer forming the upper part of the **lithosphere**. The lithosphere is split into **tectonic plates**. These plates move very slowly, at 2-5 cm per year, on a layer called the **asthenosphere**.

The **lithosphere** is the uppermost layer of the Earth. It is cool and brittle. It includes the very top of the mantle and, above this, the crust.

Did you know?

The Earth is 4600 million years old. But just how old is this? Compress it into 24 hours and humans have only occupied the last few seconds.

There are two types of crust:

- **Continental crust** forms the land. This is made mostly of granite, shown in Figure 2, which is a low density igneous rock. Continental crust is on average 30–50 km thick.
- Under the oceans is **oceanic crust**. This is much thinner, usually 6–8 km thick. It is also denser and made of an igneous rock called basalt.

◀ *Figure 2 Geologists collect samples of the crust and compare the two types: polished granite from continental crust (left) and basalt from oceanic crust (right)*

The asthenosphere and mantle

The movement of the tectonic plates is evidence that there is a 'lubricating' layer underneath the lithosphere. This is the asthenosphere. You might think that this layer would be a liquid. But if it was a liquid, the heavy tectonic plates would sink into it. Geologists think the asthenosphere is partly molten rock and partly solid rock. It may be like very thick, dense, hot porridge! We need to learn more from research like that done on the drilling ship in Figure 3.

The asthenosphere is in the top layer of the **mantle**. The mantle is the largest of the Earth's layers by volume, and is mostly solid rock. We know this because sometimes you can see the top of the mantle attached to an overturned piece of crust.

Earthquake waves tell us about the physical state of the Earth. They speed up, change direction or stop when they meet a new layer in the Earth. It's mainly due to this that we know anything about the **core**. Some waves travel easily through the crust, mantle and inner core, but not through the outer core. This suggests that the outer core has a different physical state and may be liquid, not solid. What we do know is shown in Figure 4.

▲ ***Figure 3*** *Japan built a 57 000 tonne international scientific drilling ship, the Chikyu, which drills 7 km through the oceanic crust to reach the mantle – further than anyone has ever gone before*

Layer		Density (grams/cm³)	Physical state	Composition	Temp (°C)
Lithosphere	Continental crust	2.7	Solid	Granite	Air temp – 900°C
	Oceanic crust	3.3	Solid	Basalt	Air temp – 900°C
Mantle	Asthenosphere	3.4–4.4	Partially molten	Peridotite	900–1600°C
	Lower mantle	4.4–5.6	Solid		1600–4000°C
Core	Outer core	9.9–12.2	Liquid	Iron and nickel	4000–5000°C
	Inner core	12.6–13.0	Solid	Iron and nickel	5400°C

◀ ***Figure 4*** *The properties of the Earth's layers*

Clues from space

At a depth of 2900 km, we can't sample the Earth's core. Geologists think that it is metal – mostly nickel and iron. Evidence for this comes from **meteorites**, which are fragments of rock and metal that fall to Earth from space. Most come from the asteroid belt between Mars and Jupiter.

Meteorites come in several types:

- Stony meteorites, with a similar composition to basalt.
- Stony-iron meteorites, containing a lot of the mineral olivine.
- Iron meteorites, which are solid lumps of iron and nickel.

These meteorites may be fragments of the lithosphere, mantle and core of a shattered planet. Iron meteorites may show that the Earth's core is made up of iron and nickel.

Your questions

1 Draw a cross-section of the Earth. You need a large circle divided into layers.

- **a** Label the layers with details of density, temperature and physical state.
- **b** Label each layer with a text box labelled 'Evidence'. In each one, say *how* we know about the different layers.

2 Explain the main differences between the lithosphere and asthenosphere.

Exam-style questions

3 Define the term 'tectonic plates'. (1 mark)

4 Explain **one** difference between oceanic and continental crust. (3 marks)

5 Explain **two** ways in which we know about the composition and physical properties of the Earth's structure. (4 marks)

1.15 The Earth's heat engine

In this section, you'll understand how the core's internal heat source generates convection, the foundation for plate tectonics.

Hot rocks

Inside the Earth it is hot. We know this because of:

- molten lava spewing from active volcanoes
- hot springs and geysers.

Heat from inside the Earth is called **geothermal** ('earth-heat'). The heat is produced by the **radioactive decay** of elements such as uranium and thorium in the core and mantle. This raises the core's temperature to over 5000 °C.

Some elements are naturally unstable and radioactive. Atoms of these elements release particles from their nuclei and give off heat. This is called **radioactive decay**.

Did you know?

In a few billion years the core and mantle will stop convecting because the radioactive heat will run out. This will shut down our magnetic field, and life on Earth will be destroyed by radiation from space.

Did you know?

Did you know that the Earth's magnetic field sometimes 'flips', so north becomes south and south becomes north?

The inner core is so deep and is under such huge pressure that it stays solid. The outer core is liquid because it is under lower pressure. As heat rises from the core, it creates **convection currents** in the liquid outer core and mantle (see Figure 1). These vast mantle convection currents are strong enough to move the tectonic plates on the Earth's surface. Earth scientists now think that the convection currents that move the Earth's plates are not as deep-seated as shown in Figure 1. However, they are still forceful enough to move very large 'slabs' of crust. The convection currents move about as fast as your fingernails grow. Radioactivity in the core and mantle is the engine of plate tectonics.

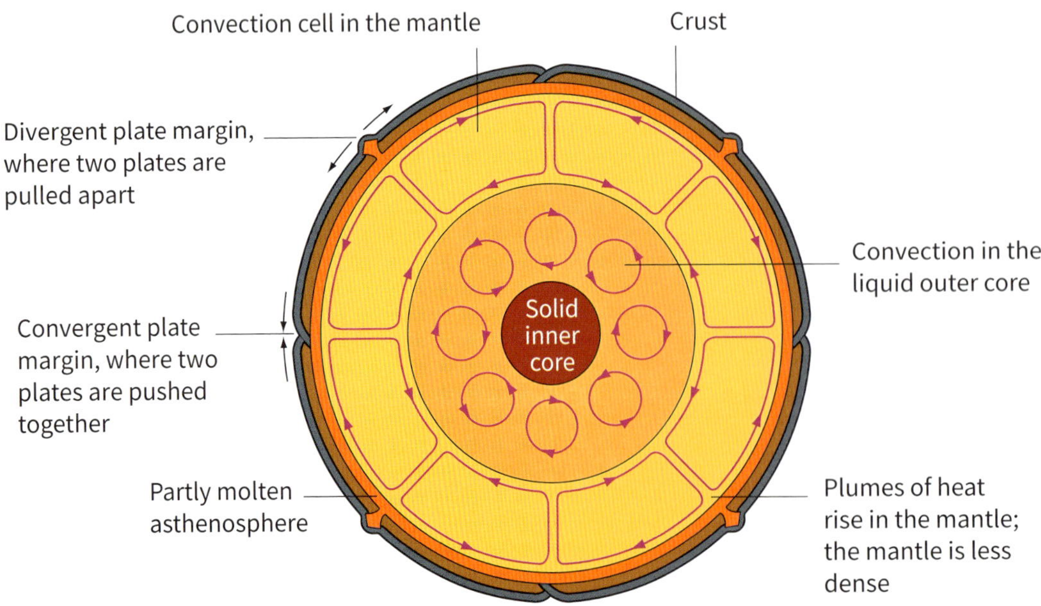

▲ ***Figure 1*** *How plate tectonics is driven by convection currents*

Plumes

The parts of convection cells where heat moves towards the surface are called **plumes**. These are concentrated zones of heat. In a plume, the mantle is less dense. Plumes bring **magma** (molten rock) to the surface. If magma breaks through the crust, it erupts as **lava** in a volcano.

- Some plumes rise like long sheets of heat. These form **divergent plate boundaries** at the surface.
- Other plumes are like columns of heat. These form **hot spots**. Hot spots can be in the middle of a tectonic plate, like Hawaii and Yellowstone in the USA.

▲ ***Figure 2*** *The northern lights, or aurora borealis*

Magnetic field

The Earth is surrounded by a huge invisible magnetic field called the magnetosphere. This is a force field which you can sometimes see, better known as the northern lights or ***aurora borealis***, shown in Figure 2. These form when radiation from space hits the magnetosphere and lights up the sky. The magnetosphere protects the Earth from harmful radiation from space and the sun (see Figure 3).

The Earth's magnetic field is made by the outer core. As liquid iron in the outer core flows, it works like an electrical dynamo. This produces the magnetic field.

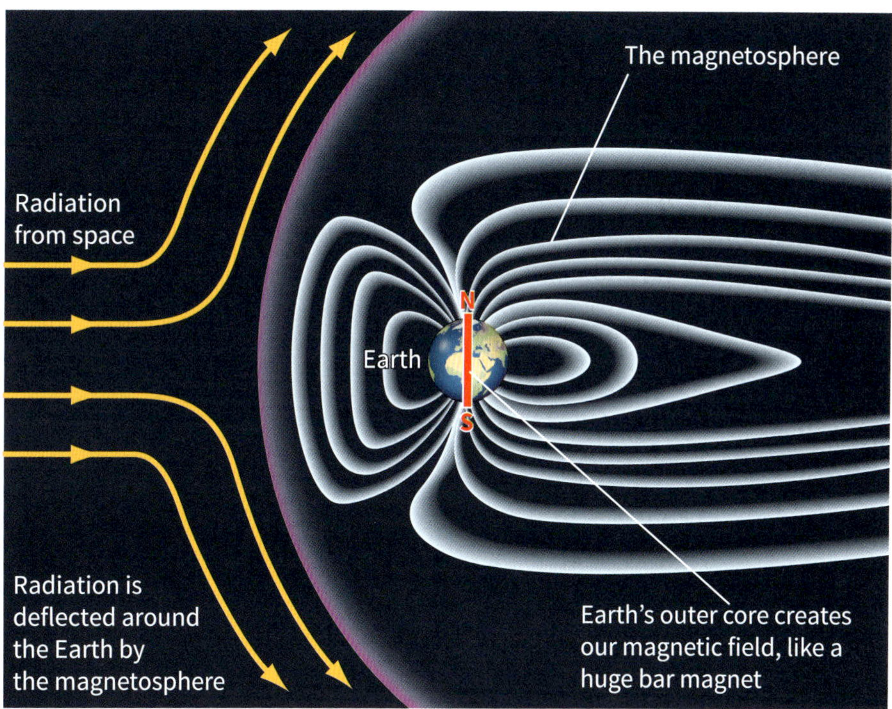

▲ ***Figure 3*** *How the magnetosphere protects the Earth from harmful radiation*

Your questions

1. Explain why the centre of the Earth is over 5000°C.
2. Look at the diagram of the Earth's convection currents. What happens to the crust at the top of one of the convection currents in the mantle?
3. Each set of words below has an odd one out. For each, explain which is the odd one out:
   - inner core, outer core, mantle, crust
   - convection, northern lights, plume, cell, current

Exam-style questions

4. Define the term 'geothermal'. (1 mark)
5. Explain the difference between magma and lava. (2 marks)
6. Other than the mantle, explain the properties of **two** of Earth's internal layers. (4 marks)

1.16 Plate tectonics

In this section, you'll understand how the Earth's tectonic plates have moved in the past, and about plate boundaries.

Pangea, the supercontinent

Scientists know that the continents were once all joined together. They formed a supercontinent called Pangea. Figure 1 shows the position of the continents 250 million years ago. Identical rocks and fossils dating from this time have been found in West Africa and eastern South America. This tells us that Africa and South America were once joined. Pangea started to split apart about 200 million years ago. Since then, plate tectonics has moved the continents to the positions they are in today.

▲ ***Figure 1*** *Pangea, the original supercontinent*

Moving plates

Today, the Earth's lithosphere is split into 15 large **tectonic plates** and over 20 small ones. These are like the patches that make up a football. The plates move very slowly on the asthenosphere. Where two plates meet, there is a **plate boundary**. There are three types of plate boundary, as shown in Figure 3:

- Divergent plate boundaries – formed when two plates move apart.
- Convergent plate boundaries – formed when two plates collide (Figure 2).
- Conservative plate boundaries – formed when two plates slide past each other.

Plate boundaries are where the 'action' is. Most earthquakes and volcanoes are found on plate boundaries.

Did you know?

Every year, the distance between the UK and the USA grows by about 2 cm. This is because the mid-Atlantic divergent plate boundary creates new oceanic crust.

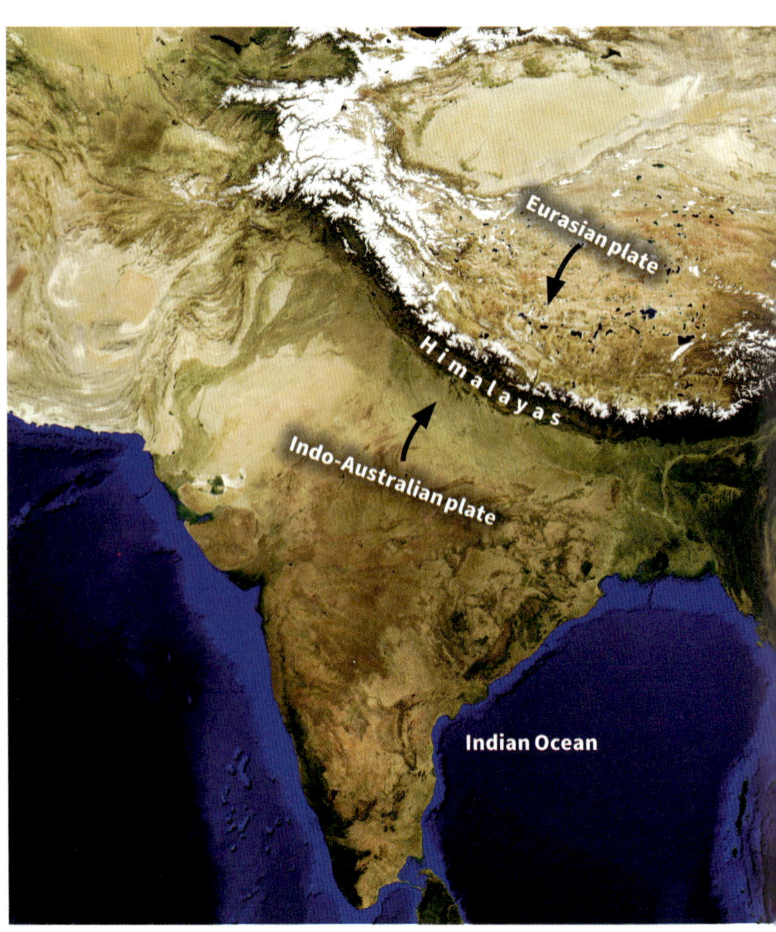

▶ ***Figure 2*** *The Himalayas – a collision between the Indo-Australian and Eurasian plates*

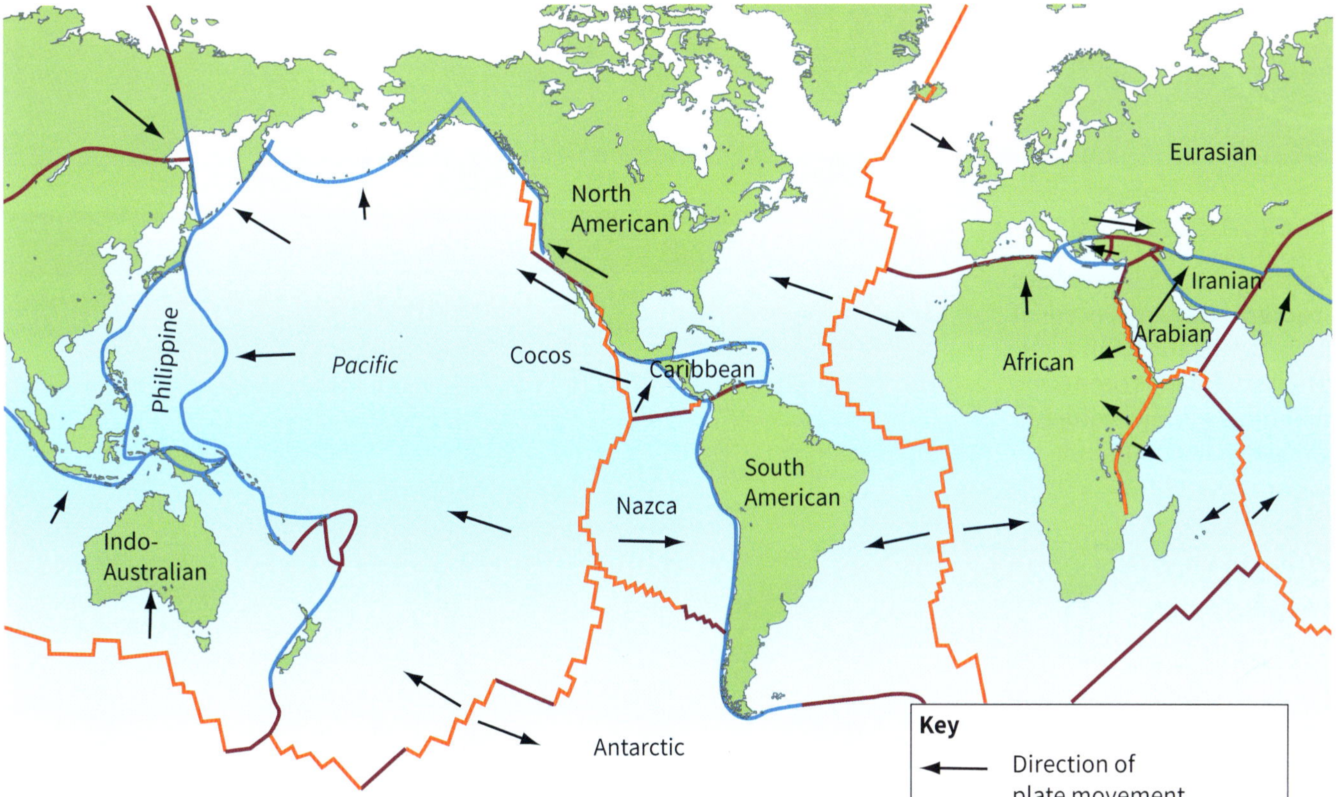

▲ ***Figure 3*** *The world's tectonic plates*

Crust: old and new

Most continental crust is 3–4 billion years old. The oldest oceanic crust is only 180 million years old. Why the age difference?

New oceanic crust forms constantly at divergent plate boundaries:

- Convection currents bring magma up from the mantle.
- The magma is injected between the separating plates.
- As the magma cools, it forms new oceanic crust.
- The plates continue to move apart, allowing more magma to be injected.

Old oceanic crust is destroyed by **subduction** at convergent plate boundaries – it is 'recycled' by the Earth. Continental crust was formed billions of years ago, and has not formed since. It is less dense than oceanic crust, so can't be subducted and destroyed.

Subduction describes oceanic crust sinking into the mantle at a convergent plate boundary. As the crust subducts, it melts back into the mantle.

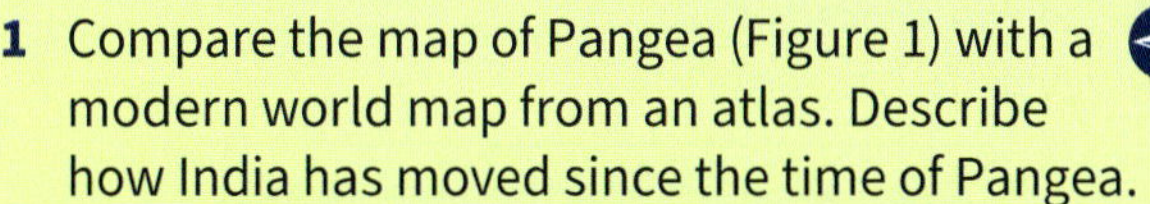

Your questions

1 Compare the map of Pangea (Figure 1) with a modern world map from an atlas. Describe how India has moved since the time of Pangea. 

2 Look at the map of tectonic plates in Figure 3.
   - a Which plate is the UK on?
   - b Name a country which is split by two plates. 
   - c Name two plates that are moving apart. 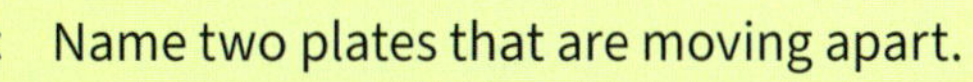
   - d Name two plates that are colliding. 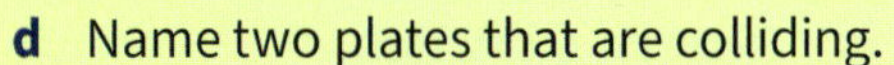

Exam-style questions

3 Define the term 'plate boundary'. (1 mark)

4 Study Figure 3. Name **one** example of: 
   - a a divergent plate boundary
   - b a convergent plate boundary. (2 marks)

5 Study Figure 3. Explain the location of divergent plate boundaries. (4 marks)

6 Explain the process of subduction along convergent plate boundaries. (4 marks)

1.17 Boundary hazards

In this section, you'll understand which hazards happen at different plate boundaries.

Tectonic hazards

Earthquakes and volcanoes (**tectonic hazards**) occur at plate boundaries. Different plate boundaries produce different tectonic features (shown in Figure 1) and hazards.

Conservative boundaries

As plates slide past each other, friction between them causes earthquakes. These are rare but very destructive, because they are shallow (close to the surface). The San Andreas fault is shown in Figure 2.

▶ ***Figure 1*** *Features of plate boundaries*

Plate boundary	Example	Earthquakes	Volcanoes
Conservative	San Andreas fault in California, USA. North American and Pacific plates sliding past each other.	• Destructive earthquakes up to magnitude 8.5. • Small earth tremors almost daily.	No volcanoes.
Divergent	Iceland, on the mid-Atlantic ridge. The Eurasian and North American oceanic plates pulling apart.	• Small earthquakes up to 5.0–6.0 on the Richter scale.	• Not very explosive or dangerous. • Occur in fissures (cracks in the crust). • Erupt basalt lava at 1200 °C.
Convergent	Andes mountains in Peru and Chile. Nazca oceanic plate is subducted under the South American continental plate.	• Very destructive, up to magnitude 9.5. • Tsunami can form.	• Very explosive, destructive volcanoes. • Steep sided, cone-shaped. • Erupt andesite lava at 900–1000 °C.
Collision zone	Himalayas. Formed as the Indian and Eurasian continental plates push into each other.	• Destructive earthquakes, up to magnitude 9.0. • Landslides are triggered.	Volcanoes are very rare.

▼ ***Figure 2*** *The San Andreas fault – a conservative plate boundary*

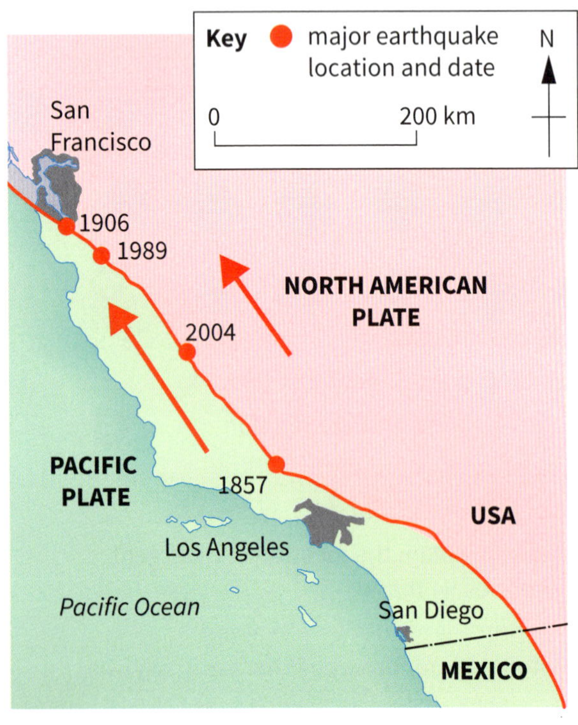

Collision zones

Collision zones are a type of convergent boundary. They form mountain ranges like the Himalayas (see Figure 3). Two continental plates of low-density granite collide, pushing up mountains. Earthquakes happen on faults (huge cracks in the crust) in collision zones. An example is the 2015 earthquake in Nepal (see Section 1.21).

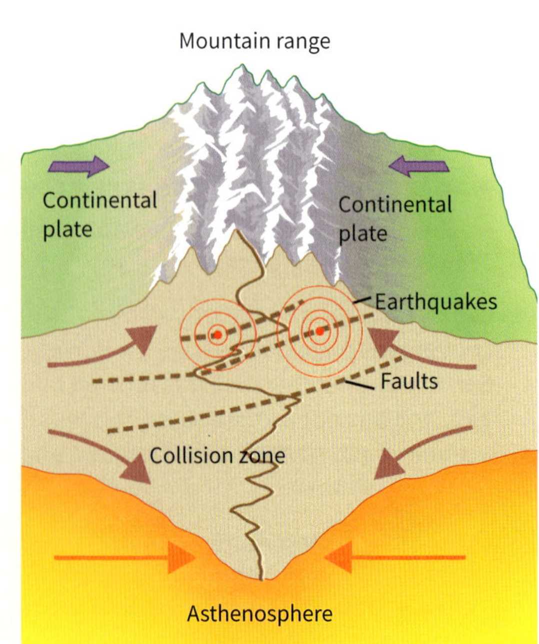

▶ ***Figure 3*** *A collision zone boundary*

Divergent boundaries

As plates move apart, magma rises up through the gap, as Figure 4 shows. The magma is **basalt** and is very hot and runny. It forms **lava flows** and shallow sided volcanoes. Earthquakes are caused by **friction** as the plates tear apart. These earthquakes are small. They don't cause much damage.

Tectonic hazards are natural events caused by movement of the Earth's plates that affect people and property.

Convergent boundaries

As the plates push together (see Figure 4), oceanic plate is **subducted**. As it sinks, it melts and creates magma called **andesite** (after the Andes). Sea water is dragged down with it. This makes the magma less dense so it rises in **plumes** (or 'blobs') upwards through the continental crust. The water erupts as steam, making volcanoes very explosive. Volcanic ash and 'bombs' are blasted up and outwards as **pyroclasts**.

Did you know?

Every year there are about 100 000 earthquakes strong enough to be felt. The largest earthquake recorded was a magnitude 9.5 in Chile in 1960.

Sinking oceanic plate can stick to the continental plate. Pressure builds up against the friction. When the plates finally snap, energy is released as a violent earthquake. These earthquakes can be devastating, especially if they are shallow.

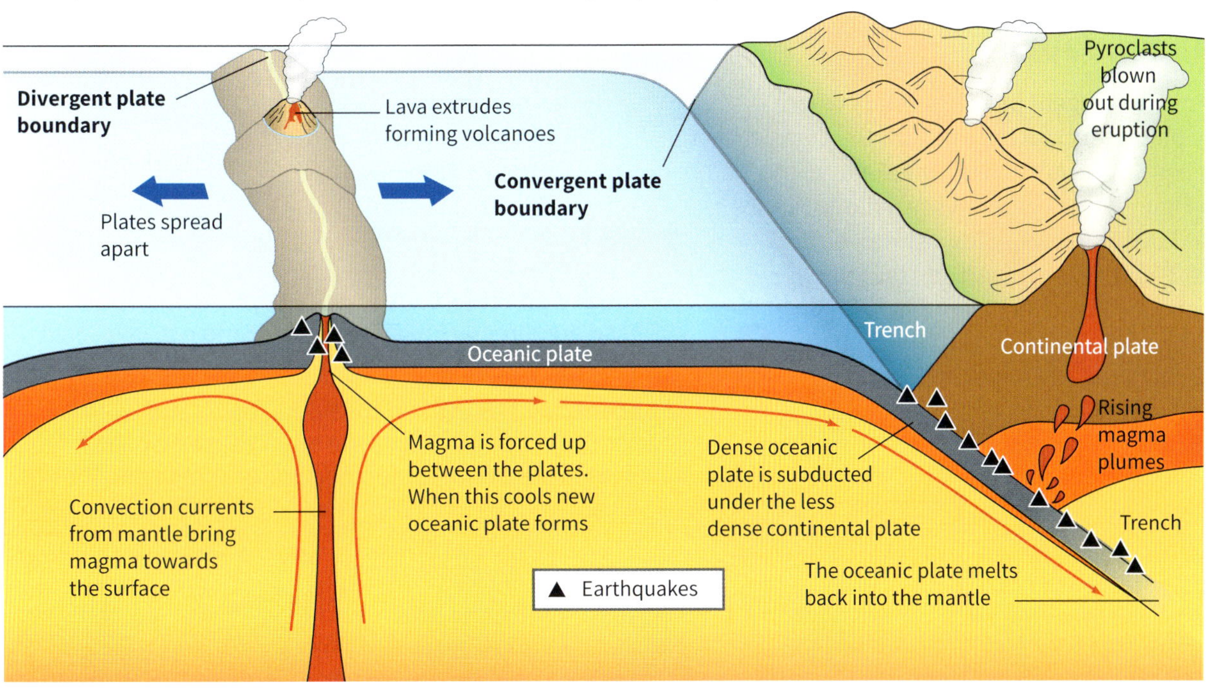

▲ ***Figure 4*** *Comparing the hazards along divergent and convergent plate boundaries*

Your questions

1 Which type of plate boundary is most dangerous for humans to live on? Give your reasons.

2 Match the words below into pairs: divergent, fault, collision, fissures, landslides, explosive, convergent, conservative
Write a brief explanation of your pairs.

3 Look at Figure 2, the San Andreas fault. The plates can 'lock', preventing sliding. Suggest **a** why this occurs, **b** why earthquakes here can be severe.

Exam-style questions

4 Explain the term 'collision zone'. (2 marks)

5 Explain why some volcanoes are more steep-sided than others. (4 marks)

6 Explain how volcanoes may form along divergent plate boundaries. (4 marks)

7 Suggest **two** reasons why some plate boundaries are more hazardous than others. (4 marks)

1.18 Volcanoes in the developed world

In this section, you'll assess the impact of volcanoes on developed countries.

Destructive power

The most devastating volcanoes are the most explosive ones. The **Volcanic Explosivity Index** (VEI) measures destructive power on a scale from 1 to 8. Mount St Helens, which erupted in May 1980, measured 4. Modern humans have never experienced an eruption measuring 8.

Volcanoes produce many hazards (see Figure 1). Some are from the **primary effects** of the volcano, whereas others are **secondary**. Some benefit the areas affected, others cause problems.

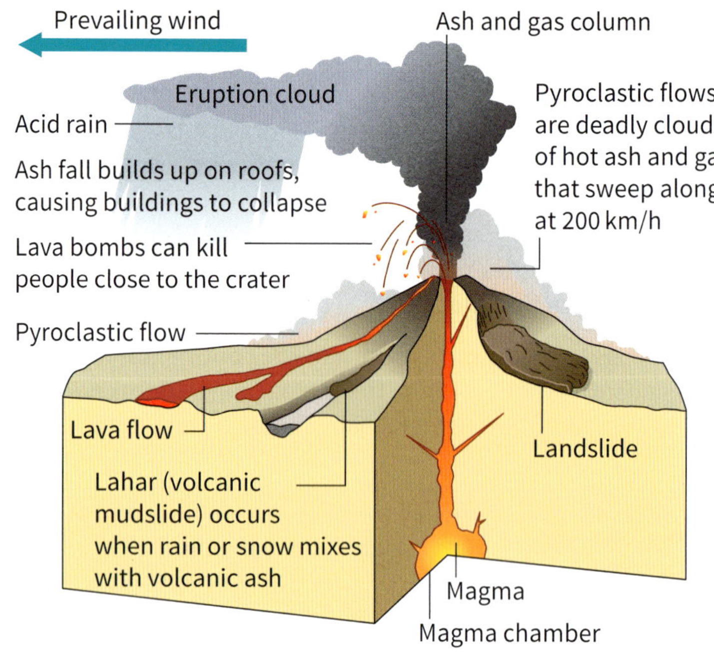

▲ ***Figure 1*** *Volcanic hazards*

Primary effects – caused instantly by the eruption. These are directly linked to the volcano e.g. lava, acid rain, gases and earthquakes.

Secondary effects – in the hours, days, and weeks after the eruption. These are often caused by the volcano e.g. disease, food and water shortages.

Sakurajima, Japan

Japan is on a convergent plate boundary where the Pacific plate is subducted beneath the Eurasian plate, causing active volcanoes. One, Sakurajima (shown in Figure 2), has erupted since the 1950s, sometimes 200 times a year. It's a stratovolcano (see Figure 3) with layers of ash and lava burying buildings and farmland. Poisonous gases cause alerts locally and bring acid rain, killing plants.

▼ ***Figure 2*** *Sakurajima – some benefits and problems caused by the volcano*

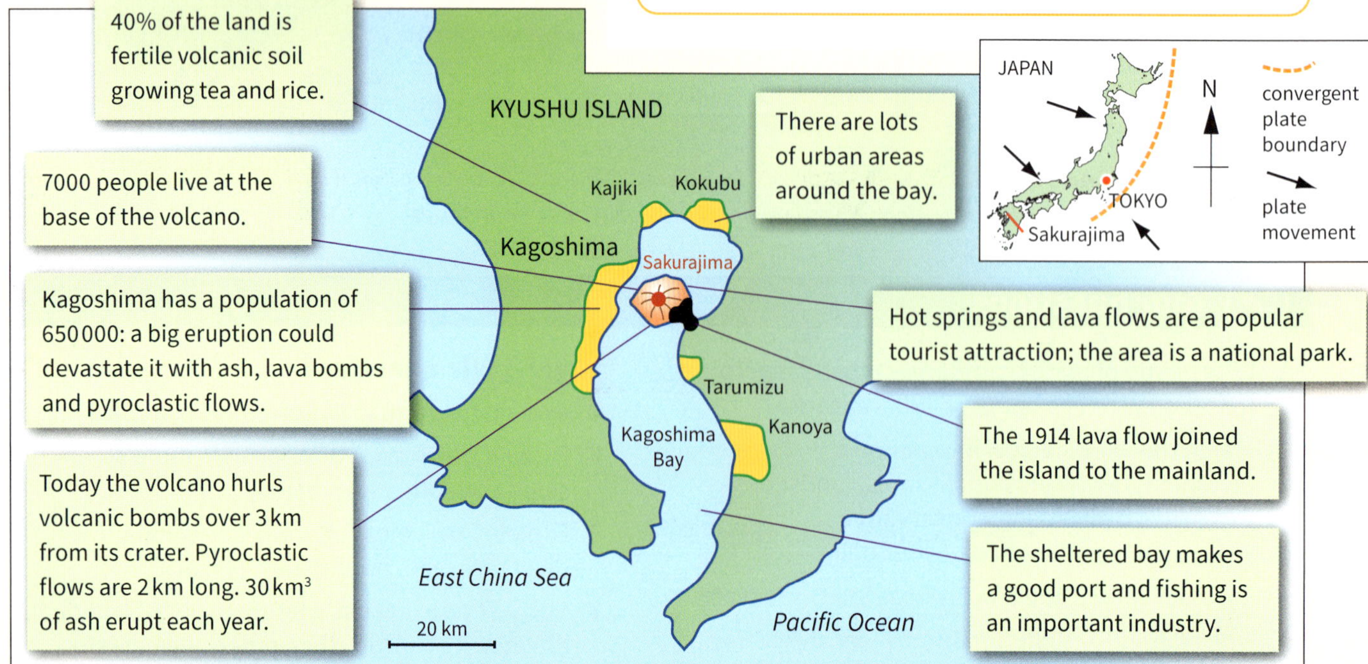

◀ Its eruptions can be **predicted** – scientists can say when it will erupt, then warn people to take shelter or **evacuate**. Figure 4 shows how Sakurajima is monitored and its evacuation procedures.

Japan is a developed country. It can afford to spend money on monitoring, protection (like the shelter in Figure 5) and evacuation. When Sakurajima erupts, it rarely causes deaths. Homes, crops and industries are destroyed, but most people have insurance and the Government helps to repair damage. In developed countries, tectonic hazards damage property (economic costs) but cause less harm to people (social costs).

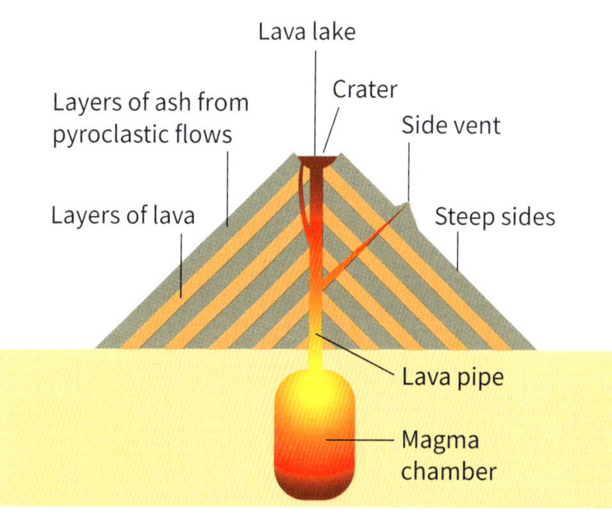

▶ ***Figure 3*** *A stratovolcano, also called a composite cone*

Aircraft are used to measure the amount of gas the volcano gives off

A tunnel in the volcano has seismometers. These monitor earthquakes which increase as magma rises so scientists can predict an eruption

Tiltmeters detect when the volcano swells up as it fills with magma

Concrete lahar channels divert dangerous mudflows

Boreholes measure water temperature as magma heats it up

Hot springs are monitored

Concrete shelters protect against volcanic bombs and ash

Evacuation routes clearly sign posted; regular evacuation drills

◀ ***Figure 4*** *How Sakurajima is monitored, and its evacuation procedures*

▲ ***Figure 5*** *An emergency shelter in Kagoshima, Japan. Sakurajima is in the background.*

Your questions

1 Copy and complete the following table to show the effects of Sakurajima:

	Benefits	Problems
Primary effects		
Secondary effects		

2 Make a table like the one below:

Protection	Prediction

Use it to list methods used to protect people from Sakurajima, and how scientists predict an eruption.

Exam-style questions

3 Explain the difference between 'primary' and 'secondary' effects of a volcanic eruption. (3 marks)

4 Explain **two** ways in which volcanic eruptions can be predicted. (4 marks)

5 Explain **two** ways in which people affected by volcanic eruptions can prepare for this hazard. (4 marks)

6 Assess the impact of volcanic eruptions on **one** developed country. (8 marks)

1.19 Developing world volcanic hazards

In this section, you'll assess the impact of volcanic eruptions for people in the developing world.

At risk

Most volcanic eruptions with high death tolls occur in poorer countries, where people are at greater risk.

- Building often takes place in risky locations, because there are few affordable places to live.
- Safe, well-built houses are expensive, so the poor live in cheaper buildings which might collapse.
- Insurance is unaffordable for many.
- Governments rely on overseas **aid** funding for warning systems.
- Road and tele-communication systems are poor, so evacuation warnings may not be heard.

Eruptions are infrequent, but may last over several years. The following example of Nyiragongo began in 2002, but continued into 2021.

▲ ***Figure 1*** *Lava destroyed 40% of Goma, including 45 schools, and covered half the airport. Water and electricity supplies were also cut off by the lava.*

Mount Nyiragongo, Democratic Republic of the Congo

In January 2002, a fast flowing river of basalt lava, 1000 km wide, poured out of Mount Nyiragongo (a large stratovolcano) and into the city of Goma (see Figure 1). 100 people died, mostly from poisonous gas and getting trapped in lava. A number of **social impacts** also resulted.

- 12 500 homes were destroyed by lava flows and earthquakes. As the eruption was predicted, 400 000 people were evacuated. Many people had to move to overcrowded **refugee** camps.
- Disruption to the mains water supplies caused concern about the spread of diseases.

There were many **economic impacts**:

- Poisonous gases caused acid rain which affected farmland and cattle – many farmers lost income.
- Due to poverty, most people could not afford to rebuild their homes.

Within days, people began returning to Goma. The main problem was poverty. Having fled with nothing, over 120 000 were homeless and needed help; with little clean water, food or shelter,

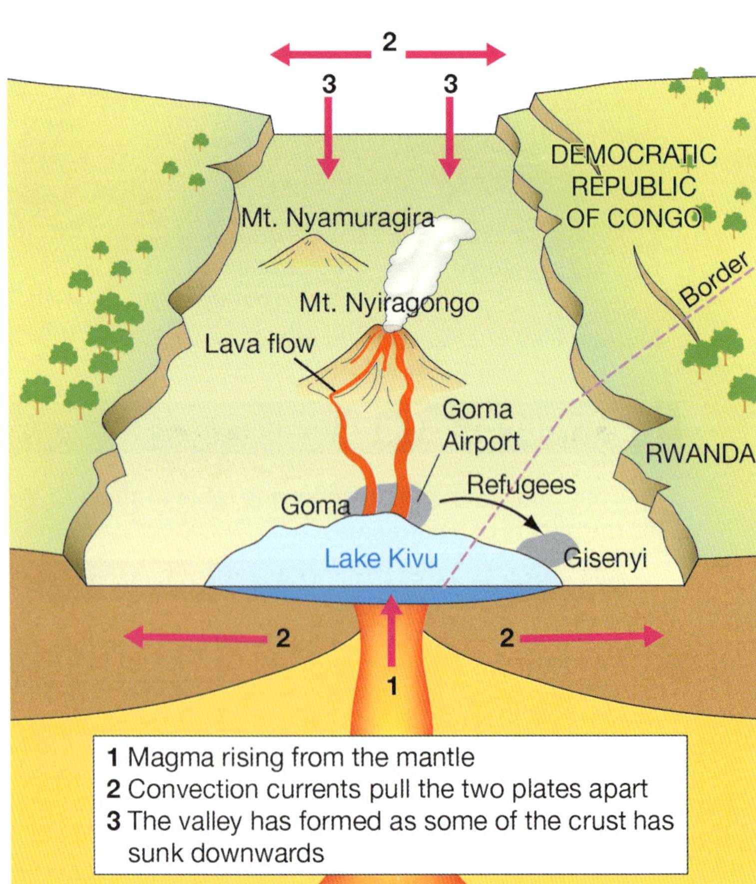

▲ ***Figure 2*** *The threats from Mt Nyiragongo and Lake Kivu*

diseases like cholera could spread. The United Nations (UN) and Oxfam began a **relief effort**.

- The UN sent 260 tonnes of food in the first week. Families got 26kg of rations.
- In the UK, a TV appeal raised money, and governments around the world gave US$35 million to supply aid.

Future threats

The World Bank funded a monitoring programme by the Goma **Volcano Observatory** to warn people nearby. The population living near the volcano has doubled to 1.5 million since 2002, and the volcano is still active.

- In 2016, the observatory discovered an active vent, following rumblings from the volcano.
- In 2020, a rise in the lava lake led the observatory to predict an eruption between 2024 and 2027.
- The volcano erupted in 2021. 32 people died – mostly in traffic accidents as people rushed to evacuate the area following predictions.

Now the World Bank has withdrawn funding, threatening the observatory.

There is also volcanic activity under Lake Kivu (Figure 2), though its threats differ from Sakurajima (see Figure 3). Gases like carbon dioxide and sulphur dioxide rise and get trapped in lake muds, which earthquakes shake free. In 2005, this happened to volcanic Lake Nyos in Cameroon. 1700 people suffocated from breathing in too much carbon dioxide.

	Mt Sakurajima, Japan	Mt Nyiragongo, DRC
Volcano type	Steep-sided stratovolcano (or composite cone) over 1000 m high.	Stratovolcano over 3400 m, high but less steep than Mt Sakurajima.
Magma type	Andesite. High gas content, high viscosity.	Basalt. Low gas content, very low viscosity.
Explosivity	VEI 4-5	VEI 1
Hazards	Lava flows, volcanic bombs, pyroclastic flows, ash fall. Erupts almost continually, but with major eruptions once every 200–300 years.	Lava flows and gas emissions. Contains a lava lake within its crater, which can drain causing huge, fast-moving lava flows.

▲ ***Figure 3*** *Comparing Mt Sakurajima and Mt Nyiragongo*

Aid is help. It can be short-term such as food given in emergency, or long-term such as training in health care.

Social impacts are impacts on people.

Refugees are people who are forced to move due to natural hazards or war.

Economic impacts are impacts on the wealth of an area.

A **relief effort** is like aid. It is help given by organisations or countries to help those facing an emergency.

A **volcano observatory** is a centre in which Earth scientists use monitoring equipment (e.g. seismographs) to try to predict further eruptions from tectonic activity.

Your questions

1. Explain in your own words what we mean by aid and relief effort.
2. **a** Draw a table identical to the one for Question 1 on page 43, and complete it to show the effects of Mount Nyiragongo.

   b Which volcano seems to have the greatest effects – Nyiragongo or Sakurajima? Explain your answer.
3. How successful was the relief effort in helping people affected by Mount Nyiragongo's eruption? Give reasons.
4. Suggest reasons why people still live around Mount Nyiragongo and Lake Kivu.

Exam-style question

5. Define the term 'refugee'. (1 mark)
6. Evaluate the view that the impacts of volcanic eruptions on developing or emerging countries are greater than those on developed countries. (8 marks)

1.20 Earthquake!

In this section, you'll understand how earthquakes are measured and assess their awesome power.

Why is the ground shaking?

Earthquakes can't be predicted. They start without warning and can be catastrophic. An earthquake is a sudden release of energy. Underground, tectonic plates try to push past each other along fractures – building up pressure which is suddenly released, sending out pulses of energy.

The power of an earthquake (how much the ground shakes) is its **magnitude**. A **seismometer** measures this using the **Richter scale** (Figure 1).

The scale is logarithmic. A magnitude 6.0 quake is 10 times more powerful than a magnitude 5.0.

Energy travels outwards from the **focus** (origin) as earthquake waves, see Figure 2. The shallower the focus, the more destructive the earthquake. The location on the earth's surface above the focus is the **epicentre**, which experiences most shaking.

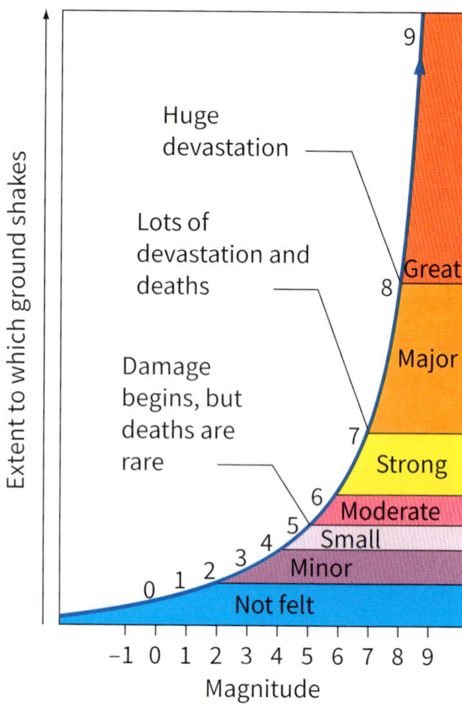

▲ ***Figure 1*** *The Richter scale*

Tsunami

Earthquakes beneath the sea bed can generate a tsunami (see Figure 3). Tsunamis are waves that travel at up to 900 km/h, with wavelengths of over 200 km. In the open ocean, wave height is less than 1 m, but as the waves approach the coast they slow down, bunch up and wave height increases up to 30 m. When a tsunami hits, it causes a very powerful flood, pushing several kilometres inland destroying homes, bridges and infrastructure. Warning systems in the ocean can detect tsunamis and set off sirens and alarms but this is only useful if the epicentre is some distance from the coast. All countries affected by the 2011 tsunami which struck Japan were warned by early warning systems in the Pacific Ocean.

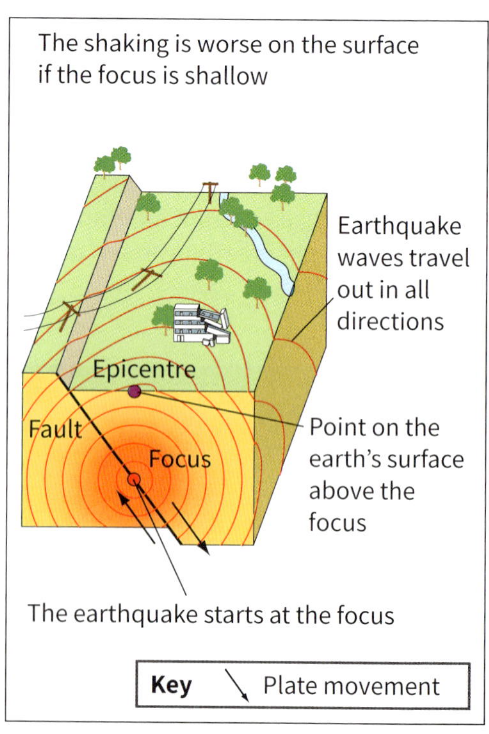

▲ ***Figure 2*** *The focus and epicentre of an earthquake*

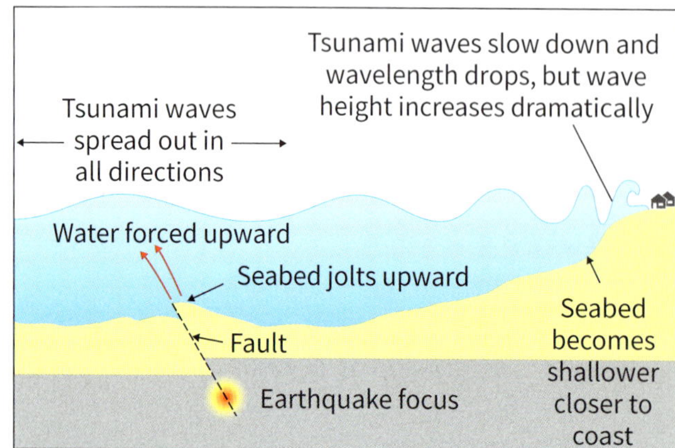

▲ ***Figure 3*** *How a tsunami forms*

Haiti and Japan

On 12 January 2010, a magnitude 7.0 earthquake struck Haiti. On 11 March 2011, a much larger earthquake (9.0) hit the region of Sendai in Japan. However, the earthquakes had very different causes and effects.

Sendai, Japan 2011	Port-au-Prince, Haiti 2010
• Magnitude – 9.0 • Focus – 30 km deep on a convergent plate boundary • Epicentre – 70 km from the coast in Sendai Bay	• Magnitude – 7.0 • Focus – 13 km deep on a conservative plate boundary • Epicentre – 25 km from Port-au-Prince (population 2.5 million)
Primary effects • 1 dam collapsed, 2 nuclear power stations fractured, and an oil refinery set on fire by damaged gas pipes. • Tohoku motorway badly damaged in northern Japan. Sendai airport closed by the tsunami. One rail link near Sendai badly damaged. • US$ 235 billion of damage caused by earthquake and tsunami combined – the costliest disaster in history.	**Primary effects** • 316 000 people died and a further 300 000 were injured, though the data are contested. Different sources show different totals. • Many houses were poorly built and collapsed instantly. 1 million people made homeless. • The port, communication links and major roads were damaged beyond repair. Rubble from collapsed buildings blocked road and rail links.
Secondary effects (caused by the tsunami) • 15 900 people died, 2600 missing, 6150 injured, 350 000 homeless. Most were at work or school when it struck at 2:46 pm. • 93% of deaths caused by drowning. • Two nuclear reactors went into meltdown because flooding damaged the cooling systems. Most people displaced by the earthquake were rehoused away from the area. • Businesses disrupted by damage, clearance and rebuilding. • Homelessness, disrupted schooling, unemployment and increased stress lasted for years as the authorities struggled to cope with damage.	**Secondary effects** • The water supply system was destroyed – a cholera outbreak killed over 8000 people. • The port was destroyed — making it hard to get aid to the area. • Haiti's important clothing factories were damaged. These provided over 60% of Haiti's exports. 1 in 5 jobs were lost. • By 2015 most people displaced by the earthquake had been re-housed.

Long-term planning

The secret of survival is planning. Japan is a developed country, so it can afford to do this. There is a 70% **probability** (chance) of a magnitude 7.2 earthquake hitting Toyko in the next 30 years. There is no way of **predicting** when this might happen, but planning helps.

- Every year Japan has earthquake drills. Emergency services practise rescuing people. People keep emergency kits (water, food, a torch and radio) at home.
- Many buildings are earthquake-proof (see Figure 4). Gas supplies shut off automatically, reducing fire risk.
- Tsunami walls have been designed to protect the coast.

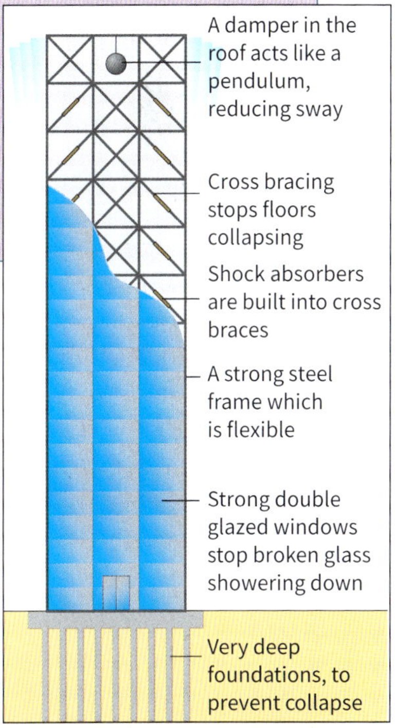

▶ ***Figure 4*** *Earthquake-proofed buildings*

Your questions

1 a Draw a large version of the Richter scale diagram.
b Label the strength of the Haiti and Sendai earthquakes on the diagram.
c How many times greater was the Sendai earthquake than the one in Haiti?

2 a Use a website, like Wikipedia, to find a list of earthquakes since 2015. Mark their magnitudes and death tolls on your diagram of the Richter scale.
b Is there a link between magnitude and death toll?

3 Using Figure 3, list the stages in the formation of a tsunami.

4 a Classify earthquake impacts in Sendai, Japan into social and economic.
b Which are greater? Explain.

Exam-style question

5 Assess the impacts of earthquakes on **one** developed country. (8 marks)

1.21 Earthquakes in the developing world

In this section, you'll assess the impacts of earthquakes on developing countries, and understand how people respond to them.

Death and destruction

Earthquakes in developing or emerging countries often have higher death tolls than volcanic eruptions. Some twenty-first century earthquakes have proven highly destructive (see Figure 1).

Location	Year	Deaths	Magnitude	Key facts
Sendai, Japan	2011	15900	9.0	The tsunami (secondary effect) caused most of the deaths. The economic costs were over US$235 billion.
Kashmir, South Asia	2005	86000	7.6	One-third of the deaths were due to landslides (secondary effect). Many children died in poorly built collapsed schools.
Aceh, Indonesia	2004	280000	9.3	Most deaths were caused by a tsunami (secondary effect) that hit 14 countries around the Indian Ocean.
Bam, Iran	2003	30000	6.6	Many people were trapped when their poorly built, mud brick homes collapsed in the densely populated city.

▲ ***Figure 1*** *Some of the most severe earthquakes of the 21st century*

The 2015 earthquakes in Nepal

Nepal is one of Asia's poorest countries. It is rural, much of it is isolated, and it's landlocked. Its landscape is almost entirely surrounded by the Himalayas, the world's biggest mountain range (see Figure 2). On 25 April and 12 May 2015, it was hit by two earthquakes, magnitude 8.1 and 7.3 respectively. The aftershocks of the first earthquake registered up to 6.7 intensity – enough to cause major damage on their own. Although the second earthquake in May was less intense, its impacts were great because of the damage from the first earthquake. In total, almost 400 aftershocks were recorded within four months.

The earthquake was caused by the collision between two plates – Nepal sits where the Indian Plate collides with and thrusts upwards the Eurasian Plate (see section 1.17). The epicentre occurred along what geologists call the main frontal thrust of the plate boundary. Big earthquakes are rare – about every 750 years. The epicentre was shallow – about 6 km deep.

The social impacts were devastating. In total:

- 9107 people died. It might have been more, but the earthquake happened at noon when most rural workers were in the fields. The figure includes 329 killed in a landslide in the Langtang Valley, caused by the earthquake, and 19 killed by avalanches on Mount Everest.
- 23000 people were injured, many seriously. Over 6000 were still being treated one month later.
- Several hundred thousand (there are no exact data) were made homeless (Figure 3).
- Drinking water supplies and sanitation drains were destroyed, increasing health risks.

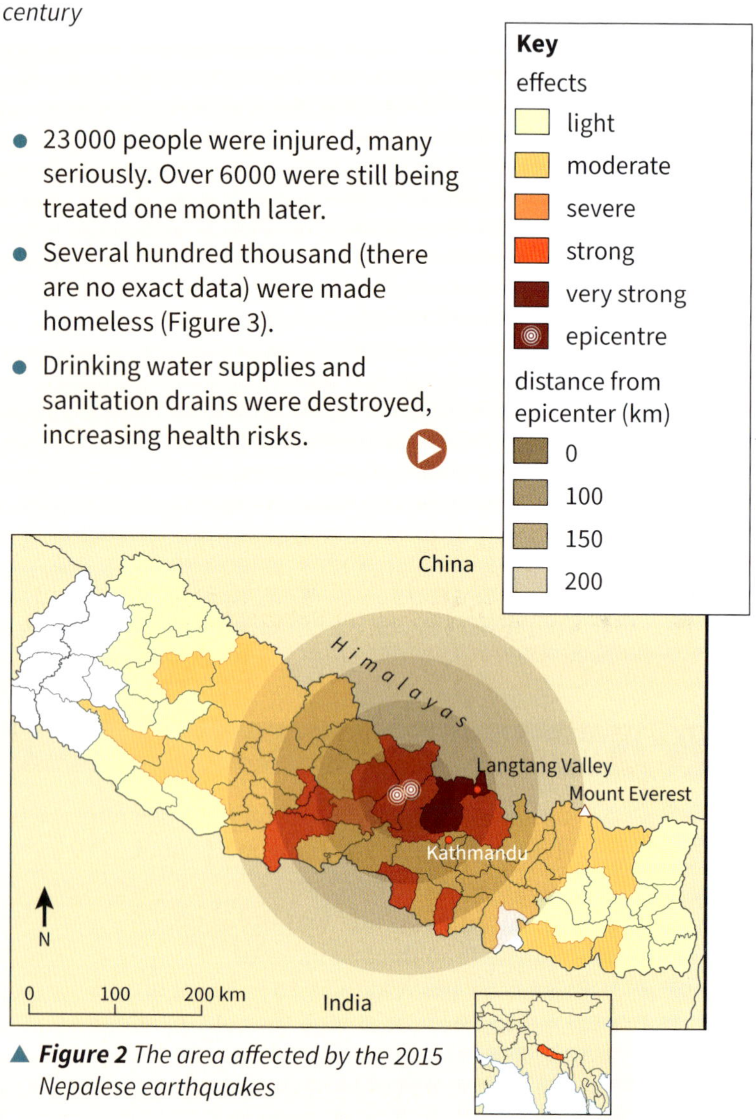

▲ ***Figure 2*** *The area affected by the 2015 Nepalese earthquakes*

Aftershocks often occur as the fault 'settles' into its new position. They can injure or kill rescuers. Many aftershocks destroy buildings that were weakened by the first earthquake.

The economic impacts were large too.

- Tribhuvan International Airport in Kathmandu, the capital, was closed, hindering the delivery of international relief supplies.
- The US Geological Survey (USGS) estimated rebuilding costs at US$7 billion – that's 35% of Nepal's GDP. It estimated that Nepal could never repay loans, and would have to rely on aid gifts. Half the money was given by the Asia Development Bank.
- Crop planting was delayed or prevented by the earthquake – April-May is the key planting season before the monsoon. Many rural families lost their income for the year as a result.

Environmental impacts were considerable. Many ancient temples were destroyed beyond repair, including 30 which were part of the World Heritage Sites in the Kathmandu region. In total, over 1000 temples were damaged across the whole country.

International responses

Nepal quickly asked the world to help:

- The UN provided blankets, tents, water and hygiene kits.
- India provided troops to search for people and clear rubble.
- The UK was the largest donor of financial aid – US$51 million from the government and a huge US$80 million from the public.
- The US and China provided helicopters to access isolated mountain areas.

▼ *Figure 3* *Damage caused by the earthquake*

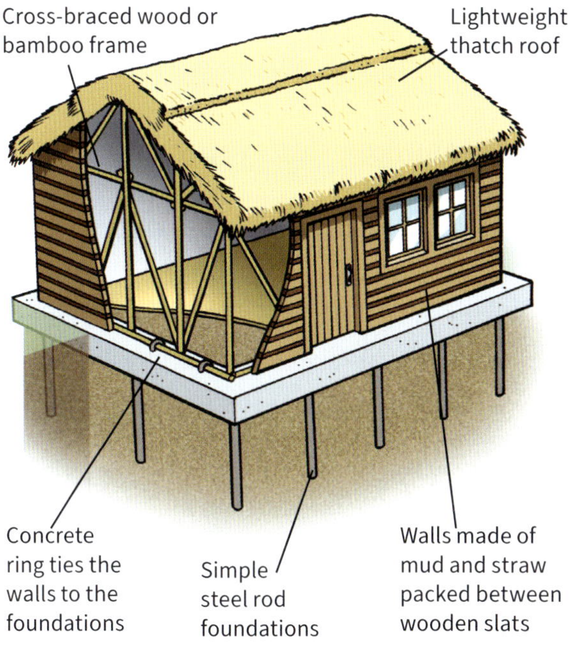

▲ *Figure 4* *Making cheaper houses safer*

Several organisations are involved in rebuilding. Buildings can be made cheaply to withstand earthquakes. Figure 4 shows how inexpensive buildings in China were made safer after earthquakes there.

Your questions

1 Explain why the Nepalese earthquake happened. Include plate names and boundary type in your answer.

2 Describe the region affected by the earthquake using Figure 2.

3 In a table, list and classify the impacts of the Nepalese earthquakes showing social (people), economic (to do with money), and environmental.

4 Research the 2023 earthquake in Türkiye and Syria. You can use the USGS website.

   a Create a fact file of the earthquake including its locations, time of day or night, strength, and timing of any secondary quakes.

   b Research the impacts on (**i**) people, and (**ii**) property.

   c Research the reasons why damage was so extensive in (**i**) urban areas of Türkiye, (**ii**) Syria. 

Exam-style question

5 Evaluate the view that the impacts of earthquakes on developing or emerging countries are greater than those on developed countries. (8 marks)

2.1 Measuring development

In this section, you'll understand different ways of measuring development.

Development in Malawi

Every day in Malawi, eastern Africa (see Figure 1), farmers work on their land. 50% of them have about one hectare of land (the size of a rugby pitch), where they grow their families' food. They are **subsistence farmers** – producing enough to feed themselves, but little extra. Though skilled, most use hand tools and rely on family labour. With more to sell, they'd have a bigger income, and could afford education or better tools. Until this happens, Malawi's **level of development** will remain low.

▲ ***Figure 1*** *Malawi (capital Lilongwe) is about two-thirds the size of England, but with less than a third of England's population*

Level of development means a country's wealth (measured by its GDP), and its social and political progress (e.g. its education, health care or democratic process in which everyone can vote freely).

Measuring development

The United Nations (UN) claims that Malawi is one of the world's 25 poorest countries. The UN uses **development indicators** – or measures of how a country is improving – to work this out (see Figure 2). Looking at them together gives a picture of a nation's level of development.

Economic development indicators include:

- **GDP (Gross Domestic Product)**: The total of goods and services produced by a country in a year (in US$). Dividing GDP by population gives GDP per person (or *per capita*). GDP is now measured in PPP (**Purchasing Power Parity**) which shows what it will buy in each country. Low-income countries usually have low prices, so $1 will buy more there.
- **Poverty line**: The minimum required to meet someone's basic needs – the World Bank uses $1.90 per person, per day.
- **Measures of inequality**: These show how equally wealth is shared among the population. It includes the percentage of GDP owned by the wealthiest 10% of the population, and by the poorest 10%.

However, development can also be defined using **social indicators**, for example:

- **Access to safe drinking water**: The percentage of the population with access to an improved (piped) water supply within 1 km.
- **Literacy rate**: The percentage of the population, aged over 15, who can read and write.

▼ ***Figure 2*** *Socio-economic development indicators for Malawi in 2019, compared to a middle-income country (Brazil) and high-income country (the UK)*

	Malawi	Brazil	UK
GDP per capita (US$) measured in PPP	1060	14 652	46 659
Living below the poverty line %	51.5	4.2	18.6
Access to safe drinking water %	88.7	98.2	100
Literacy rate %	62	93.2	99
HDI	0.483	0.765	0.932
% Income share of the highest 10% earners	37.5	43.4	31.1
% Income share of the lowest 10% earners	2.2	0.8	1.7

The Human Development Index (HDI)

Although GDP is commonly used to show level of development it is rarely the best indicator. Wealth can be very unevenly distributed. In some countries, wealth goes to a few, whereas others spend their limited wealth on health and education, benefiting everyone.

▼ **Figure 3a** *HDI – the top five*

HDI rank	Country	HDI figure	GDP rank	Free elections?
1	Norway	0.957	10	Yes
2=	Ireland	0.955	5	Yes
2=	Switzerland	0.955	8	Yes
4=	Hong Kong, China	0.949	12	No
4=	Iceland	0.949	17	Yes

▼ **Figure 3b** *HDI – the bottom five*

HDI rank	Country	HDI figure	GDP rank	Free elections?
185	Burundi	0.433	197	Yes but unstable
186	South Sudan	0.433	193	In effect, no
187	Chad	0.398	188	A one-party state
188	Central African Republic	0.397	196	Unstable, corrupt elections
189	Niger	0.394	192	Yes but unstable

The UN developed the HDI as a better way to measure development. The HDI consists of one figure per country, between 0 and 1 (the higher the better). It is calculated using an average of four indicators: life expectancy, literacy, average length of schooling, and GDP per capita (using PPP$).

HDI gives a different rank order compared with GDP (Figure 3), though there is a close link. The poorest countries using GDP also have the lowest HDI. Governments don't have money for health care and education.

Corruption and development

One problem with trying to achieve economic development relates to **systems of governance** and corruption. The Corruption Perceptions Index (see Figure 4) was devised to help investors work out where their money is safe. It uses a scale from 0 (highly corrupt) to 100 (honest). In corrupt countries, invested money is likely to be used to bribe officials, or purchase weapons.

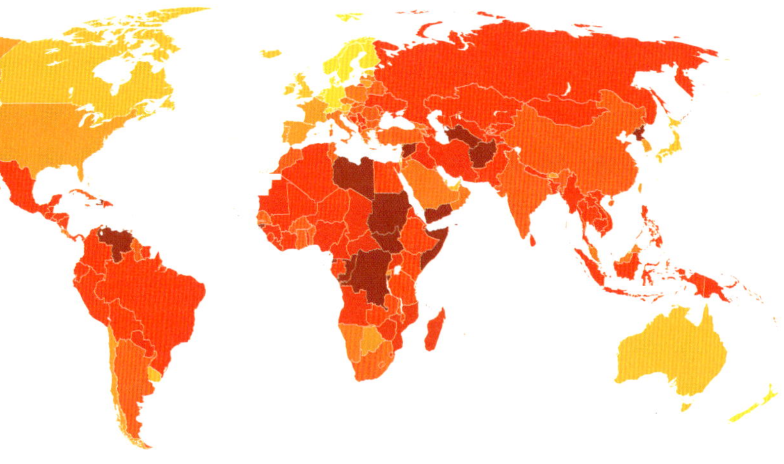

SCORE

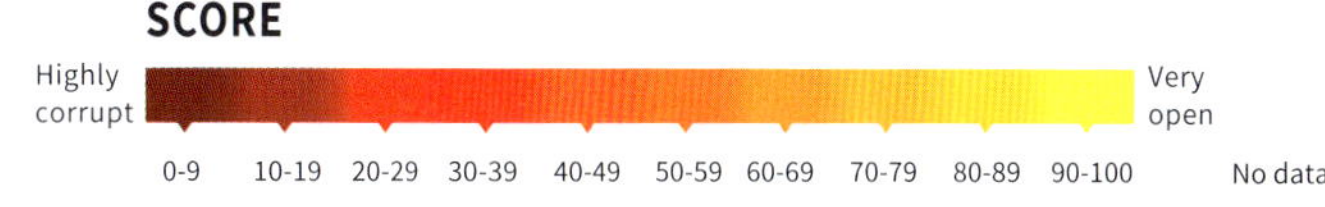

▲ **Figure 4** *The Corruption Perceptions Index shows countries graded according to how corrupt they are judged to be*

Your questions

1 a Choose one country in sub-Saharan Africa, and one in Europe. Use the CIA World Factbook online to research development data and compile a table like Figure 2 for your countries.

b How similar are these to Malawi and the UK?

2 Copy and complete the table opposite to explain reasons and effects for the world's poorest countries.

3 Explain why countries are often ranked in a different order using GDP and HDI.

4 Using Figures 2 and 3, identify things that improve in a country as GDP increases.

	Reasons for this	The effect of this on the country
a They have low life expectancy.		
b They have high maternal mortality.		
c They have low literacy rates.		
d They rarely have free elections.		

Exam-style question

5 Explain how **two** named economic indicators help to show a country's level of development. (4 marks)

2.2 Development and population

In this section, you'll understand how populations change as countries develop, particularly in Malawi.

Malawi's population

Malawi has the world's most **youthful population**. In 2020, 43.5% of its population was aged 0-14 – 8.3 million out of a total of 19.1 million. Its **median age** was 17. The young dominate its **population structure**. Age-sex structure tells us a lot about a country's level of development, as Figure 1 shows. The **demographic data** for Malawi are typical of many of the world's poorest countries.

Population structure – the number of each sex in each age group (e.g. 10–14).

Demographic data – all data linked to population e.g. birth rate, death rate etc.

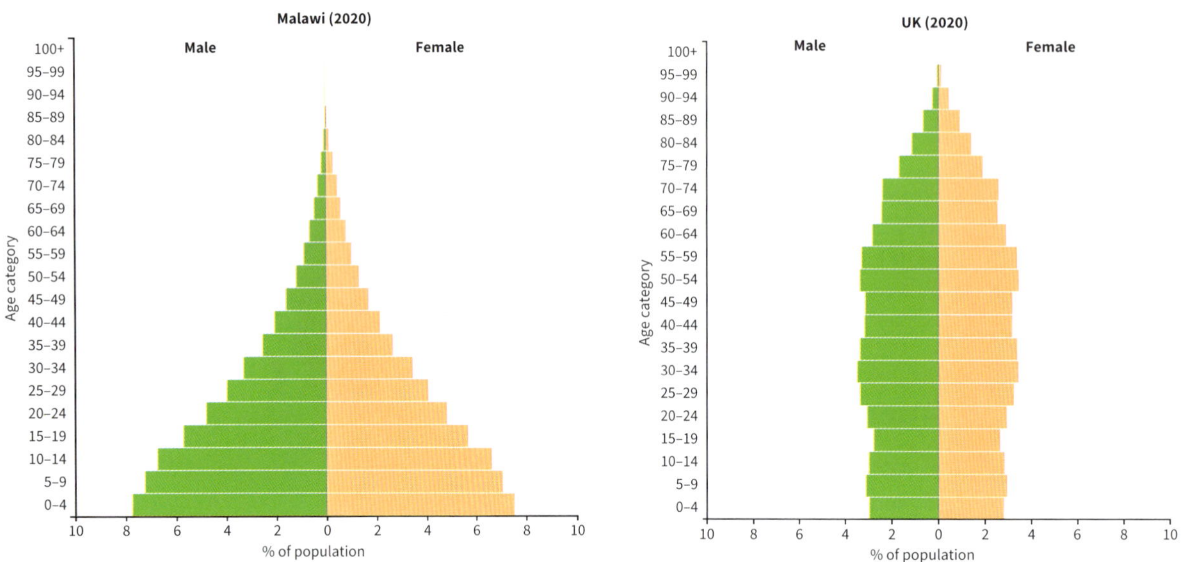

▲ ***Figures 1a and 1b*** *Comparing Malawi's (1a) and the UK's (1b) population structure*

Population change

Development brings change. As countries develop, GDP per capita increases. This increased wealth means more money to spend on health, education and piped water supplies. Every demographic indicator changes – look at the data for Brazil and the UK in Figure 2, compared with Malawi. Rapidly falling death rates are now typical of most developing countries. Over time:

- birth and death rates, dependency ratios, fertility rates, infant mortality rates, and maternal mortality rates all fall
- life expectancy, years of schooling and literacy rates all increase.

As birth rates fall and life expectancy increases, so eventually populations age, as in the UK.

Know your demographic indicators

- **Birth rate:** number of live births per 1000 people per year.
- **Death rate:** number of deaths per 1000 people per year.
- **Dependency ratio:** numbers of those aged 0–14 and over 65 who are outside working age as a proportion of those working aged 15–64, times 100. The lower the ratio, the greater the number who work and are less dependent. It's not a reliable measure as many children and elderly work in Malawi.
- **Fertility rate:** average number of births per woman.
- **Infant mortality:** number of children per 1000 live births who die before their first birthday.
- **Life expectancy:** average number of years a person can expect to live.
- **Maternal mortality:** number of mothers per 100 000 who die in childbirth.

Women's health and education

As in many developing countries Malawian women are more likely to be poor than men. Many work as landless labourers, and 25% of mothers are undernourished. As a result, Malawi's infant and maternal mortality rates are among the world's worst, even though they have halved since 2000.

- Skilled health workers attend only 20% of births, so maternal mortality is high.
- Babies of unhealthy mothers are more likely to die in their first five years.

Population and education are closely linked. Primary education is free in Malawi, but there are few state secondary schools – most charge fees which subsistence families cannot afford. So, few girls attend secondary school beyond the age of 13. In rural areas, many girls marry at 13 or 14, and have their first child soon after, so Malawi has a high fertility rate. It is caused by poverty.

As countries develop, educated women are more likely to develop a career, marry and have children later. Fertility and infant mortality rates tumble. The same happens even in countries like Malawi, among professional women – their fertility and infant mortality rates are almost as low as in high-income countries. As countries develop, the lives of most people are improved.

Data	Malawi	Brazil	UK
Population			
Birth rate (per 1000 people)	32.8	13.8	10.7
Death rate (per 1000 people)	6.3	6.3	9.1
Fertility rate	4.2	1.75	1.65
% population aged 0–14	43.5	24.1	17.6
% population aged 65 and over	2.7	8.6	18.2
Dependency ratio %	91	43.8	55.5
Health			
Life expectancy (years)	62.2	75.1	80.8
Infant mortality per 1000 live births	30.9	16	4.3
Maternal mortality per 100 000 births (2010)	349	60	7
Number of doctors per 100 000 population	0.04	2.16	2.8
Education			
Average number of years in school	11	14	17
Literacy rate %	62	93.2	99
Average age of first marriage for women	20.1	30	32.5

▲ ***Figure 2*** *Comparing population and social development indicators in 2020, for Malawi, Brazil and the UK*

Your questions

1. Using Figures 1a and 1b, compare the population structure in Malawi and the UK as follows: a the general shape of each population pyramid, b the two largest age groups in each country. 
2. Give reasons for the differences you notice.
3. Compare the main social differences between Malawi, Brazil and the UK, using Figure 2.
4. If Malawi develops in future, explain what changes you would expect to occur in: **a** its birth rate, **b** its infant mortality rate, **c** its life expectancy.
5. Explain why economic and social development can only take place if a country improves life for women and children.

Exam-style questions

6. Define the term 'birth rate'. (1 mark)
7. Explain the difference between the terms 'birth rate' and 'fertility rate'. (3 marks)
8. Study Figure 1a. Describe the population structure of Malawi. (3 marks) 
9. Explain how changes to GDP per capita might result in changes to life expectancy. (4 marks)

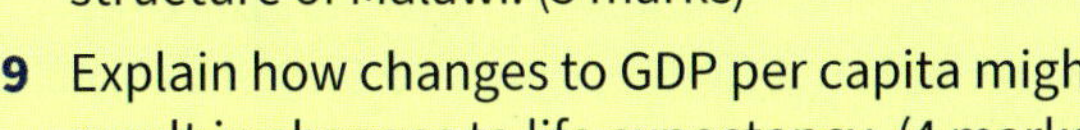

2.3 Global inequality

In this section, you'll understand the reasons for global inequality and whether it has changed in recent years.

Mind the gap

In 1980, a report was published which highlighted global differences between rich and poor, the poverty of many developing countries, and the vast wealth of a few developed countries. It was known as the Brandt Report, written by the then German Chancellor, Willie Brandt. It identified two groups of countries:

- A group of wealthy countries in North America, Western Europe, Japan and Australasia. Most were in the northern hemisphere, so were called the '**global north**'. These were the world's **High Income Countries** (HICs).
- A second group of much poorer countries in Latin America, Africa, and Asia. Most were in the southern hemisphere, so were called the '**global south**'. These were the world's **Low Income Countries** (LICs).

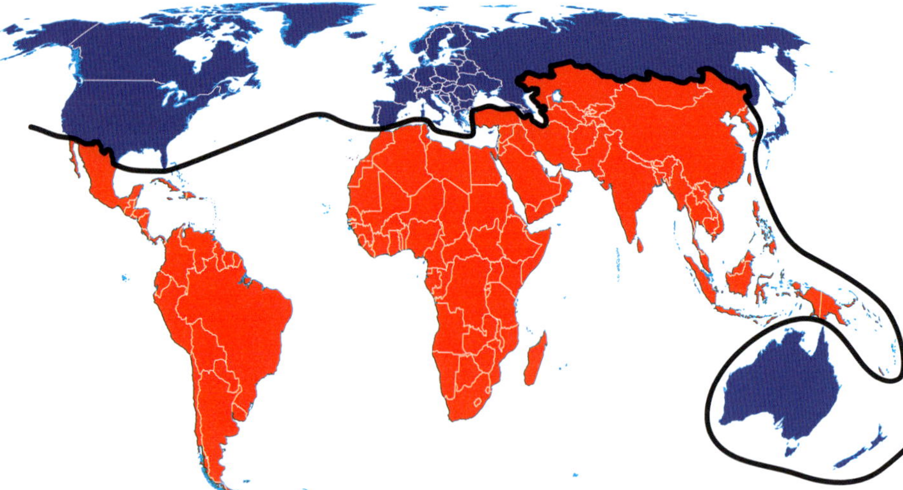

▲ ***Figure 1*** *The 'North-South Divide', from the Brandt Report of 1980*

The inequalities between these two groups became known as the 'North-South Divide', or the 'Development Gap', shown in Figure 1.

Global shares of wealth

If the world's wealth were evenly shared, there would be no 'gap' between richer and poorer countries. However, this is not so. Every year, every country submits its GDP data to the World Bank. The world's wealth is estimated at US$78 trillion! But how is it shared out?

To work this out, the World Bank ranks countries by total GDP, from 1 (the USA, with US$21.5 trillion), to 230 (Tuvalu US$47 million). They are then grouped into five equal bands of 46 countries, known as **quintiles** (or fifths), shown in Figure 2. The top quintile – 1 to 46 – consists of the world's richest countries.

On a diagram, the results look like a champagne glass:

- the top quintile – that is, the richest fifth or 20% – of the world's countries owns 83% of the world's wealth
- the bottom two quintiles – that is, the poorest 40% – own just 3%. However, there are some very wealthy people living in poor countries – Malawi's richest person in 2020 was worth US$220 million.

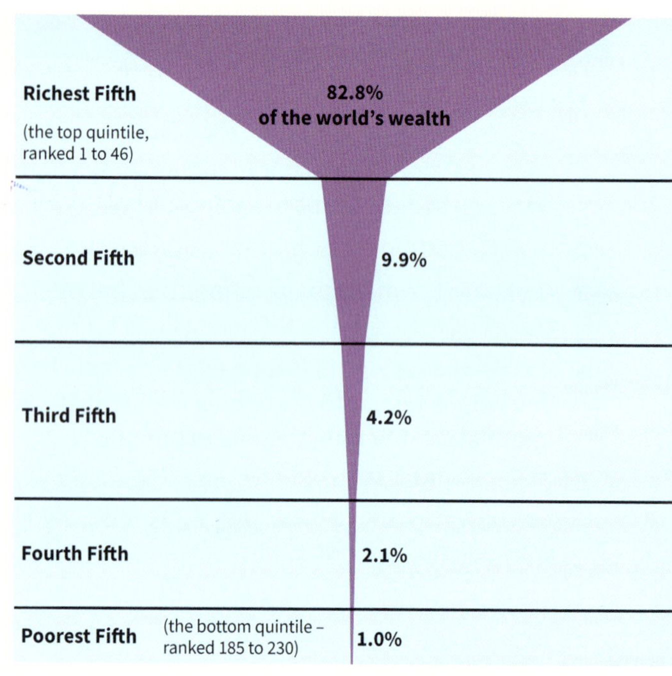

▲ ***Figure 2*** *GDP per capita (PPP), divided into quintiles*

What's changed?

Four decades after the Brandt Report many **emerging economies** have become wealthier:

- In the 1980s, rapid development took place in Latin America. This created a group of countries now known as **Middle Income Countries** (MICs) such as Brazil and Chile. Large reserves of raw materials in these countries (e.g. iron ore) encouraged investment and growth, and their cities experienced big increases in population (see Section 3.1).
- In the 1990s, equally rapid development took place in countries of south-east Asia, such as Hong Kong, Singapore, and Malaysia. Their growth was sometimes due to the relocation of manufacturing overseas by US and European TNCs. Growth was so aggressive that these countries became known as the 'Asian Tigers'. Most of this region is now classed as **Newly Industrialising Countries** (NICs).
- Even more rapid has been recent industrialisation of China and India. They are referred to as **Recently Industrialising Countries** (RICs). Together with Brazil and Russia, these countries are also known as the BRICs (after their initials).

▲ Rising ▼ Falling	1980 rank	2019 rank	The 10 largest economies (total GDP in US$)
	1	1	USA (US$21.5 trillion)
▲	9	2	China (US$14.3 trillion)
▼	2	3	Japan (US$5 trillion)
	4	4	Germany (US$3.9 trillion)
▲	13	5	India (US$2.87 trillion)
	6	6	United Kingdom (US$2.83 trillion)
▼	5	7	France (US 2.7 trillion)
▼	7	8	Italy (US$2 trillion)
▲	12	9	Brazil (US$1.84 trillion)
▼	8	10	Canada (US$1.7 trillion)

▲ ***Figure 3*** *The changing rank order of the world's ten largest economies 1980–2019*

A new world order

Has Brandt's North-South Divide and its gap survived? Figure 3 shows that the rank order of the world's ten largest economies in 2019 has changed since 1980. But two things have stayed the same:

- the USA's GDP still far exceeds the rest
- in 1980, the world's ten poorest countries were in sub-Saharan Africa – Figure 4 shows this is still true.

▶ ***Figure 4*** *The ten poorest countries in the world in 2020, all of which were in Africa*

Your questions

1 Explain the meaning of each of the following groups of terms:
  a 'global north', 'global south', 'emerging economies'
  b HICs, MICs, LICs
  c NICs, RICs, BRICs

2 Describe three ways in which the data in Figure 2 show evidence of global inequality.

3 Study Figures 2, 3 and 4.
  a Explain ways in which the world has changed since 1980.
  b Explain how far there is still a 'North-South Divide'.

Exam-style questions

4 Define the term 'development gap'. (1 mark)

5 Study Figure 3. Explain the changing rank order of the world's ten largest economies between 1980 and 2019. (4 marks)

2.4 What's holding Malawi back? 1

In this section, you'll understand how Malawi faces physical and environmental barriers to its development.

Malawi's uphill task

Since his election in 2020, Malawi's President, Lazarus Chakwera, has had his work cut out. Malawi faces increasing food prices and fuel shortages, and millions are underfed. He believes that Malawi faces an uphill task in joining a globalised world, but that globalisation could bring investment, jobs and development. Without investment, GDP and incomes remain low, and there is little tax to spend on education or health care. If people are sick, they are unable to earn. It's a vicious cycle (see Figure 1). So what's holding Malawi's development back?

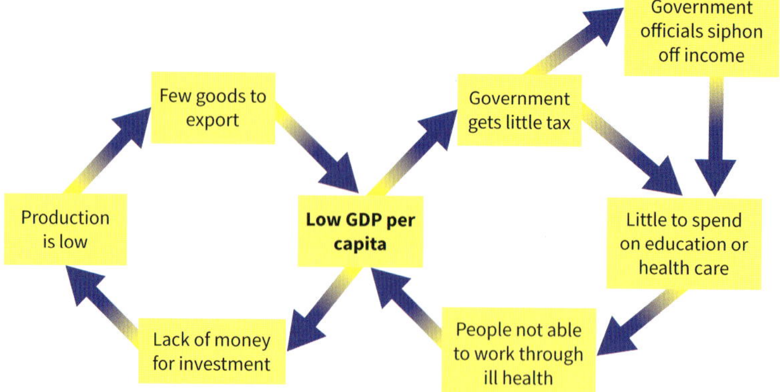

▲ ***Figure 1*** *The economic cycle of poverty*

1 It's landlocked

Malawi has no coastline – it's landlocked – so it has no port from which to export or import goods. This need not be a problem if a country is wealthy (e.g. Switzerland). Reaching the coast involves going to Nacala in Mozambique along one slow, 800 km, single track railway (see Figure 2). The line is narrow gauge (see Figure 3), which limits the speed and weight on each train. Nearly all of Malawi's exports (e.g. tobacco, sugar, tea) go this way. The trains return with imports (e.g. fertiliser, fuel, manufactured goods). But it's a slow, expensive process.

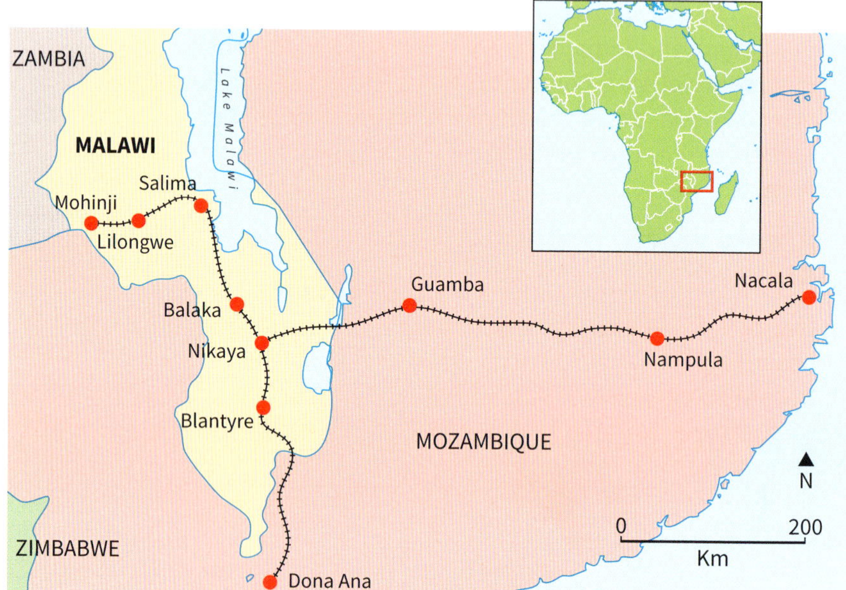

▲ ***Figure 2*** *Malawi is dependent upon a limited railway network to export its products*

2 Rural isolation

85% of Malawi's population is rural (it has the highest percentage rural population in the world). Much of rural Malawi is isolated, with poor infrastructure. Roads are mostly dirt, so it takes several hours to travel 20 km to local markets during the rainy season. When they flood, farmers can be cut off.

▶ ***Figure 3*** *Malawi's single track rail link is susceptible to flooding and damage, as shown here*

Rural telecommunications also vary. Landline services are slow, so Malawi, like most other African countries, has leap-frogged to using mobile technology. By 2019, Malawi had only 14 350 telephone landlines (1 per 1400 people), but 2.7 million internet users (1 per 8 people). Mobile phone ownership is growing rapidly – from 1 million in 2007 to 10 million by 2019 (out of a population of 20.4 million) – but rural coverage can be poor. Nevertheless, mobile technology is helping rural areas to develop.

3 Living with a changing climate

Climate change is affecting Malawi. Oxfam published a report called *Africa - up in smoke*. It showed that climate change would affect Africa more than anywhere, causing:

- water shortages as temperatures rise (increasing evaporation)
- food shortages caused by variable rainfall and increased drought.

Rainfall in Malawi has been much lower since 2000, compared to the twentieth century. The rainy season (November–April) has been shorter, so rivers have dried up, and crop yields have fallen. But when rains have arrived, they have been intense. In 2012, heavy rains damaged and reduced Malawi's maize harvest by 7%, and 10 000 families were made homeless by flooding.

▲ ***Figure 4*** *River pollution in Blantyre, Malawi*

4 Health problems

Health problems hold back Malawi's people and economy. In Lilongwe, the capital, water supplies become contaminated during the rainy season by surface run-off from built-up areas. Squatter settlements on the city edge have no sanitation or waste management. Rivers and local dams become contaminated with waste and bacteria, causing risks to human health (Figure 4). Air quality is poor from dust, industrial smoke, and exhaust from badly maintained lorries.

Covid-19 brought a crisis to Malawi, with hospitals stretched to capacity – the hospital in Blantyre, the second city, has just 80 beds. The virus spread most in the cities, peaking in January 2021 with 1300 cases daily. Prompt and effective action by the government (such as banning travel and mixing at work) brought case numbers down, but longer-term problems lie in obtaining vaccines.

In the meantime, other health problems remain serious, such as childhood diarrhoea, typhoid, malaria, and dengue fever.

Your questions

1 a In pairs, draw a labelled spider diagram to show problems facing Malawi caused by **i)** being landlocked, **ii)** rural isolation, **iii)** changing climate, and **iv)** health problems.

b Decide which is the greatest problem. Justify your answer.

2 a In pairs, write down the advantages of future investment in the following:

- building a new rail link to the coast
- improving rural roads
- building dams for water storage
- investment in hospitals and health clinics.

b Which one of these would you choose? Justify your choice in 250 words.

Exam-style questions

3 Define the term 'landlocked'. (1 mark)

4 Explain how physical factors can be a barrier to development in a developing country. (3 marks)

2.5 What's holding Malawi back? 2

In this section, you'll understand the economic and political barriers that Malawi faces in trying to develop.

The economic barriers ahead

How should Malawi develop in future? Despite its physical and environmental barriers (see section 2.4), it can only develop by increasing trade with other, wealthier countries. But this is not easy. Three problems stand in the way:

- its terms of trade, and its debt
- the tea and coffee trade
- global trade and the World Trade Organisation.

Terms of trade means the value of a country's exports relative to that of its imports.

1 Terms of trade

The value of Malawi's exports every year is less than its imports, so every year, it earns less than it spends – its **terms of trade** are stacked against it (see Figure 1). Exporting more would help. But what, and how?

One of the reasons for such poor terms of trade is that Malawi exports largely raw materials, known as **primary products.** It has traditionally sold these to developed countries, and in return bought manufactured goods that it does not make itself. This was typical of trade in the 1980s, as Figure 2 shows.

Exports	Imports	Debt total
Value: \$10.7 billion (2019) **Goods sold** Tobacco, tea, raw sugar, beans, soybean products, clothing **Sold mainly to** Developed and emerging countries: *Belgium 16%, USA 8%, South Africa 6%, Germany 6%, United Arab Emirates 5%* Developing countries: *Egypt 7%, Kenya 5%*	**Value:** \$12.8 billion (2019) **Goods bought** Postage stamps, refined petroleum, medicines, fertilisers, office machinery and parts **Bought from** Developed and emerging countries: *South Africa 17%, China 16%, United Arab Emirates 9%, India 9%, United Kingdom 8%*	US\$2.1 billion (2017)

▲ ***Figure 1** Malawi's terms of trade, and debt*

Trade has changed in recent years as the 2020 diagram in Figure 2 shows, and now a third type of trade is important.

▼ ***Figure 2** Changes in trade flows since the 1980s*

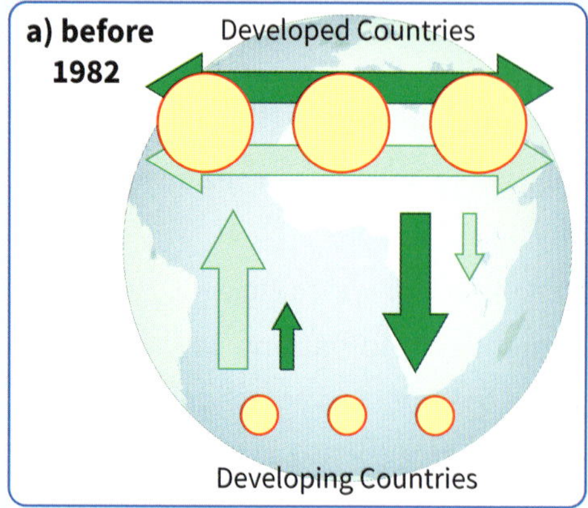

▲ Until the 1980s, the flow of trade was either between the world's developed and developing countries, or between the developed countries themselves.

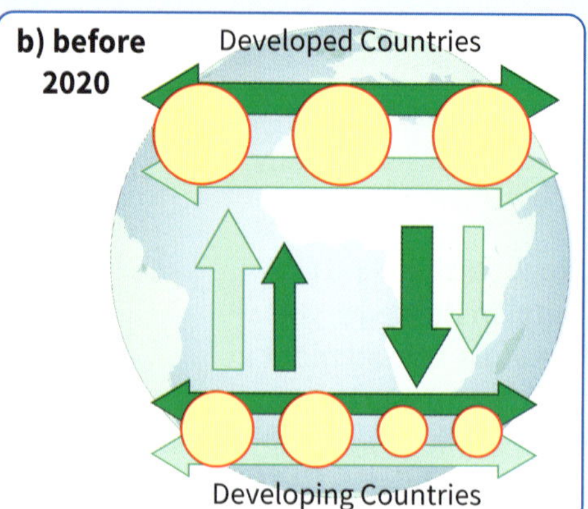

▲ Now there is a third type of trade – between developing countries like Malawi and the emerging economies of India and China. China now buys food and raw materials from African countries, and Chinese goods are exported to developing economies of Africa and Latin America.

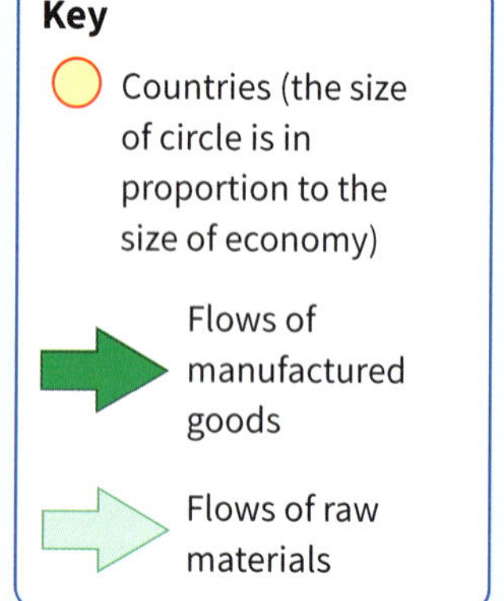

2 Colonisation and cash crops

In the nineteenth century, the British **colonised** (took over) Malawi and its land. They developed plantations to grow coffee and tea for export. Plantations still remain in British ownership, some by large TNCs (e.g. Unilever, producing PG Tips tea). These plantations hire local landless labourers, or subsistence farmers seeking extra income. But workers only get paid 1p per kg of tea leaves or coffee cherries picked, though plantation owners argue that they also get housing, water, firewood and daily lunch.

Farming is critical to Malawi. Over 80% of its population works in farming, and the country still depends on **cash crops** (sold for cash) for exports. These are known as **commodities**, and are traded on global markets. Malawi's farmers find this tough, because global prices constantly change. As Figure 3 shows; they never know what price they'll get.

Estate workers lose out. Tea is sold to the global market at US$2.75 (£1.75) per kg on average. Profits go to companies in developed countries – an example of **neo** (new) **colonialism**. In UK supermarkets, it's sold at around £20 per kg – 10 times the price paid to farmers, with most profits going to supermarkets.

▲ ***Figure 3*** *East African tea prices 2014–2019*

▲ ***Figure 4*** *A large tea plantation in Malawi*

3 Global trade and international relations

The **World Trade Organisation** (WTO) is a global organisation which aims to make trade easier. It tries to help developing countries trade with wealthier countries so they can increase wealth, jobs and investment. It also aims to get countries to agree that goods will be free of duties, or **tariffs**, which are added on to the price of goods, making them more expensive.

It doesn't always work. Malawi exports raw coffee beans, instead of roasting them which would get a higher price. The reason is that the EU and USA charge tariffs of 7.5% on imported roasted beans, but nothing on raw beans. It's cheaper for European or American coffee companies (e.g. Costa or Starbucks) to roast beans themselves rather than buy ready-roasted from Malawi.

Your questions

1 Describe how global trade has changed between 1982 and the present.

2 In what ways is Malawi's trade now **a** similar to, **b** different from diagram b in Figure 2? 

3 Draw a spider diagram to show the reasons why **a** Malawi's tea trade and **b** the WTO are a barrier to Malawi's economic development.

Exam-style questions

4 Define the term 'terms of trade'. (1 mark)

5 Study Figure 1. Describe **two** features of Malawi's trade. (2 marks) 

6 Explain **two** reasons why Malawi's debt is increasing. (4 marks)

7 For a named developing country, assess the importance of trade in its economic development. (8 marks).

2.6 How do countries develop?

In this section, you'll understand two theories of development regarding how countries develop over time.

Why are some countries poor?

Malawi illustrates how a low-income country faces barriers to economic development. These barriers may be:

- economic, e.g. terms of trade (explained in section 2.5)
- social, e.g. the impact of little education (explained in section 2.2)
- environmental, e.g. changing climate (explained in section 2.4).

However, there are different theories about the causes of poverty in countries like Malawi. The first theory, by Rostow, sees a path to progress that Malawi simply has to follow. The second, by Frank (a German sociologist), believes that Malawi's poverty is due to past relationships with other countries.

▲ ***Figure 1*** *A market stall in Uganda - subsistence farmers selling cash crops. Rostow would say this is typical of a Stage 2 economy.*

Rostow's theory

Walt Rostow was an American economist who worked in the US government after the end of the Second World War. In 1960, he published his theory (shown in Figure 2) – usually called Rostow's Model. He believed that manufacturing was central to development, and that law courts, a central currency, and policed property rights were essential to its emergence. Without these, it could never grow. He believed that countries should pass through five stages of development:

1 *Traditional society* – Most people work in agriculture, but produce little surplus (extra food which they could sell). This is a 'subsistence economy'.

2 *Pre-conditions for take-off* – there's a shift from farming to manufacturing. Trade increases profits, which are invested into new industries and infrastructure. Agriculture produces cash crops for sale (like Uganda in Figure 1).

3 *Take-off* – growth is rapid. Investment and technology create new manufacturing industries. Take-off requires investment from profits earned from overseas trade.

4 *Drive to maturity* – a period of growth. Technology is used to produce consumer goods cheaply.

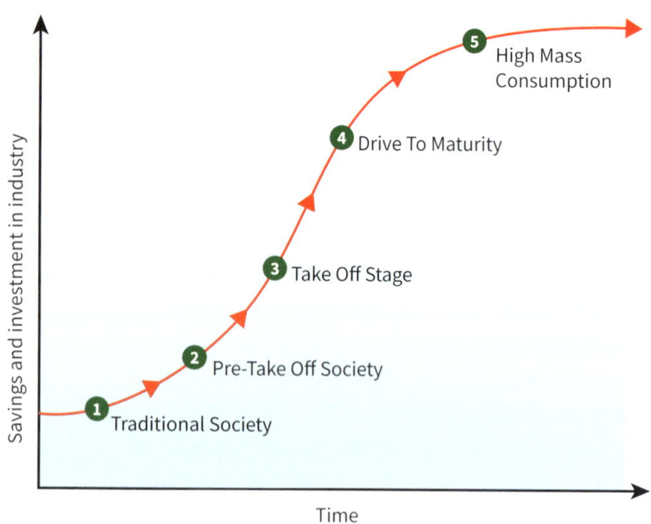

▲ ***Figure 2*** *Rostow's five-stage model of economic development*

Communism is a political and economic system without private ownership of property. Communist countries include China and Cuba, and once included the USSR and most of Eastern Europe until 1991.

5 Age of high mass consumption – a period of comfort. Consumers enjoy a wide range of goods. Societies choose how to spend wealth, either on military strength, on education and welfare, or on luxuries for the wealthy.

Rostow's theory became known as modernisation theory, because he believed that a country's institutions (such as finance or law) would have to modernise before development could take place.

▼ ***Figure 3*** *Five countries – but at which stage of Rostow's model are they?*

Country	GDP in US$ PPP	GDP growth rate %	Literacy %	Doctors per 1000 people	People working in farming %	People working in industry %
Australia	49 854	1.84	99	3.68	3.6	21.1
Mali	2 322	5.4	35.5	0.13	80	Very few
Singapore	97 341	0.73	97.3	2.29	0.7	25.6
Chile	24 226	1.03	96.4	2.59	9.2	23.7
Sri Lanka	13 078	2.29	91.9	1	27	26

Frank's dependency theory

In 1967, dependency theory was developed by the sociologist André Frank – in opposition to Rostow's ideas. Frank believed that development was about two types of global region – core and periphery (shown in Figure 4). The core represents the developed, powerful nations of the world (i.e. North America, Europe and Australasia), and the periphery consists of 'other' areas, which produce raw materials to sell to the core. The periphery therefore depends on the core for its market.

In Frank's theory, low-value raw materials are traded between periphery and core. The core processes these into higher-value products, becoming wealthy. Frank's objection to Rostow was that modernising institutions wouldn't work if countries were locked into dependency on core regions to sell their products. Poorer countries, he argued, weren't just simpler versions of wealthy countries; their poverty resulted from decisions made by wealthy elites who profited from plantations or mines.

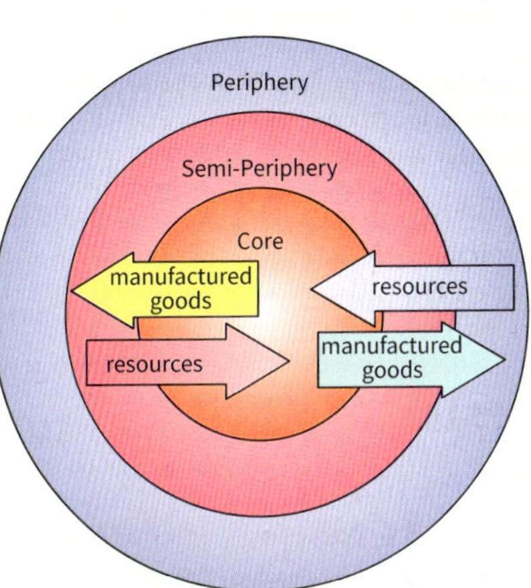

▲ ***Figure 4*** *Frank's Core-Periphery model*

Your questions

1 a Make a large copy of Rostow's model (Figure 2). 

b Using the data in Figure 3, decide which countries fit which stage. Write this on your copy and annotate it with your reasons.

2 a Make a large copy of Frank's Core-Periphery model (Figure 4).

b Decide which of the five countries listed in Figure 3 are 'core' and which are 'periphery'. Write this on your copy and annotate it with your reasons.

c Research the CIA World Factbook online to find the exports and imports of the five countries.

d Label your diagram with the goods and products traded by the five countries. 

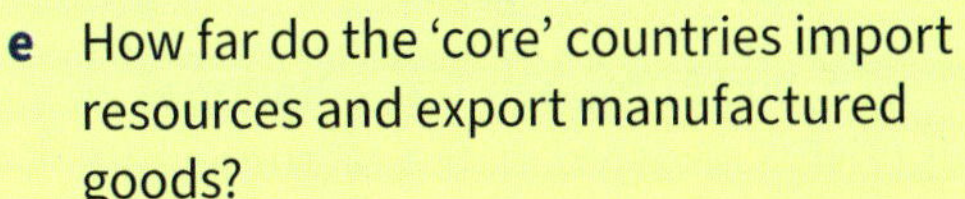

e How far do the 'core' countries import resources and export manufactured goods?

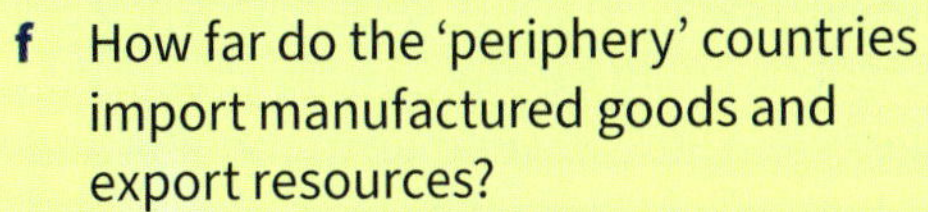

f How far do the 'periphery' countries import manufactured goods and export resources? 

Exam-style questions

3 Explain how Rostow's model can be used to explain the development of a country. (4 marks)

4 Evaluate the Rostow model as a means of explaining a country's development. (8 marks)

2.7 Development in a globalised world

In this section, you'll be able to assess the reasons why some countries benefit from globalisation more than others.

Think global!

Does it matter if Malawi has fewer phone lines or internet users (section 2.4) than wealthier countries? Or that it is landlocked? Many think that it does. They believe that, without technology and access to the coast, Malawi can never play a part in the global economy. Figure 1 shows that GDP in sub-Saharan Africa lags behind. While the NICs of south-east Asia, such as Singapore, have raced ahead, Malawi struggles. Could **globalisation** help Malawi too?

What is globalisation?

Globalisation involves a set of government policies by which countries become increasingly connected to each other. These policies include:

- economic **inter-dependence** between countries – national borders have become less important
- increasing volumes and variety of **trade** in goods and services, shown in Figure 2
- increased spread of **technology**
- international **flows** of investment into other countries
- using people in other countries to provide services if they can do so more cheaply – known as **outsourcing** (e.g. call centres)
- **culture**, where global media companies spread 24-hour news, TV, film, and music. You're as likely to hear Dua Lipa in Shanghai as in Southampton.

Changes in investment

In the 1990s, US and European Trans-National Companies (TNCs) invested in new factories and transport infrastructure in south-east Asia. They did this because goods could be made more cheaply there; Chinese wages were 90% lower than in the USA and Europe. This type of investment by one country into another is known as **Foreign Direct Investment** (FDI). TNCs could manufacture goods cheaply, then export them. The result was a huge growth in Chinese exports. It caused a **global shift** – changing where manufactured goods were made, from developed to developing countries.

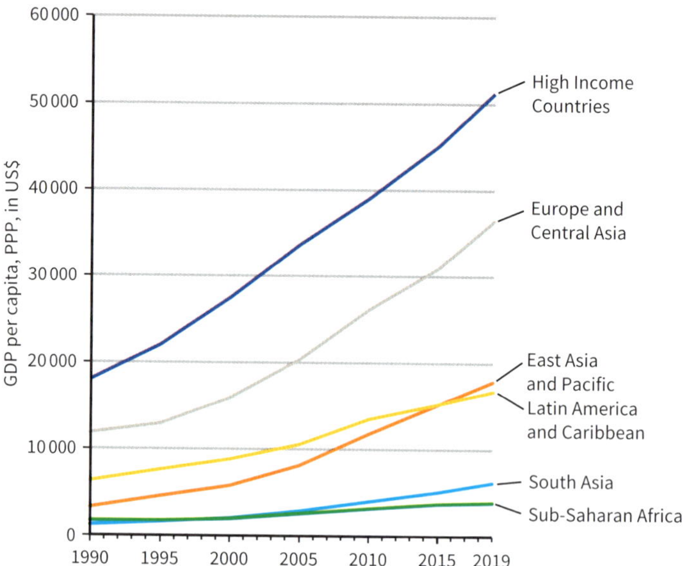

▲ ***Figure 1*** *GDP for different regions of the world, 1990–2019*

Top 5 exporters of goods 1980 (US$ billions)		Top 5 exporters of goods 2020 (US$ billions)	
USA	225	China	2591
Germany	192	USA	1431
Japan	130	Germany	1380
France	116	Netherlands	674
UK	110	Japan	641

▲ ***Figure 2*** *The top 5 global exporters, 1980 and 2020*

The Clark Fisher model

Two economists, Clark and Fisher, produced a theory – or model – that explains changes in employment structure as countries develop their economies. They identified three stages (see Figure 3):

A **Low-income countries** whose employment is dominated by the primary sector (e.g. farming and mining).

B **Middle-income countries** are dominated by the secondary sector (**manufacturing**). As economies develop and incomes rise, the demand for agricultural and manufactured goods increases.

C **High-income countries** are dominated by the tertiary and quaternary sectors. As incomes continue to rise the tertiary (services) sector develops as people start to use more services – banking, insurance, legal services and leisure. Finally, the quaternary sector develops as tertiary services become more specialised, such as IT, medical services, the media. Some of these services earn very high fees. Many countries in this category have lost their manufacturing to countries like Vietnam, where wages are lower.

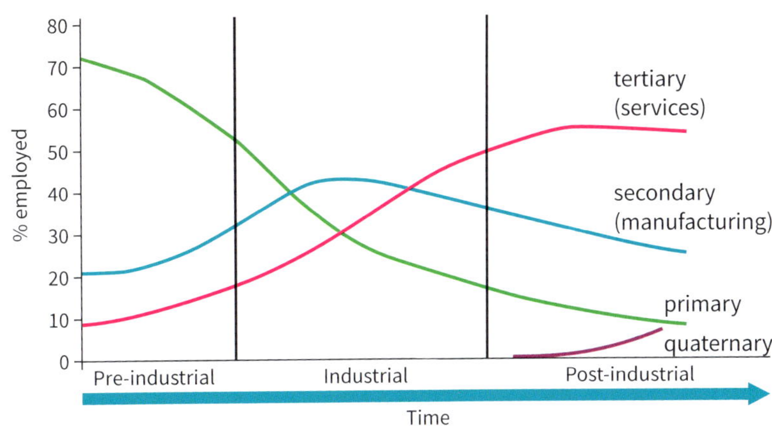

▲ ***Figure 3*** *The Clark-Fisher Model*

Development in Vietnam

Since 1990, Vietnam has seen rapid industrial growth, known as **industrialisation**, which has led to rapid changes. It's an economic and social process:

- economically, money (or capital) has been invested in factories, to turn raw materials (**primary products**) into manufactured goods (**secondary products**). This adds value. A piece of furniture, for example, is worth more than a block of wood.
- socially, it means workers moving from rural areas (the countryside) to urban (or cities) as most secondary jobs are in urban factories. As manufactured goods are worth more than raw materials, urban wages are higher than rural wages.

	Malawi	Vietnam	UK
GDP per person (in US$ PPP)	1060	8041	46 659
Where GDP comes from (%)	Agriculture: 28.6% Industry: 15.4% Services: 56%	Agriculture: 15.3% Industry: 33.3% Services: 51.3%	Agriculture: 0.7% Industry: 20.2% Services: 79.2%
Percentage of people by occupation	Agriculture: 77% Industry: 4.9% Services: 18.1%	Agriculture: 40.3% Industry: 25.7% Services: 34%	Agriculture: 1.3% Industry: 15.2% Services: 83.5%
Value of Exports (US$)	10.7 billion	249 billion	902 billion
Export goods	Tobacco, tea, raw sugar, beans, soybean products, clothing	Broadcasting equipment, telephones, integrated circuits, footwear, furniture	Banking, insurance, business services, cars, gas turbines, gold, crude petroleum, packaged medicines

▲ ***Figure 4*** *Comparing the economies of Malawi, Vietnam and the UK in 2020*

While Vietnam has benefitted from globalisation (many factory goods are exported, adding to its GDP), Malawi struggles to industrialise, and its GDP remains low.

Your questions

1. Using Figures 1 and 2, describe how GDP and exports have changed in different regions of the world since 1980.
2. Explain why:
   - **a** industrialisation has given Vietnam a higher GDP than Malawi
   - **b** the UK has a much higher GDP than Vietnam, even though its manufacturing (industry) is lower.
3. Copy the Clark Fisher model and annotate it with where **a** Malawi, **b** Vietnam, and **c** the UK each belong. Give evidence to explain your decision.

Exam-style question

4. For a named emerging country, assess how far it has benefited from globalisation. (8 marks)

2.8 Introducing India

In this section, you'll understand about India's growth and its significance as a country.

Arrival in India

It's 6 a.m. Bangalore airport in southern India hums with people as flights arrive from Europe. In the arrivals hall, taxi drivers noisily bid and out-bid each other to take passengers to office blocks in the city. Weary travellers step out, unaware of the imminent assault on their senses – the din of traffic, the heat, humidity, and torrential rain of the monsoon season. On the journey, there are three lanes of traffic marked out, but five lanes of vehicles squeeze in – the drivers know they can do it. It seems like chaos, but somehow it works.

▲ ***Figure 1*** *Bangalore's traffic congestion*

It's all about growth!

India is growing, fast. It's one of the world's **emerging countries**, and, as Figure 2 shows, it is predicted to become the world's second largest economy by 2050. Since 1997, its economy has grown by 7% annually, septupling in size!

India's geographical location (Figure 3) encourages its growth. With the Himalayas as a physical barrier to the north, its geographical location between the Middle East (then on to Europe) and south-east Asia enables it to trade easily by sea with some of the world's biggest and fastest-growing economies.

An **emerging country** is one with high-to-medium human development, and recent economic growth.

- Oil imports are speedily obtained from the Middle East into Mumbai on the west coast.
- Trading with Singapore and south-east Asia is speedier via Kolkata on the east coast.

Position in 2015	Likely position by 2050
2	1 China
10	2 India
1	3 USA
20	4 Indonesia
21	5 Nigeria
7	6 Brazil
9	7 Russia
3	8 Japan
39	9 Philippines
6	10 UK

▲ ***Figure 2*** *The world's ten biggest economies by GDP by 2050, compared to 2015*

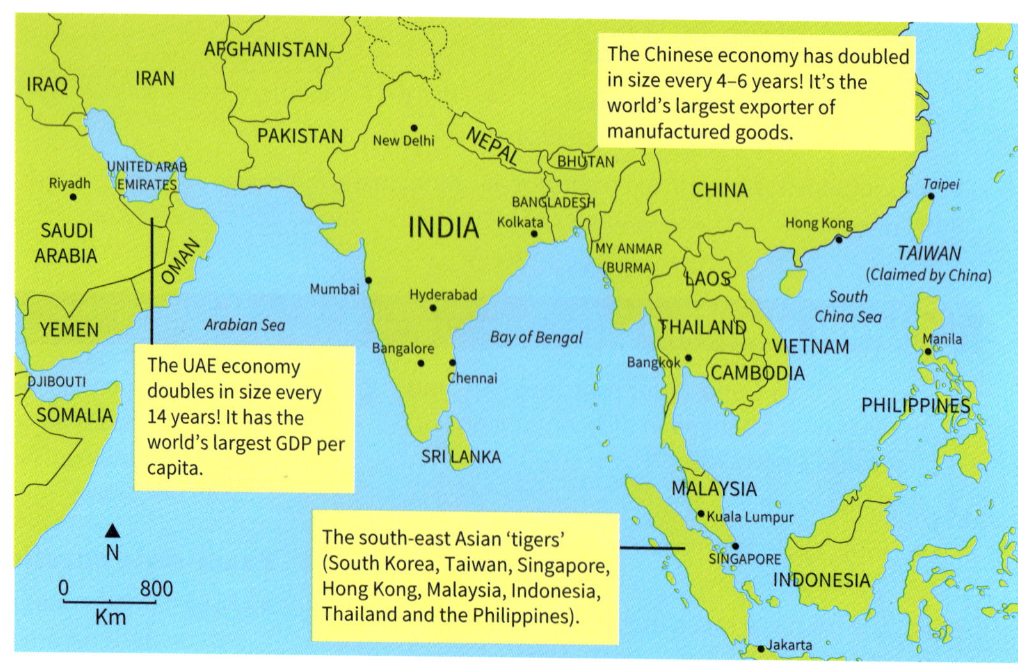

▲ ***Figure 3*** *The significance of India's location, politically and economically*

Half of the world's population lives in China, Southern Asia and south-east Asia, providing a huge market for goods and services. India's largest export customers include the Middle East, China, Singapore, and Hong Kong. The whole region is an economic powerhouse! Its time differences (+4 hours from the UK and Europe) mean that financial trading is easy, though less so with the USA (-9 hours on the east coast and -13 on the west).

Understanding India's significance

India is large! With 3.3 million sq km, it's 13 times larger than the UK, though it's only a third the size of the USA. But as a country, it's much more significant than size alone!

Socially, India has:

- the world's second largest population, 1.42 billion in 2023, and is likely to overtake that of China by 2024.
- the world's 4th and 5th largest cities – Mumbai (population 20 million) and Kolkata (15 million).

Politically India:

- is the world's largest democracy – in 2019, 911 million people were registered to vote.
- was the largest and most significant British colony, with direct British rule, known as the Raj, between 1857 and its independence in 1947.
- has a growing global influence. It was one of the founding members of United Nations and of the G20 industrial nations. It takes part in UN peacekeeping missions and contributes the second-largest number of troops to the UN.

Culturally India:

- is the birthplace of four of the world's religions, Hinduism, Buddhism, Jainism and Sikhism. Today 78% of the population practices Hinduism, 15% Islam, 2.5% Christianity, and 2% Sikhism.
- has some of the world's ancient cultures. Hindu civilisations are over 5000 years old.
- has the world's largest film industry, Bollywood, producing over 1200 films each year!

Environmentally, India has some of the world's:

- richest biodiversity. While elephants and tigers (Figure 4) are well known, its diversity includes 6% of the world's bird and plant species. But population and economic growth threaten them.
- worst environmental problems, with land, air and water pollution. India is the world's third greatest emitter of greenhouse gases and has 13 of the world's top 20 cities for air pollution.

▲ ***Figure 4*** *India's Bengal tigers are a threatened species as India's population and economy grow*

Your questions

1 Study Figure 3 and explain how India's geographical location could help it to continue to grow rapidly in future.

2 Draw a spider diagram to show India's significance in the world.

3 Based on this section, draw and complete a SWOT analysis about India to show its **S**trengths, **W**eaknesses, future **O**pportunities, and **T**hreats.

Exam-style questions

4 Study Figure 2. Suggest **two** reasons for the projected changes in GDP rank order by 2050. (4 marks)

5 Assess the influence of India's geographical location on its development. (8 marks)

2.9

India's place in a globalised world

In this section, you'll assess the reasons for rapid globalisation in India, and its impacts.

Globalisation and economic growth

The theory of globalisation is easy – if a country can't produce everything it wants, then it must trade with others. So many countries and companies make products that prices should fall. Companies always try to produce goods wherever it is cheapest, so countries such as India gain, with cheap labour. Trade from those countries increases hugely, as goods are made and sold to the world's wealthier countries.

The effects of globalisation on India have been startling, as Figure 1 shows. Look at the following changes:

- the massive increase in exports – up by over 33 times in 28 years!
- the increase in output, as measured by total GDP.

The impact on India's people has also been significant:

- a 600% increase in GDP per capita
- reduced unemployment and poverty.

▼ ***Figure 1*** *Economic change in India 1991–2019*

	India 1991	India 2019
GDP total, (US$) in PPP	1.2 trillion	9.2 trillion
GDP per capita (US$) in PPP	1150	6700
Exports value (US$)	17.2 billion	572 billion
Imports value (US$)	24.7 billion	624 billion
Unemployment rate %	20	8.5
Living in poverty %	36	21.9
Main exports	Commodities – tea, coffee, iron ore, fish products	Refined petroleum, diamonds, packaged medicines, jewellery, cars, clothing
Main imports	Petroleum products, textiles, clothing, machinery	Crude petroleum, gold, coal, diamonds, natural gas

Economic liberalisation in India

India's globalised economy began in 1991, when a programme of **economic liberalisation** began. Before 1991, the government decided which industries produced what and where. Liberalisation (or freedom) changed it to a 'market economy', where consumers, known as the 'market', and individual companies, decide:

- what people will buy, based on demand
- where goods can be made most cheaply
- where investment in products will make most profits.

Governments supporting a market economy encourage foreign investment, and reduce or abolish:

- import tariffs (see section 2.5)
- controls on how much money is brought into, or out of, a country
- taxes, especially on company profits.

The importance of transport

How can goods produced in India be cheaper in the UK than those made in Europe? Surely transport costs wipe out any advantages of cheaper labour? Transporting goods is actually cheaper today than it was in 1980, even though fuel is more expensive. Three changes have helped to reduce costs: shipping, containerization, and aircraft technology.

- **Shipping**. Ships transport 90% of goods traded between countries, and are becoming larger. A ship only needs one crew, and improvements in fuel efficiency mean that a large ship costs only slightly more to run than a small one. Global shipping has increased hugely – Figure 2 shows sea trade in 2014.

- **Containerization**. The UK imports textiles, clothing, and footwear from India. These arrive in containers on ships, like those in Figure 3, which are cheaper and quicker to transport to ports, load onto ships, and unload at the other end.
- **Aircraft technology**. Air transport is more expensive than sea, so it's had to become more efficient. Only 0.2% of UK imports arrive by air, but these make up over 20% by value. Imports from India by air are 70 times more valuable (e.g. jewellery, fresh fruit) than those transported by sea.

▼ ***Figure 2*** *Sea trade flows connecting producers and consumers. The width of the line is proportional to the amount of trade.*

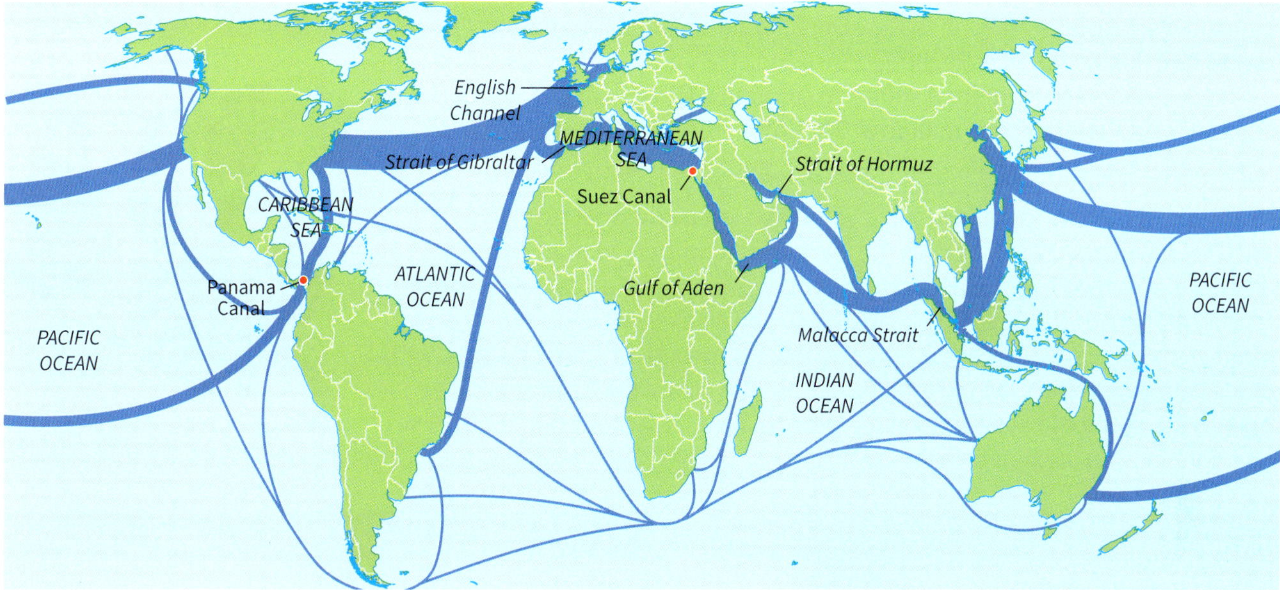

Foreign Direct Investment

Much of India's economic growth has come from Foreign Direct Investment (FDI – see Section 2.7) which the government encouraged after 1991. By 2005, India was one of the largest recipients of overseas investment. Most of it came from major **Trans National Companies** (TNCs) such as Oracle, and international banks (such as Merrill Lynch) who have invested in telecoms and services. The service economy has grown most, with TNCs investing in IT, research and development, and call centres, all providing cheaper services.

▲ ***Figure 3*** *Containers in Mumbai, India's largest container port*

Your questions

1 a Describe the changes to India's economy from 1991–2019 shown in Figure 1.

b In pairs, create a mind map to show the reasons for these changes.

2 Describe the flows of shipping trade shown in Figure 2.

3 Explain how economic liberalisation, foreign direct investment, and TNCs have all been linked in India's economic growth.

4 Explain why FDI is important in developing an economy.

Exam-style questions

5 Define the term 'foreign direct investment'. (1 mark)

6 Explain why the size of global trade flows varies on Figure 2. (4 marks)

7 Assess the economic impacts of rapid globalisation on **one** named emerging or developing country. (8 marks)

2.10 How TNCs operate in India

In this section, you'll understand how one TNC (BT) operates in India, including outsourcing.

It's all change!

It's August in the monsoon season in New Delhi, India's capital. That means a hot, humid bus ride to work for Rashmi, one of the city's many IT workers. She manages a development team for a software company. New Delhi is booming – its estimated population of 31.5 million in 2023 makes it three times the size of London. In the monsoon season, traffic gets clogged up as the roads flood.

Rashmi is typical of many people in India who have left traditional rural lives behind. They have moved to cities for the increasing numbers of manufacturing and service jobs in the IT sector. Her employer pays her £5000 a year – four times an average salary in India. She works mostly six-day weeks, but it's easier than in the countryside. In many rural areas there is now a shortage of young people, with only older family members left to farm. Fewer people now work as farmers (see Figure 2a) while services have seen a huge growth in value, accounting for 61.5% of India's GDP in 2019, as shown in Figure 2b.

▲ ***Figure 1*** *A call centre in India*

	1991	2019
Agriculture % of labour force	62	47
Industry % of labour force	11	22
Services % of labour force	27	31

▲ ***Figure 2a*** *Changing employment in India*

	1991	2019
Agriculture % of GDP	31	15.4
Industry % of GDP	11	23
Services % of GDP	27	61.5

▲ ***Figure 2b*** *How contributions to India's GDP have changed*

The part played by TNCs in India's growth

The annual turnover (or earnings) of the world's largest TNCs is higher than the GDP of most countries! Of the 100 largest global economies, 69 are corporations! In India, companies such as BT have led a process called **outsourcing** – where a company moves services overseas, such as software development or call centres, because labour is cheaper (Figure 3). India benefits from the fact that a high proportion of its qualified population speak English.

In India, three types of outsourcing have occurred:

- **Call centres.** Most Indian call centre employees (see Figure 1) are graduates earning £3000 a year (20% of what BT has to pay in the UK).
- **Software development**. Universities such as Bangalore provide technically-qualified graduates who enable BT to develop and support its broadband product out of India.
- **Company administration**, e,g. accounting.

Rank	Region	Country	City
1	Asia Pacific	India	Bangalore
2	Asia Pacific	Philippines	Manila
3	Asia Pacific	India	Mumbai
4	Asia Pacific	India	Delhi
5	Asia Pacific	India	Chennai
6	Asia Pacific	India	Hyderabad
7	Asia Pacific	India	Pune
8	Asia Pacific	Philippines	Cebu City
9	Europe	Poland	Kraków
10	Asia Pacific	China	Shanghai

▲ ***Figure 3*** *Top global outsourcing locations. Six out of the top ten cities are located in India.*

BT – a global company, shown in Figure 4 – has Indian headquarters in New Delhi, though its software development takes place in Bangalore. Bangalore is India's 'Silicon Valley', and has experienced an IT boom. The Indian government offers reduced taxes to attract companies there. Changes in communications technology help companies like BT. Instead of flying managers to meetings around the world, conference calls (using Zoom, Teams or similar) make it easy to 'meet'. People are as likely to discuss work with someone overseas via Zoom or Teams, as with the person at the next desk.

Figure 4 *Countries where BT has offices*

BT and the 'new' economy

BT is typical of TNCs that are part of the 'new economy' – based on the sale of services, rather than manufactured products. Most high-income countries now earn most of their GDP from the new economy, and India's growth has taken advantage of this. Unlike manufacturing companies, companies in the new economy are not tied to locations where raw materials are available. They are 'footloose' – they can locate anywhere – as long as high-quality communication links are available e.g. superfast broadband.

▲ ***Figure 5*** *A software company in Bangalore's 'Technology City'– an area which has attracted major IT companies*

The new economy is sometimes referred to as the 'knowledge economy' because it relies on skilled, well-qualified people with high levels of technical knowledge. TNCs usually locate close to university cities and research centres in science parks to take advantage of this. They can as easily locate in Bangalore (see Figure 5) as Birmingham – it all depends on skill and wage costs. However, most decisions are made at global headquarters, in cities like London, with other operations in lower-wage countries.

Your questions

1. In pairs, use the internet to find different images of New Delhi. Produce a presentation which tries to 'sell' India as a country where BT should outsource.
2. Explain the effects of the growth of TNCs on the following groups of people in India: **a** graduates from India's universities, **b** construction workers, **c** people living in rural villages.
3. In pairs, suggest possible reasons why services have grown so much in value (Figure 2b) but with a lower % of people (Figure 2a).
4. How much is India **a** gaining, and **b** losing from globalisation of companies such as BT?

Exam-style questions

5. Define the term 'outsourcing'. (1 mark)
6. Explain **two** reasons why large TNCs outsource their operations to emerging countries. (4 marks)
7. Explain why the governments of emerging countries are often keen to attract large TNCs. (4 marks)
8. Assess the economic and social impacts of TNCs on emerging countries. (8 marks)

2.11 The impacts of change in India

In this section, you'll assess the social and economic impacts of change in India.

Head for the city!

By moving to a city, Rashmi (in section 2.10) is part of a big change in India. In 1990, 25% of India's population was **urban** (lived in towns and cities) – 200 million people. By 2019, this had risen to 34% – over 400 million people! The change results from rural poverty (push factors) and the increase in urban jobs (pull factors), leading to **rural-urban migration** to cities like New Delhi (Figure 1).

▲ ***Figure 1*** *New apartment blocks on the outskirts of New Delhi*

A time of social change!

Economic development has social impacts, including urbanisation.

- **Urban expansion**, particularly the construction of new apartments for single professionals, like those in Figure 1.
- For educated urban women, developing a career results in **later marriage**, and fewer children. Birth and fertility rates fall, as Figure 2 shows.
- **Population structure** changes, with reduced younger age groups and a lower dependency ratio.
- **Social customs** change, especially in cities. In traditional rural villages, the Hindu caste system determines a person's status (and job), and a person marries within their caste. In cities, away from their families, young urban Hindus are freer to marry outside their caste.

Changes in health and education have also followed economic growth. India's infant mortality rate has fallen by over 55% since 1991, shown in Figure 3. This is due to:

- increased access to safe water supplies, as waterborne diseases (e.g. diarrhoea) are one of the biggest child killers
- rapid expansion of hospitals in rural areas.

Although the data in Figure 3 show improvements, women's literacy remains 17% lower than men's. However, there is evidence that the gender gap has narrowed since 2011.

Population Indicators	1991	2021
Birth rate (per 1000 people)	30	17.5
Death rate (per 1000 people)	10	7.1
Fertility rate	4.0	2.2
% population aged 0–14	37.7	26.3
% population aged 65 and over	3.8	6.7
Dependency ratio %	70.4	48.7

▲ ***Figure 2*** *Changes in India's population data, 1991–2021*

HDI, Health and Education Indicators	1991	2019
HDI	0.38	0.645
Life expectancy (years)	59.7	70
Infant mortality per 1000 live births	89	39.6
Maternal mortality per 100 000 births	550	145
Number of doctors per 1000 population	0.41	0.86
Average number of years in school	2.4	12
Literacy rate %	50	74
Average age of first marriage for women	18.7	21

▲ ***Figure 3*** *Changes in HDI, health (yellow) and education (blue) data for India, 1991–2019*

Economic change – Winners and losers

India's middle class – its managers, university lecturers, or well-paid IT workers like Rashmi – is growing. There will be 250 million middle class people by 2025! They benefit from good salaries and employment opportunities in India's cities. But that leaves a billion who are not well paid. Most have jobs, but these are mostly low paid.

Garment workers

After 1991, India's government allowed large TNCs to set up factories, like the one in Figure 4, taking advantage of low wages. By 2015, clothing was India's largest manufacturing industry, employing 80 million people and earned US$ 300 billion in GDP. For TNCs, India's minimum wage is 87% lower than in the UK, and many clothing companies don't even pay that much. European and American clothing retailers which buy Indian clothing include Walmart and Zara.

▲ ***Figure 4*** *One of many thousands of clothing factories in India*

This employment trend is controversial:

- Poverty is widespread in India. There is no shortage of people willing to work 100 hours a week in factories for an average of £35, including overtime.
- Most textile jobs are unskilled, and there's no equal pay agreement, so 70% of employees are young women on lowest pay.
- Many sweatshops discriminate against older women returning to work after raising children.

Did you know?

By 2030, European and American middle classes will shrink from 50% to 22% of the world's total, while Asia's share will more than double to 64%.

Looking forward – can it last?

Some believe that India's economic growth might soon peak. Between 2000 and 2050, all those born during the 1990s (when birth rates were higher) will be working adults. Economists call this a 'window of opportunity' – when the economically active population is at its highest. This only lasts so long – by 2050, India's population will be ageing. If birth rates fall further, its dependency ratio will rise. But for now, India's economy booms!

Your questions

1 Explain why economic growth leads to rural-urban migration.

2 a Using Figures 2 and 3, describe recent changes in population, health and education in India.

b Explain how economic growth has contributed to **any three** of these changes.

3 Draw and complete a table to show the winners and losers from economic changes in India since 1991. Consider different age and gender groups.

4 In pairs, discuss whether you think companies such as Walmart, Gap and Zara should use cheap labour in India.

Exam-style questions

5 Define the term 'rural-urban migration'. (1 mark)

6 Explain **one** positive and **one** negative impact of the rapid growth of manufacturing industry upon a named emerging or developing country. (4 marks)

7 Using examples, assess the social impacts of globalisation upon **one** named emerging or developing country. (8 marks)

2.12 Unequal development

In this section, you'll assess how change has affected parts of India differently.

Booming India – but for whom?

India's wealth creates some problems.

- Its economic growth is mainly urban, and wealth is concentrated in cities. People migrate to cities for work and spend money earned on housing and services, which creates more jobs. This causes an upward spiral that economists call the **multiplier effect,** shown in Figure 1. Over time, this effect can develop over a whole region, which becomes known as a **core region**. But other regions, which are further from the core, don't reap the same benefits.
- Environmental problems can result from rapid economic growth, unless it is regulated.

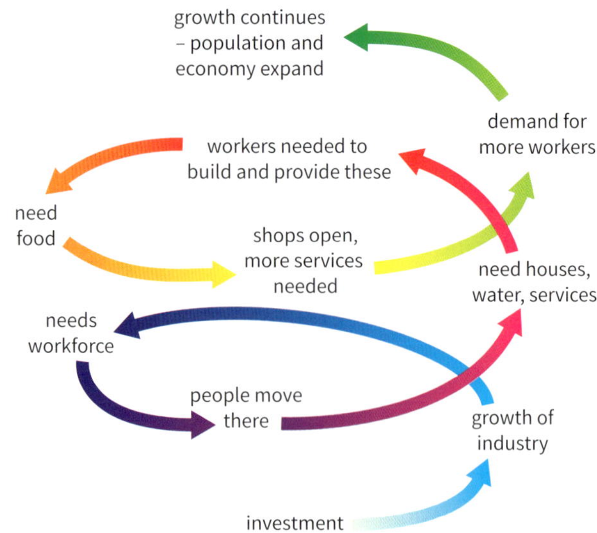

▲ ***Figure 1*** *The upward spiral caused by the multiplier effect*

Maharashtra – an urban core region

India's wealth is unequally shared. It varies between states, and between urban and rural areas. Figure 2 shows total GDP for each state. Maharashtra has by far the biggest GDP. It's India's richest core region. It contains India's largest city, Mumbai, in which live some of India's most talented, well-qualified and well-paid people, as well as many poor, often landless farmers who fuel rural-urban migration.

Maharashtra's economic growth has come from:

- **service industries** – e.g. banking, IT, call centres (see section 2.10).
- **manufacturing** – half of Mumbai's factory workers make clothing. Other industries include food processing, steel, and engineering.
- its **port,** which is the second largest in the country.
- a booming **construction** industry, building factories and offices.
- **entertainment**. Mumbai hosts the world's largest film industry, Bollywood.

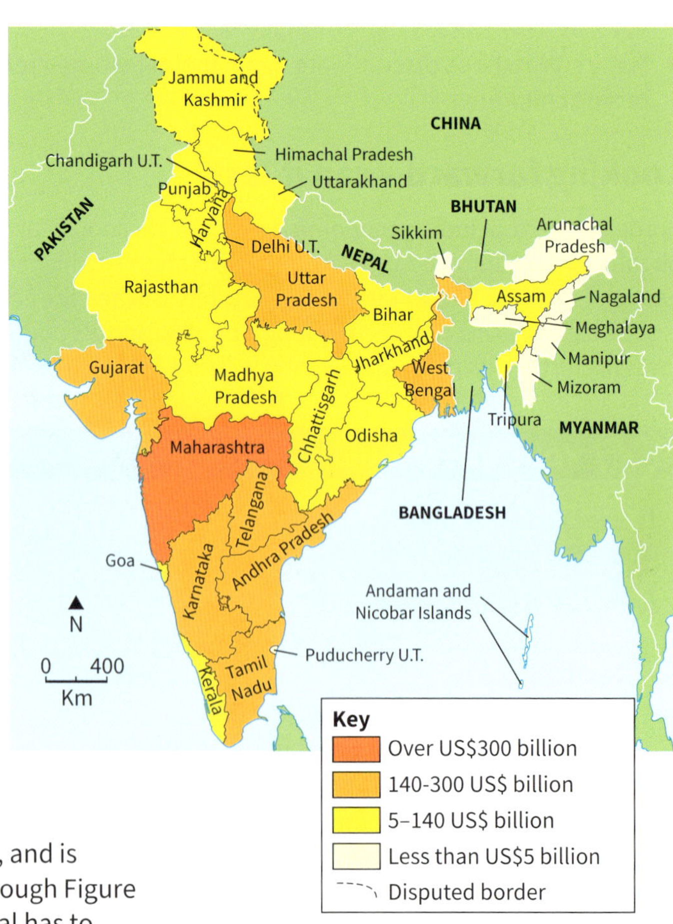

▲ ***Figure 2*** *The total GDP of India's states and territories (in US$) in 2019–2020*

Bihar – the rural periphery

By contrast, the state of Bihar receives little investment, and is distant from cities. It's a part of India's '**periphery**'. Although Figure 2 shows that Bihar's *total* GDP is not the lowest, that total has to be shared between 100 million people. It's the *per capita* income shown in Figure 3 that makes it India's poorest state.

- 86% of its population is rural. Many are subsistence farmers, trapped in a cycle of poverty, shown in Figure 4.
- Half of its households earn less than 80p a day, and 80% work in low-skilled jobs.

Even with 100 million people, Bihar gets little investment, because people can't afford basic services – only 59% of its population has electricity.

- As in Maharashtra, Bihar is also a traditional caste-based society. Those in higher castes are literate, whereas those in lowest castes are mostly illiterate. It is difficult to marry outside one's caste, so those who are poor stay poor.
- School attendance is low. Only a third of children complete primary school, and 2% reach Years 12 and 13. Overall, literacy in Bihar is 47%.
- Women are poorest in Bihar, and have India's lowest literacy rates (33%). They rarely own land, and most are low-wage labourers.

State	GDP per capita in $US PPP 2019–2020
Goa	25044
Delhi	19974
Maharastra	10477
India average	**6907**
Manipur	3834
Uttar Pradesh	3635
Bihar	2395

▲ ***Figure 3*** *The per capita income of India's core and periphery states*

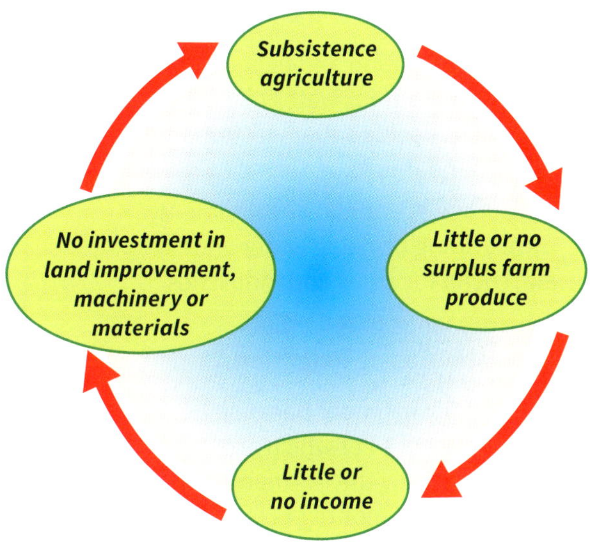

▲ ***Figure 4*** *The cycle of poverty*

The environmental impacts of growth

With little regulation, India's economic growth has made several environmental issues much worse including:

- **Water pollution** from poorly managed garbage and waste removal services (much consumer waste is dumped in rivers), poor or variable street drainage, cremations near rivers, and lack of sewage treatment.
- **Air pollution** caused by old public transport, urban traffic, and emissions from coal-fired power stations. Even in rural areas, burning fuelwood and biomass causes localised pollution. Each contributes to India's increasing emissions of greenhouse gases.
- **Loss of biodiversity** and land degradation, as more land is needed for food, cities and industry.

Your questions

1. Using Figure 3, calculate by how much in US$ **a** Maharastra's GDP per capita is above India's mean, **b** Bihar's is below it.
2. Using data, make up four statements to show that:
   - **a** Maharashtra is part of India's 'core',
   - **b** Bihar is part of its 'periphery'.
3. Copy the cycle of poverty in Figure 4. Explain what could happen if a family member **a** becomes ill, **b** gets into debt, **c** goes to Mumbai and sends wages home.

Exam-style questions

4. Define the term 'core region'. (1 mark)
5. Explain **one** reason why levels of development vary within a country. (4 marks)
6. Assess the extent to which India has an economic 'core' and 'periphery'. (8 marks)

2.13 A top-down project: The Narmada River Scheme

In this section, you'll evaluate top-down development in India.

Top-down – the government decides!

Over much of India, rainfall is seasonal and unevenly spread, shown in Figure 1. Between November and March, almost no rain falls across much of India. Parts of the north-west are so dry that it's semi-desert. Then, between May and September, the **monsoon** brings heavy rain across India. Dams make it possible to store these rains for the dry season.

As India's population and economy increase, demand for water rises. The Government decided that western India needs super dams to:

- encourage economic development, by providing drinking water and electricity for cities and industries
- farm dry lands to feed the population, using **irrigation**.

The Indian Government has built over 4500 dams, 14 of which are huge **super dams**. Now the Narmada, one of western India's major rivers (shown in Figure 2), is being tackled by building 3000 dams, of which 30 will be super dams, taking 100 years to complete! But how well will this work?

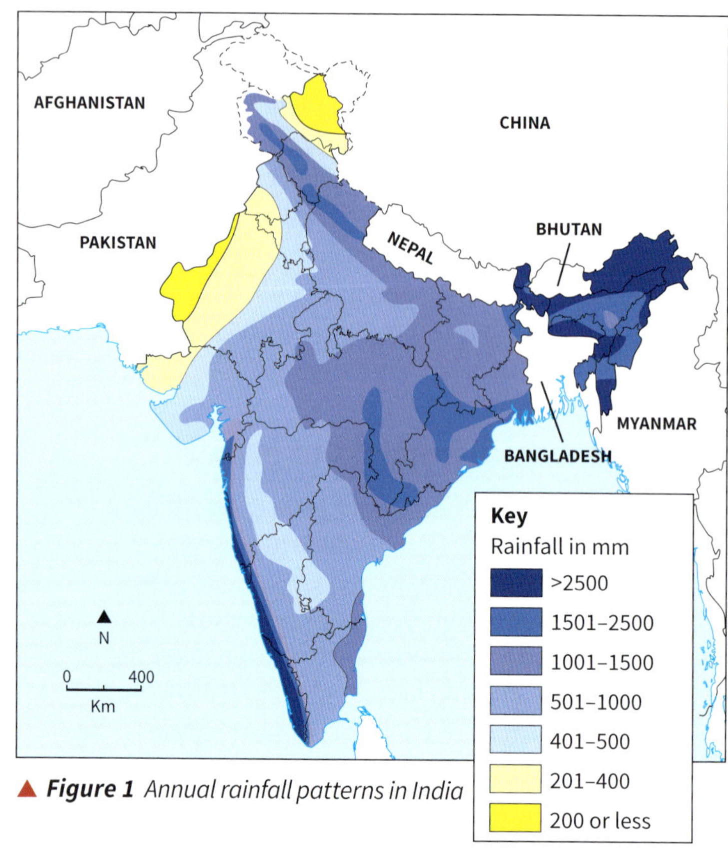

▲ ***Figure 1*** *Annual rainfall patterns in India*

Irrigation is the artificial watering of land that allows farming to take place.

Top-down development and IGO funding

'Top-down development' occurs when decisions about development are made by governments or large companies. Those decisions are imposed on people, in theory because of their intended benefits. 'Top down' involves:

- decision-makers identifying a country's needs or opportunities, e.g. to develop energy sources and transport, or improve food security
- experts planning changes
- local people being told about plans, but having no say in whether, or how, they happen.

Top-down schemes are usually large and expensive, so they often involve overseas loans from **Inter-Governmental Organisations** (IGOs) – that's government banks and agencies working together. The Sardar Sarovar Dam has been funded partly by the World Bank, partly by Japanese banks, and partly by state governments in India who stand to benefit from the dam.

The argument says everyone benefits from 'top-down' because of a process called 'trickle down', where jobs (and therefore wealth) 'trickle down' to the poor. Some 'top-down' schemes have bad reputations because they bring more problems than gains, while others bring huge benefits.

The Sardar Sarovar Dam

Figure 3 shows the Sardar Sarovar Dam on the Narmada River. It is one of the world's largest dams. When complete, it will store monsoon rains for use during the dry season. Originally 80 metres high, the government plans to raise it to 163 metres to increase its capacity.

Who benefits?

- **India's cities.** The dam is multipurpose, providing 3.5 billion litres of drinking water daily, and hydroelectric power (HEP).
- **Farmers** in western India. A network of canals will irrigate 1.8 million hectares of farmland in Gujarat, Maharashtra, Rajasthan and Madhya Pradesh (see Figure 2). These states suffer drought causing loss of crops and animals each year.

Who loses?

- **Local residents.** 234 villages have been flooded by the dam, forcing 320 000 people out. Few rural families can afford electricity from the scheme – only cities benefit.
- **Local farmers.** Good quality farmland has been flooded. Damming the river means that fertile sediment, deposited on flood plains each year, is also lost.
- **Western India.** Religious and historic sites have been flooded.
- **People downstream.** The region has a history of earthquake activity. Seismologists believe that the weight of large dams can trigger earthquakes, which could destroy the dam and cause massive loss of life.

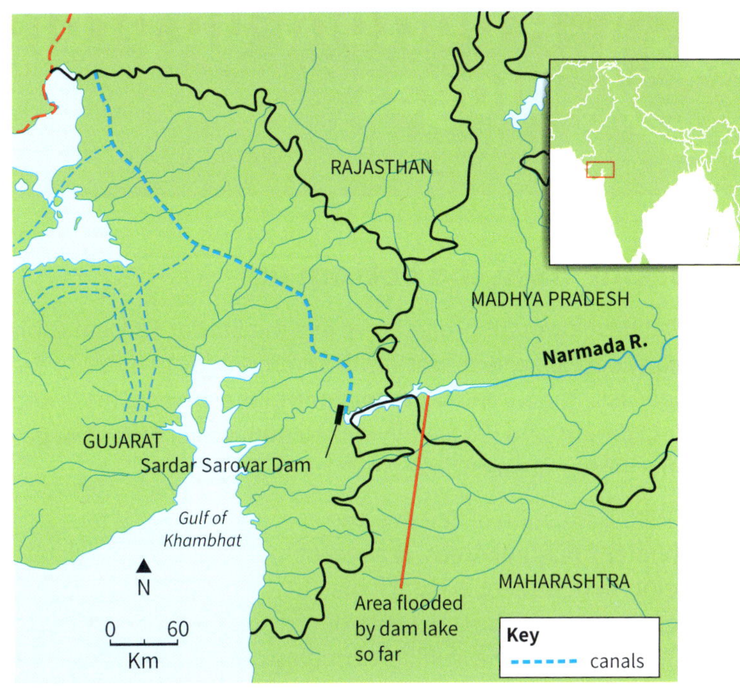

▲ ***Figure 2*** *The Narmada River in western India, and the site of the largest dam – the Sardar Sarovar*

▲ ***Figure 3*** *The Sardar Sarovar dam*

Your questions

1 a Complete a large copy of the table about the economic, social and environmental benefits and problems of the Sardar Sarovar dam.

	Benefits	Problems
Economic		
Social		
Environmental		

b Now highlight or underline in one colour those benefits or problems which are local, and in another those which are further away.

c Which are the greatest benefits – economic, social or environmental? Are they local or further away?

d Which are the greatest problems?

e Explain whether you think top-down schemes like this should be built if they cause such problems.

Exam-style question

2 For a named top-down development project, evaluate its benefts and problems. (8 marks)

2.14 A bottom-up project: biogas

In this section, you'll evaluate bottom-up development in rural India.

Working from the bottom up

Not all development projects are like the Sardar Sarovar Dam (in section 2.13). In many cases, experts work with communities to identify their needs, offer people assistance, and let them control their lives. This is known as **bottom-up development** and it's usually run by **non-government organisations** (NGOs) such as charities or universities.

ASTRA (Application of Science and Technology in Rural Areas) is a recent development project in rural India. Researchers from the University of Bangalore spent time in villages finding out about people's lives. They talked to families, recorded how they spent their time, and listened to their problems. This is how a bottom-up development project begins.

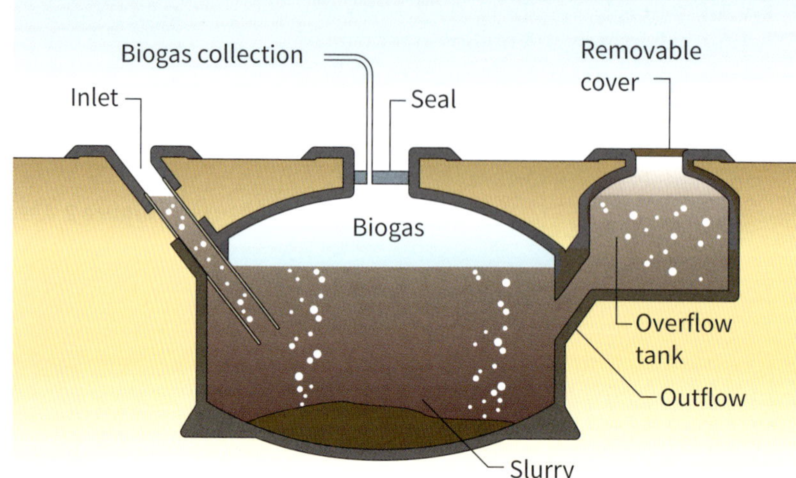

▲ ***Figure 1*** *A biogas plant*

The problem of time

ASTRA found that, for most rural families, daily routine takes time – especially for women and girls. Cleaning, collecting fuel, preparing and cooking food, fetching water, tending sacred cows, looking after the vegetable patch – all before any paid work is done! Rural girls have little education and few complete primary school. The biggest job is collecting fuelwood. Every family needs 25–30 kg of it every week, and it takes hours to collect. As population grows, it's in increasingly short supply.

Intermediate technology uses low-tech solutions using local materials, labour and expertise to solve problems.

Solution – think cow dung!

The answer to the fuel (and time) problem was under their noses – cow dung! Cow dung is a valued resource, because it produces gas, called **biogas**. The gas is used for cooking by day, and powering electricity generators at night. The dung is fed into a brick, clay or concrete-lined pit that forms part of a biogas plant (see Figure 1). The pit is sealed with a metal dome and the dung ferments to produce methane. As pressure builds, methane is piped into homes.

It's simple, uses local materials, and is an example of **intermediate technology**. Figure 2 shows how it uses little space, uses materials available in India, and can be located in a village without impact.

▲ ***Figure 2*** *A village biogas plant*

The benefits of biogas

By 2010, four million cattle dung biogas plants had been built in India. These created 200 000 permanent jobs, mostly in rural areas, as well as other benefits:

- Unlike firewood, cooking with gas produces smoke-free kitchens (shown in Figure 3), so there are fewer lung infections.
- Heat is instant, so cooking is quicker.
- There's no ash, so there's less cleaning.
- No longer is time spent gathering wood or dung, so girls now have more time to go to school.
- Cattle are now kept in the family compound, making dung collection easier. Previously, cattle would graze local woodland, eating saplings and preventing trees from regenerating.
- When cattle dung is fed into the digester, micro-organisms that cause disease are destroyed as the dung ferments.
- After digestion, the sludge is richer in nutrients than raw dung, so it makes a better fertiliser.
- Many villages now use biogas to power electricity generators that provide light at night and pump drinking and irrigation water from underground. Farmers can now get three crops of vegetables a year using pumped water.

▲ ***Figure 3*** *Cooking indoors using a new, clean, biogas stove*

Your questions

Impact	Short-term (immediate or in a few months)	Medium-term (up to a year)	Long-term (over a few years)
Social			
Economic			
Environmental			

1 a Copy and complete the table above to show the benefits of biogas for communities, and whether these are short-, medium- or long-term.

b Which benefits are greatest? Over which time period?

c How suitable does biogas seem to be for rural communities in India? Explain your answer.

2 a Draw a table to show the overall benefits and problems of top-down (section 2.13) and bottom-up development.

b Which seems better for **a** national, **b** local needs?

Exam-style questions

3 Define the term 'intermediate technology'. (1 mark)

4 Explain **two** problems faced by girls living in many rural communities in developing or emerging countries. (4 marks)

5 Explain **two** benefits of a named bottom-up development project. (4 marks)

6 For a named bottom-up development project, evaluate its benefits and problems. (8 marks)

2.15 India – which way next?

In this section, you'll evaluate the challenges that India faces and how its role is changing.

The challenges ahead

In spite of rapid economic growth, India has many problems. It has not invested enough in **infrastructure** to keep pace with economic and population growth. A quarter of its people have no electricity supply, and its per capita electricity consumption is low compared with other NICs. Its electricity network is inadequate, but demand continues to rise. The system often crashes. In July 2012, India suffered the world's biggest power cut which affected 620 million people, in areas shown in Figure 1. That's 50% of India's and 9% of the world's population! Most companies have their own generator to avoid chaos.

It's the same with water, education and wealth distribution.

- Water resources are declining and projects such as the Narmarda schemes (see section 2.13) are essential, if controversial.
- Far too few people have a good education.
- Poverty is widespread (see Figure 2). Little wealth spreads to India's rural villages, where two-thirds of its population live.

▲ ***Figure 1*** *Areas of India affected by a major power blackout in July 2012*

What's the cause of India's problems?

The cause is the rural-urban divide, political divisions, and lack of infrastructure. The reasons are simple:

- To attract investment, India's tax rates on businesses are low. As a result, TNCs pay little tax. The government therefore has relatively little money to pay for public services, e.g. water, sanitation. Private companies could invest in these services, but so few people could afford them, they do not regard them as worth doing.
- For those on high incomes, there are many ways to avoid paying tax – so the government gets little tax revenue.

It's the same in cities. Among the smart office blocks, few buildings are actually finished. It's as though builders walked away. This makes sense when you know India. Higher property taxes are due on every *completed* building – so companies never finish! The top floor of a block might be left open, used instead as an outdoor 'street' where food traders set out stalls for office workers! As long as it remains incomplete, lower property taxes are paid. This is business, Indian style.

300 million Indians earn less than US$1 a day. That compares to 85 million people in China (with its bigger population). 45% of Indian children under-five are malnourished. Two-thirds of India's homes have no toilet. Only half of rural villages have electricity.

▲ ***Figure 2*** *India's future problems – adapted from an article on the BBC World Service online*

India on the world stage

India's role is increasing in Asia, and also globally, as a major world economy.

- In **Asia**, relations with some neighbours are tense. There have been three wars with Pakistan since 1947, when the two were separated because of religious differences. Both countries have nuclear weapons, raising fears of future conflict. Kashmir, lying between them, is disputed between the two (see Figure 3 and its location on Figure 1). Water scarcity is a source of conflict, as the main rivers of both countries rise in the mountains of Kashmir. HEP projects on India's side could take irrigation from farming areas of Pakistan like the area in Figure 4.
- **Globally**, in 2021, India became a non-permanent member of the UN Security Council, after a long campaign. It belongs to the G20 group of the world's largest economies, which includes the USA and some EU members. The G20 aims to improve international cooperation in many issues. By joining, India can help resolve problems needing global action, such as climate change. However, India's government refused to phase out its use of coal in the COP26 talks on climate change in 2021. India can also support investment from the World Bank and development banks, such as the Asia Development Bank, to help the economies of developing countries.

▲ ***Figure 3*** *Kashmir, disputed territory in the Himalayas between India and Pakistan*

▲ ***Figure 4*** *Irrigation ditch in Pakistan*

Your questions

1. In pairs, brainstorm reasons why spending on infrastructure is important for emerging economies.
2. List the arguments for and against TNCs paying more tax to help India develop.
3. In pairs, research India's relationships with one of: China, Afghanistan, Pakistan, Bangladesh. Research **a** what relationships are like, **b** reasons for this, **c** where improvements could be made. Produce a six-slide PowerPoint.
4. Write a 200-word speech to explain why India should become a permanent member of global organisations such as G20 or the UN Security Council.

Exam-style questions

5. Explain **one** reason why a major power cut in an emerging country could damage its economic growth. (2 marks)
6. Explain **one** benefit and **one** problem for an emerging country if it decides to raise taxation levels on large companies. (4 marks)
7. Explain why a named emerging country might wish to extend its global influence politically. (4 marks)
8. For an emerging country, evaluate the impacts of its international relations with other countries. (8 marks)

3.1 A world of growing cities

In this section, you'll understand past, current, and likely future trends in urbanisation.

New arrivals

Every day, coaches arrive at the bus station in Kampala, Uganda's capital, shown in Figure 1. People pour off the coaches, many on their first day in the city. Most are young, hoping for a job, perhaps, or a college place, or to join a relative. The majority have left rural villages where their families live, but where employment opportunities are limited.

This is the world's biggest migration. It is called **urbanisation** and occurs as people move from rural (countryside) areas to urban (towns and cities). An increasing percentage of the world's people now live in urban areas. Most cities have younger populations than rural areas. For migrants, cities have benefits – employment is often more plentiful – but they can bring social isolation as people live without families and friends.

▲ ***Figure 1*** *The main bus station in Kampala*

Urbanisation means a rise in the percentage of people living in urban areas, compared to rural areas.

How does urbanisation vary between different regions?

In 2007, the world passed a milestone. For the first time, more people lived in urban areas than rural. The United Nations (UN) predicts that by 2050 over 68% of the world's population will be urban. The biggest increase is in Asia and Africa, shown in Figure 2.

- Asia – urban population is expected to grow to about 64% by 2050.
- Africa – urban population will grow to 58% by 2050, though this will still be the world's lowest urban percentage.

The causes of this growth are:

- mostly, migration to cities
- but also, higher natural increase, i.e. more births than deaths.

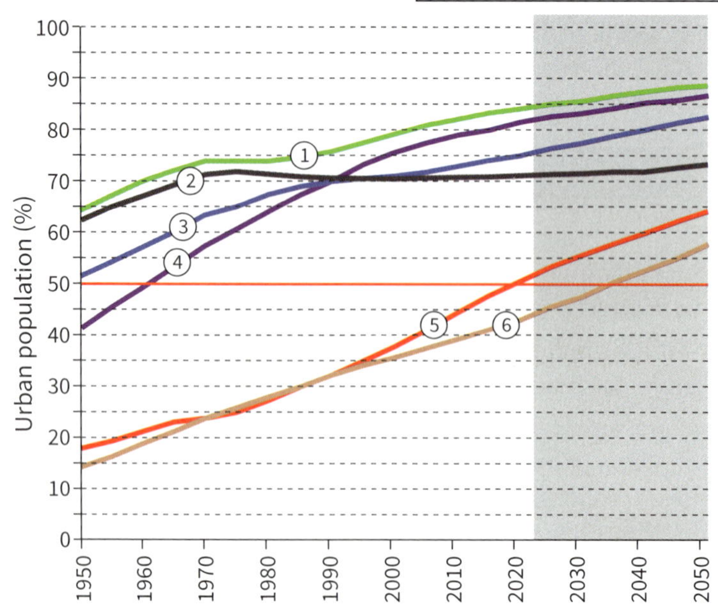

▲ ***Figure 2*** *The increasing urban population in different world regions, 1950–2050 (projected)*

The changing balance

Urbanisation isn't just a process. It can affect individual cities differently. Figure 3 shows the changing distribution of the world's ten largest cities from 1975 to the predicted population levels by 2025.

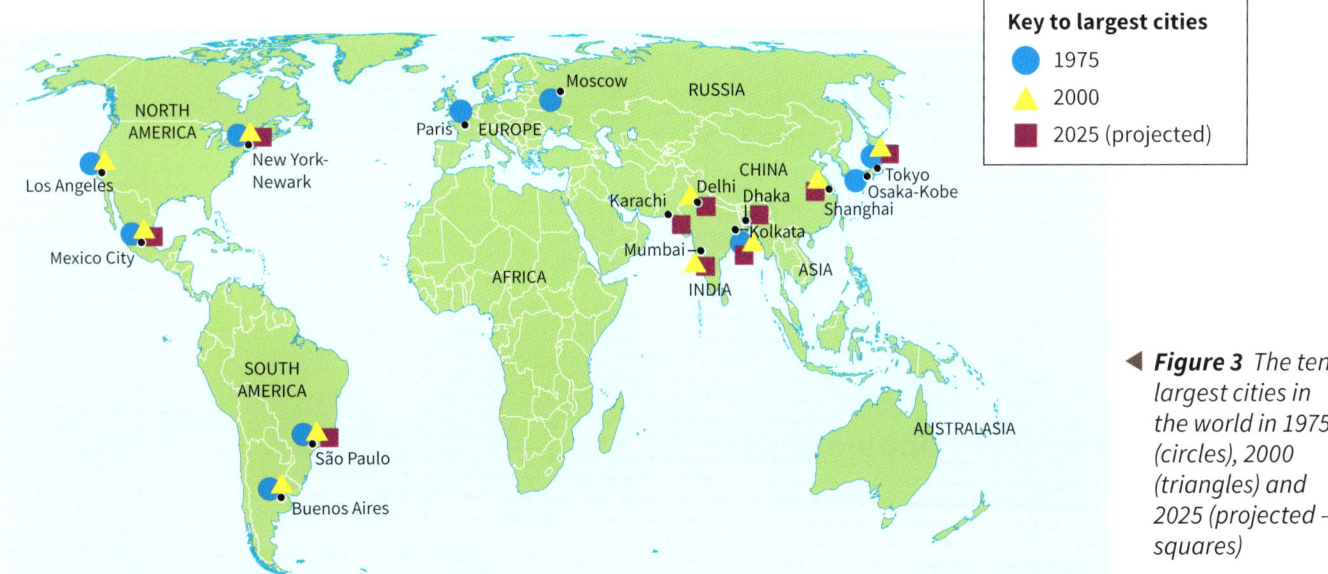

◀ ***Figure 3*** *The ten largest cities in the world in 1975 (circles), 2000 (triangles) and 2025 (projected – squares)*

The cities in Figure 3 show that big changes have taken place. In 1975, six of the world's ten largest cities were in the most developed countries. The populations of these cities were already high because they had grown during the industrial revolutions of the 18th and 19th centuries. The UK was the first country in the world to reach 50% urban population in 1861.

Urbanisation in developing countries has mostly taken place since the 1950s. Their urban populations have risen rapidly by 2.3% every year since 2000. That means urban populations in developing countries double every 30 years! Even so, less than 40% percent of people in developing countries lived in urban areas in 2015.

By 2025, only two of the world's ten largest cities will be in the most developed countries. Urban populations in these countries are now rising slowly. But there are exceptions. London's population fell between 1951 and the 1990s. Redevelopment of the city has created more jobs and housing. Now its population is growing faster than at any time – 16% during 2011–21!

Your questions

1 Explain the difference between 'urban', 'city' and 'urbanisation'.

2 a Using Figure 2, compare the differences in urban populations in different regions of the world in **i)** 1950, **ii)** 2000, **iii)** 2050.

b London's population was 9.5 million in 2021. If growth continues at 16% per decade, calculate its population **i)** by 2031, **ii)** by 2051.

3 Using Figure 3, describe changes in the distribution of the world's largest cities, 1975–2025.

Exam-style questions

4 Define the term 'urbanisation'. (1 mark)

5 Study Figure 3. Describe the projected distribution of the world's largest cities by 2025. (3 marks) 

6 Explain **two** reasons why the world is increasingly urbanised. (4 marks)

3.2 The world's megacities

In this section, you'll understand the global pattern of megacities and world cities.

A world of millionaires

As the world becomes more urbanised, so towns and cities grow in population and area. As cities grow, so do the words used to describe them. The term 'million city' is used for any city with a population of over one million.

- In 1950, there were 83 million cities.
- In 1997, there were 285 – 106 in high income countries, and 179 in emerging and less developed countries.
- By 2015, there were over 500!

Understanding urban populations

City population sizes are sometimes confusing. Search for the population of London and you'll get different answers, depending on the boundaries. For example, London's population in 2021 was either:

- 9.4 million (the **metropolitan area**, called London, with its 33 boroughs), or
- 11 million (the **built-up area**, including places such as Watford), or
- 15 million (the population within a **50 km radius**), including places such as Chelmsford.

So London is the biggest metropolitan area in Western Europe, but Paris is its largest city!

▲ ***Figure 1*** *The Greater Manchester conurbation*

As cities grow, they merge. This forms a **conurbation**, or a continuous urban area, for example, Manchester is now the Greater Manchester conurbation, including towns such as Oldham, shown in Figure 1. The population of Manchester varies, depending on whether you mean the city (553 000 in 2019), or the conurbation (2.8 million in 2019).

The growth of megacities

As cities grow, some become so large that they become **megacities** (over 10 million people). Figure 2 shows Tokyo-Yokohama, two cities which have merged and are the world's largest megacity.

Like the world's largest cities (see section 3.1), the largest megacities have changed:

- In 1980, most were in high income countries – New York, Tokyo, Paris, London. The populations of some of these have hardly changed since.
- Increasing numbers of megacities are in the emerging countries. By 2020, only two of the world's megacities (Tokyo-Yokohama and New York) were in high income countries. The rest were in emerging countries, e.g. Sao Paulo, Shanghai, and Mumbai.

▼ ***Figure 2*** *The Tokyo-Yokohama region. Areas in purple are urban areas.*

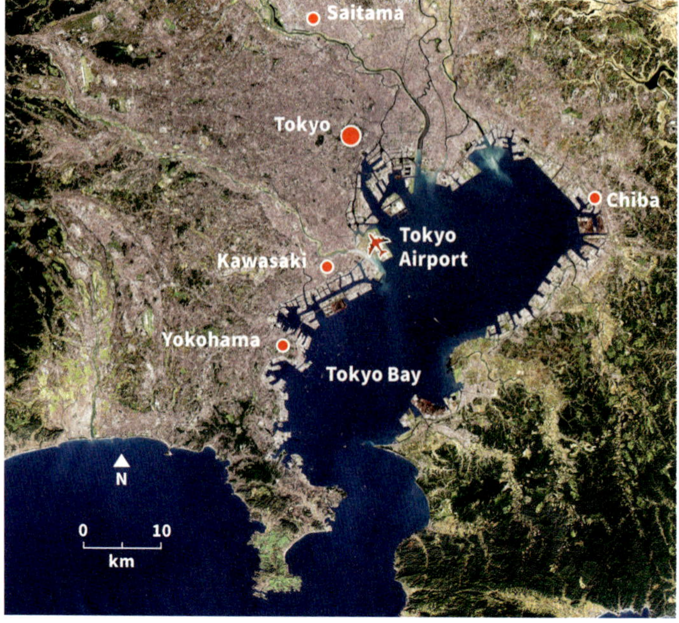

▲ ***Figure 3*** *The major 'world cities', a term used by Loughborough University for influential cities*

World cities

A few megacities play a disproportionate role in world affairs. These are called **'world cities'**. They have **urban primacy** – that means an importance and influence bigger than their size suggests. London is one of these, both in the UK (which its economy dominates), and in the world. It is not the world's largest city, but it plays a big role globally.

Think of each world city as a wheel. The cities are 'hubs' (centres), where economic activity occurs. Spokes radiate out with flows of:

- **investment.** London and New York are the world's biggest financial centres. Half the world's money – several trillions of US dollars – comes through London each year!
- **airline traffic.** In 2019, Dubai (86 million passengers) was the world's largest international airport. But add together London's two largest airports, Heathrow and Gatwick, and London is the biggest (119 million).
- **decision-makers** in TNC headquarters. They decide what to produce, what to sell, and where economic activity occurs. 80% of the world's largest companies have headquarters in cities of the USA, the EU, and Japan.
- **political decisions.** Government decisions in the UK can affect people globally, e.g. about where to invest, or trying to resolve conflicts.

In 2020, the world cities were graded based on their influence in the global economy. The cities and gradings are shown in Figure 3.

Your questions

1 Explain the difference between city, metropolitan area, conurbation, million city, megacity and world city.

2 Measure the size of Tokyo-Yokohama in Figure 2 **a** from north-south, **b** from east-west.

3 Using Figure 2, draw a series of labelled diagrams to show how the different parts of Tokyo-Yokohama (e.g. Kawasaki) grew over time until the whole area become a megacity.

4 Draw a spider diagram to explain what makes a city a world city.

Exam-style questions

5 Study Figure 3. Describe the distribution of the world cities. (3 marks)

6 Explain the term 'urban primacy'. (3 marks)

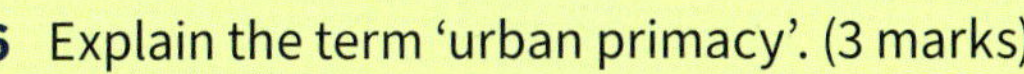

7 Explain **two** ways in which world cities influence the global economy. (4 marks)

3.3 Urban process and change

In this section, you'll assess the factors that contribute to the growth and / or decline of cities in developing, emerging and developed countries.

Urbanisation on a huge scale!

Imagine a city whose **net growth** adds a population equivalent to that of Nottingham *every year*! That's Guangzhou, China, which grew by 3.3 million people in the second decade of the 21st century! It was the world's fastest growing city. At the present time, the world's fastest-growing cities are in Asia and Africa. Urban populations there are growing by 3.0% each year, compared to the global average of 2.1%.

The main cause is economic growth, which creates new jobs.

- In emerging countries, TNCs have invested in factories, causing rapid industrialisation.
- In high income countries (HICs), some 'world cities' are growing rapidly as their service economies expand.

In each case, migration causes urbanisation as people move to find work.

Net growth means the number left after subtracting those leaving from those arriving.

Deindustrialisation – closure of industries.

Case Study 1 – Kampala, Uganda

Kampala (population 3.3 million), is the capital of Uganda, and typical of many African cities. Its growth is driven mainly by **internal migration**, but **natural increase** (an excess of births over deaths) also plays a part, as young migrants go on to have children.

Most people come from rural areas - **rural-urban migration** - shown in Figure 1. It is a result of factors which 'pull' people to Kampala, and others that 'push' them from the countryside. 'Pull' factors include:

- jobs in growing businesses. A recently opened steelworks in southern Kampala, owned by a TNC, now employs 2000 people.
- jobs in construction, building infrastructure (e.g. water, transport).
- better services (e.g. health and education) which make quality of life better than in rural areas.
- better life chances, with more opportunities.

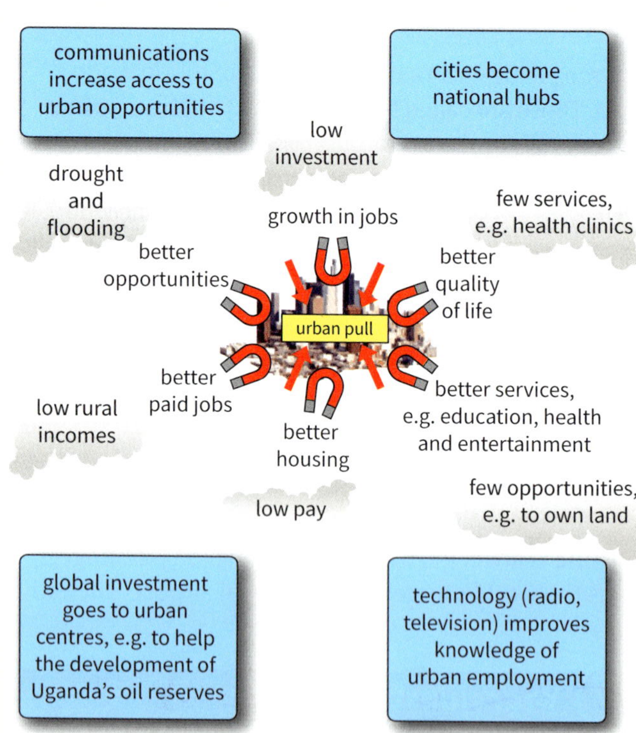

▲ ***Figure 1*** *Rural-urban migration in Uganda*

Case Study 2 – New York City, USA

While the population of many cities in high income countries has slowed, New York City's growth increased to 8.34 million by 2019 (see Figure 2). During 2010–19, its population grew by a net 161 000 people, but this masks a real churn in people arriving and leaving:

- Natural increase (the difference between births and deaths) added 565 000 people and international migration another 496 000 (total 1.061 million).
- 900 000 left the city to live elsewhere in the USA. 

A major cause of its growth is the '**knowledge economy**'. New York is one of two world cities (see section 3.2) with its focus on expertise in finance. It needs well-qualified people with university degrees and specialised training. It now has to 'import' experts from overseas, as there are not enough in the USA, as well as unskilled migrants. This has increased **international migration**. With 37% of its population foreign-born, New York has the world's largest urban immigrant population.

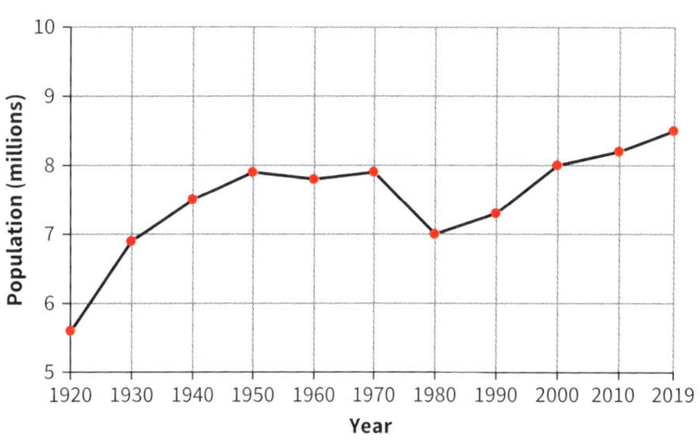

▶ ***Figure 2*** *New York's population growth since 1920*

Case Study 3 – Detroit, USA

Cities also decline. Detroit is home to General Motors (GM), the USA's largest car company. GM and its suppliers created thousands of jobs, so that by 1950 Detroit's population was 1.85 million as people moved there. But two problems led to Detroit's decline (see Figure 3).

- Between 1960 and 2000, its wealthier population left to live in suburbs outside the city. This left a poorer population, so income from local taxes fell. Unable to provide enough services, Detroit City Council went bankrupt in 2013.
- Between 2000 and 2010 GM's sales halved. It survived, but now makes cars using robotics, needing fewer people, and parts from overseas. This created unemployment and **de-industrialisation** (closure of industries). People left to find work, causing population decline.

In 2020, although unemployment remained high, it fell to 10% (from 15% in 2015). The decline in population has also slowed. Many homes in Detroit have been sold off to meet personal debts, some for just a few dollars (Figure 4). But demand is now rising as new investment is creating new jobs.

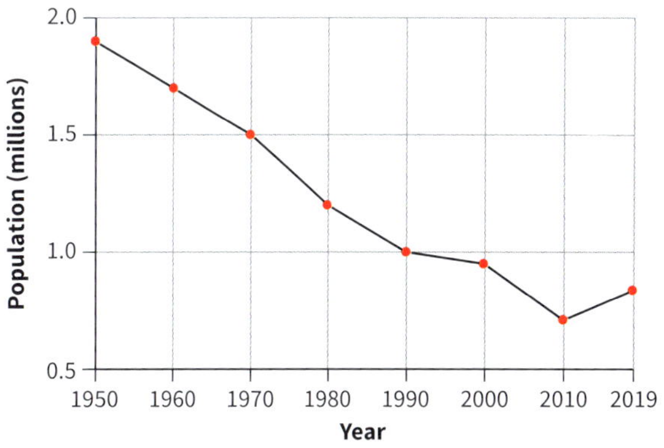

▲ ***Figure 3*** *Detroit's changing population*

▲ ***Figure 4*** *Demand in parts of Detroit is so low that houses have been auctioned for US$1*

Your questions

1. **a** Copy Figure 1 and add labels to explain the reasons for Kampala's growth.
   b Make a second copy, and label it to show the 'churn' in New York's population.
2. In pairs, design a new copy of Figure 1 so that it shows the reasons for Detroit's decline.
3. Using examples from Kampala, New York City and Detroit, explain the differences between these terms:
   a 'internal migration' and 'international migration'
   b 'rural-urban migration' and 'internal migration'
   c 'natural increase' and 'net population growth'.

Exam-style questions

4. Study Figure 2. Describe the changes in population in New York. (3 marks)
5. Assess the economic factors that can lead to changes in urban populations. (8 marks)

3.4 How urban economies differ

In this section, you'll understand how urban economies differ in developing, emerging and developed economies.

Time to get up!

It's 5 a.m. Grace has just woken ready for today's work. She sells fruit and vegetables from a street stall in Kampala, Uganda's capital. If she's early, she'll sell fruit to those catching buses to work. Later, she'll sell to passing drivers on the road leading out of the city.

Like many Kampalans, Grace works in the **informal economy.** Traders like her do not figure in most development data about countries – yet millions like her earn their living on the street:

- selling goods, e.g. clothes and groceries
- cooking or selling food, e.g. fruit and vegetables
- offering a service, e.g. mending a car tyre.

Informal economy means an unofficial economy, where no records are kept. People in the informal economy have no written contracts or employment rights.

Formal economy means one which is official, meets minimum legal standards for accounts, taxes, and workers' pay and conditions.

A developing city – Kampala, Uganda

Kampala's informal economy is large. Uganda earns half its estimated GDP from informal work, and 80% of people work in it. That makes it worth US$33 billion a year! Most workers are women and young people, and are poor. For them, it offers opportunities when there are few others.

Meanwhile, the **formal economy** is growing slowly because most Ugandans are rural subsistence farmers.

- Manufacturing is small, employing only 7% of Uganda's population.
- Services are the main part of Kampala's formal economy, e.g. shops and stalls, banks, offices of Ugandan companies (e.g. Air Uganda), and government offices.

▲ ***Figure 1*** *A fruit and vegetable stall on the street in Uganda*

An emerging city – New Delhi, India

Compared with much of India, New Delhi is wealthy. The annual World Wealth Reports rank New Delhi as one of the world's wealthiest cities, so you might think that the formal economy would be more important. Not so! India's 'Hindu' newspaper estimates 80% of workers in New Delhi are in the informal economy. Economists believe that it provides between 50% and 80% of India's GDP, worth over US$4 trillion per year! In New Delhi, selling food, cigarettes, or clothing on the street is as common as in Kampala.

Like many capital cities, most people work in services, which earned New Delhi 86% of its GDP in 2017–18. Manufacturing (e.g. clothing) contributes 12%, and is mostly part of New Delhi's informal economy. Factories would normally be classed as the formal economy, but in New Delhi there are no regulations about worker records, tax payments, or working conditions (e.g. ventilation), and no minimum wage or benefits (e.g. holidays), so many factories are part of the informal economy.

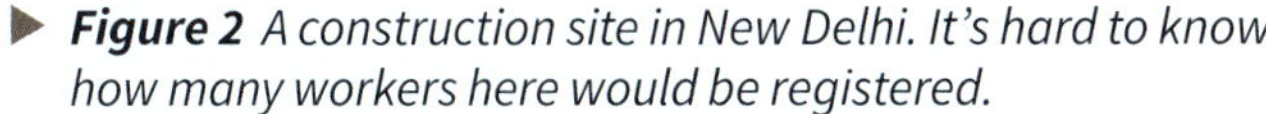

▶ ***Figure 2*** *A construction site in New Delhi. It's hard to know how many workers here would be registered.*

A developed city – New York City, USA

The city of New York is one of the USA's biggest economic assets. If it were a country, it would have the world's 12th largest economy, the size of Spain! Manufacturing is small with 10% of employment, fed by cheap migrant labour which makes up two thirds of all jobs in this sector. The most valuable part of the city's economy is its 'knowledge economy' (see section 3.3). In 2020, financial companies alone provided 10% of New York's employment, and a quarter of the city's wages.

But below the surface, the informal economy thrives. Economists claim that it earns 7% of US GDP each year, and is worth US$1 trillion! In New York, the informal economy consists of two main groups of people:

- migrants, both legal and illegal
- self-employed workers who may not declare income to tax officials.

The informal sector is greatest in construction, street selling (Figure 3) cleaning, and the hotel and catering industry. Workers have no protection, and often have to work long hours for less than minimum wage.

▲ ***Figure 3*** *Informal street selling in New York*

Your questions

1 Give examples of the **a** formal and **b** informal economy in your local area.

2 a Copy and complete the following table about the *informal* economy in the three capital cities.

City	% of GDP from Informal economy	Value in US$	Main jobs
Kampala			
New Delhi			
New York			

b Compare the *formal* economies of the three cities.

3 In pairs, design a spider diagram to show the advantages and disadvantages of working in the informal economy in these three cities.

Exam-style questions

4 Define the term 'informal employment'. (1 mark)

5 Assess the reasons why urban economies differ between cities in developed countries and those in developing or emerging countries. (8 marks)

3.5 The changing face of New York

In this section, you'll understand how urban populations change over time.

Delving into the past

August 2021. Allan, a retired teacher, is off to the Museum of Immigration on Ellis Island in New York's harbour (Figure 1). It was the registration point for 50 million 20th century immigrants to the USA. He's searching for details about his grandfather, a Jewish immigrant, who had walked from eastern Russia to Amsterdam to escape persecution and get a ship. For millions, New York offered the prospect of a new life.

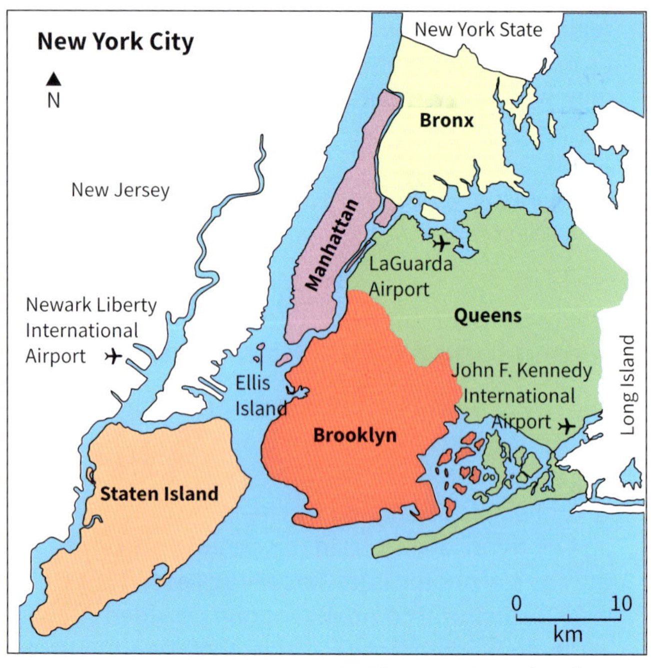

▲ *Figure 1 Map showing the five boroughs of New York city – Manhattan, Brooklyn, Staten Island, Bronx, and Queens. Ellis Island is also shown.*

Why New York grew

What is now New York City began as a fort on Manhattan Island in the 1600s. By 1860 its population had exploded to 860 000. Its deep harbour enabled it to trade with Europe. Clothing and food industries – and work – grew around the harbour.

As well as Europeans, New York was the destination for two other major groups of migrants:

- Puerto Rico, a Caribbean island, which in 1917 became a US territory, gave its citizens free movement within the USA. It brought millions of people to New York, immortalised in the song 'America' from 'West Side Story'.
- African American migrants came from the southern USA, escaping poor economic conditions, racial segregation, the spread of racist ideology and widespread discrimination. Cities such as New York offered greater freedom.

Communities developed among migrants of similar origins. These formed **ethnic enclaves** – communities whose bonds lay in family ties, food shops and places of worship. It led to districts such as Little Italy, many of which survive now.

Suburbanisation

As an island, Manhattan's rapid expansion could either grow upwards (Figure 2), or cross into Long Island or mainland New York state. Transport made outward expansion possible.

- The extensive **subway** and **rail** system expanded after the first underground line opened in 1900. From Manhattan, people could go either to The Bronx, Brooklyn or Queens (see Figure 1) or even beyond to Long Island. Greater space outside the city promised a quality life. You could buy a large house with garden 30 miles away on Long Island, but be in Manhattan in 40 minutes.
- As car ownership grew after the 1930s, **road bridges** crossing the Hudson and East Rivers fed traffic into Manhattan from new fast **freeways** from Long Island and The Bronx.

▲ *Figure 2 New York's famous skyscraper skyline – crowding as many people and businesses onto Manhattan as possible!*

Counter-urbanisation and 'white flight'

Suburbanisation had a huge effect on New York's population. Between 1950 and 1980, New York lost 12% of its population. A process took place known as **counter-urbanisation** – people leaving cities, as shown in Figure 3. Those moving tended to be white second-generation migrants who had done well enough to move out. It was known as 'white flight' and left behind poorer migrant communities and African Americans. Although rare now, there were ethnic rivalries, like those in the film 'West Side Story', which made many feel unsafe. As the wealthy left, income from business and sales taxation fell, but welfare demands on the city grew. In 1975 the city was nearly bankrupt.

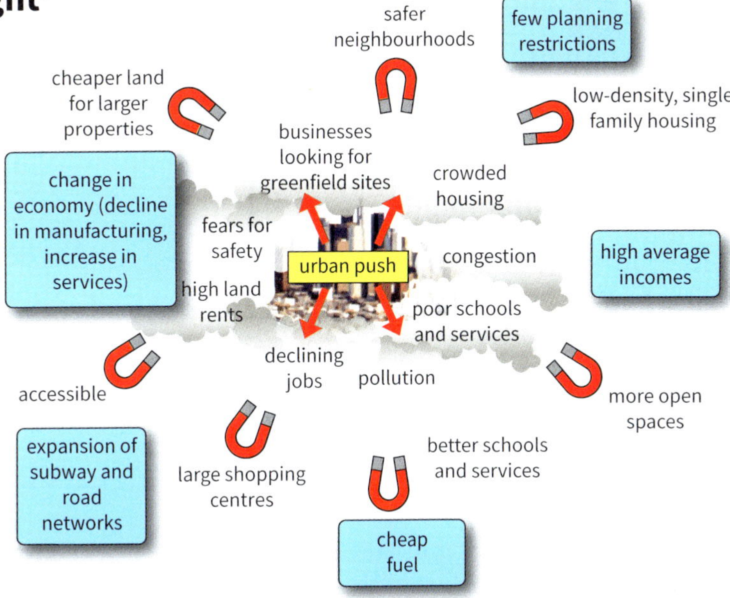

▲ ***Figure 3*** *Reasons for New York's counter-urbanisation*

Re-urbanisation

Since 1980, three changes have attracted people back into New York, known as **re-urbanisation**.

- The knowledge economy has created employment in the city.
- Closure of docks and industries has created space for **regeneration** (re-developing former industrial areas or housing to improve them). Areas such as Battery Park (see Figure 4) now contain new apartments and offices on **brownfield sites** (urban land which has been developed before).
- The city is safer, due to increased employment and 'zero tolerance' policies towards crime.

▲ ***Figure 4*** *New York's Battery Park at the southern tip of Manhattan*

Your questions

1 a Explain the differences between urbanisation, suburbanisation, counter-urbanisation, regeneration, and re-urbanisation.

b Make a sketch copy of Figure 2 on page 85 and label it with these four terms.

2 a Using Figure 1, describe the location, layout and physical size of New York.

b Explain why the subway system and road bridges were essential to the growth of New York.

3 Make a new copy of Figure 3 to show the causes of re-urbanisation since 1980.

Exam-style questions

4 Define the term 'counter-urbanisation'. (1 mark)

5 Explain **two** reasons for the re-urbanisation of many cities. (4 marks)

6 Assess the impacts of suburbanisation upon cities. (8 marks)

3.6 Land use in cities

In this section, you'll understand different urban land uses and what causes these.

Understanding cities

Cities can look complex. Their size means it's hard to see anything but buildings. Yet buildings form a pattern of **land use** (what land is used for), usually arranged as shown in Figure 1. Once you know this, it is easier to see changes between one part of a city and another.

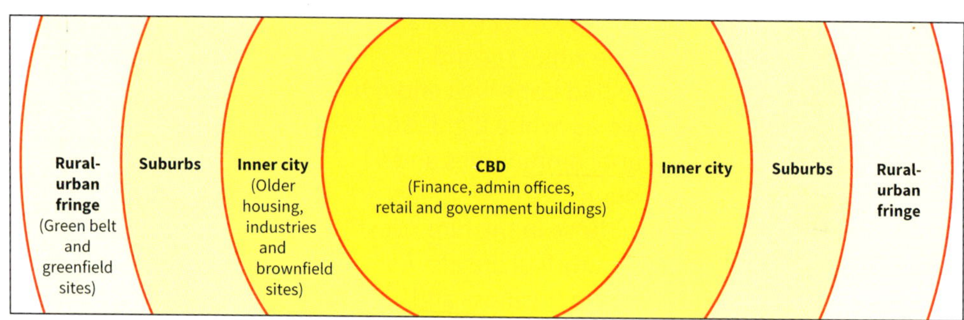

▲ ***Figure 1*** *The pattern of land use in towns and cities*

Land use in urban areas is usually easy to recognise. City centres (commercial areas) look different from residential areas (where people live). In turn, each of these looks very different from industrial areas.

The table below shows how this pattern of land use has developed. It's caused by three things – **accessibility** (how easy a place is to get to), **cost of land**, and **planning decisions** made mostly by councils.

Type of land use	Characteristics	How these areas develop
Commercial (offices and retail)	• Mostly in the CBD – **Central Business District.** • Buildings taller than the rest of the city. • Buildings are at a higher density – few open spaces. • 'Low-rise' business and retail parks on the city edge (the **rural-urban fringe**).	• This is the most accessible part of the city (the railway station is there and most roads meet there). • Demand for land is greatest, which forces prices up. Space is limited so the only way to build is up! • Land is expensive so every bit of land is used. (City parks are protected by law). • Planners allow these near main roads for retail customers or staff to reach without going into the city.
Industrial	• Away from the centre, either in the **inner city** (older 19th century industries) or on the city edge (more recent industries). • Close to transport links e.g. motorways, rail, ports.	• Most industries need space, so these are away from the CBD. New industries are built on 'industrial estates' reserved by planners to keep industry separate. • Industries need transport. Older industries relied on canals, rivers and rail. New industries rely on road.
Residential	• Usually surrounding the CBD and industries in **suburbs**. • The oldest properties are close to the centre. • Residential areas differ between one part of a city and another.	• Land is cheaper further from business and commercial areas. • Cities grow outwards in 'rings', with oldest suburbs near the centre and newest on the outskirts. • Land is expensive near the centre, so terraces and flats are common. Further away from the city, cheaper land means houses can have larger gardens.
	Housing varies between different parts of cities: • 19th century houses are **terraced**, in parallel rows. • 20th century houses are lower-density housing estates further out, with **semi-detached and detached** housing. • 21st century housing varies from apartments in the inner city to large housing estates on the outskirts.	• Industry owners built these at high densities for factory workers. • Land was cheaper further out, so houses had a garden and garage. Planners developed estates to house those moved from inner city housing in the 1950s and 60s. • In the 21st century planners prefer to allow housing on 'brownfield' land rather than use 'greenfield' sites (farmland that has never been built on) on the edge of cities.

▲ ***Figure 2*** *Land use development*

Land use in Leeds

Leeds is the largest city in West Yorkshire. It remains one of the UK's most important industrial cities, though manufacturing is a shadow of what it was until the 1970's. The satellite photo shown in Figure 3 shows central Leeds and some of the areas which have been regenerated since the 1980's. Find the following:

- The River Aire – where much manufacturing industry used to be concentrated.
- Leeds railway station – the centre of the city.
- The M621 – leading to the south-west to Bradford and Manchester. The M1 (not shown on Figure 3), comes in from Sheffield and further south.
- The following former industrial suburbs – Holbeck and Hunslet. These have experienced major redevelopment.
- Armley and Burley – suburbs which are changing because of the expansion of the two largest universities in Leeds. 'Studentification' changes communities as more houses are rented out, and bars and cafes develop.

◀ ***Figure 3*** *Pléiades satellite image of Leeds city centre and part of the inner city*

Thinking beyond

How might cities change in the future if people work increasingly from home or shop online?

Your questions

1 Copy Figure 1. Using a town or city you know well, name examples of areas you know on your diagram. Use Google Maps to help you.

2 Draw a spider diagram with three 'arms', labelled 'accessibility', 'cost of land' and 'planning decisions'. Label it to show how different land uses develop.

3 Study Figure 3 of Leeds. Identify the following:

a the CBD **b** industrial areas **c** residential areas.

4 Explain the following statements using Figure 3:

a The CBD is easy to identify.

b There are industrial areas near the River Aire. 

c There are industrial areas further out of the city.

d Residential areas are further from the CBD.

Exam-style questions

5 Define the term 'accessibility'. (1 mark)

6 State **two** land use zones across a city. (2 marks)

7 Explain why industrial areas tend to be found in certain parts of cities. (4 marks)

8 For a named city, explain **two** reasons why its inner suburbs are changing. (4 marks)

3.7 Mumbai – a growing megacity!

In this section, you'll assess the location of Mumbai and its urban structure.

Mumbai – a world city!

Figure 1 shows one of the world's largest cities, but where is it? Its city lights are typical of a developed city such as New York, Tokyo, or London. It's actually Mumbai in India, one of the world's **megacities** (see section 3.2). It is India's main commercial city, in the state of Maharashtra which is India's richest state (see section 2.12). Mumbai is India's second largest city, and in 2021 was the seventh biggest in the world. In 2020, it was also graded 'Alpha' as a world city (see section 3.2), because of its economic importance.

▲ ***Figure 1*** *Mumbai at night*

Mumbai's site and situation

The main city of Mumbai lies on an island, shown in Figure 2, by the deep-water estuary of the Ulhas River. Mumbai's port has grown round the estuary to become India's largest container port. Much of the city is low-lying, just above sea level. It lies 19°N of the Equator, so it's tropical, with a monsoon between June and September. Torrential monsoon rains flood low-lying roads and traffic can quickly come to a standstill.

In 2021 the estimated population of Mumbai was 20 million, though its metropolitan area is far bigger. As it has grown, it has spread to the mainland to form a conurbation. Its metropolitan area in Figure 2 includes Navi Mumbai, Thane, Bhiwandi, and Kalyan. In total, about 25 million people lived there in 2021.

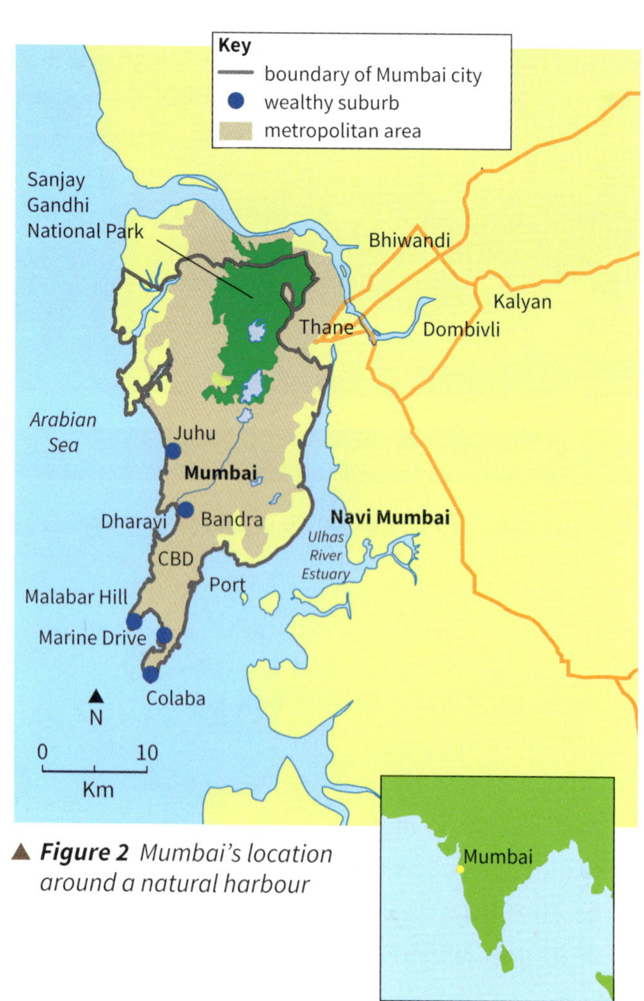

▲ ***Figure 2*** *Mumbai's location around a natural harbour*

Mumbai's national and international connections

Mumbai is well connected to other important economic locations.

- Nationally, its deep-water harbour has made it India's second biggest **port**. Large container ships can access Mumbai. Its waterfront is 10 km long, allowing huge port development, with manufacturing industries nearby.
- Internationally, Mumbai's location on India's west coast makes it closer to Europe via the Suez Canal than other Indian ports. Shipping times to Europe are five days shorter than from Kolkata on the east coast.

- By **air**, Mumbai is nine hours from UK airports. Its international airport handled 47 million passengers in 2019 (about 60% of London's Heathrow). It is a four hour flight to Singapore, and under three hours to Dubai or other Middle East destinations. Nationally, most other Indian cities are within two hours' flight time. This makes it possible to travel on business to any of these cities and back in one day.

The structure of Mumbai

Figure 3 shows a model of the structure of cities in developing countries. Mumbai is not *exactly* like this, as Figure 4 shows.

- Because it was built around the harbour, the CBD is not in the centre, but near the island tip (see Figure 4).
- Some industrial areas are near the port, but land is so expensive that many have moved out to places such as Navi Mumbai, where land is cheaper.

The structure of Mumbai has similarities, but also differences, compared to UK cities (see section 3.6).

Residential areas in Mumbai show wide inequality.

- Wealthy suburbs (see Figure 2) are all inner city areas along harbour or coastal waterfronts, close to the CBD.
- Middle-low income areas are in older parts of the city on the island, further from the CBD.
- Low-income groups live in **'chawls'** – these are low quality multi-storey buildings. 80% of homes are single rooms.
- The poorest 60% of people live in **informal** housing (mostly squatter shacks) on the outskirts, although there is some informal housing in Dharavi, near the centre.
- There are also thousands of people living on Mumbai's streets.

▼ ***Figure 3*** *A model of the structure of developing cities*

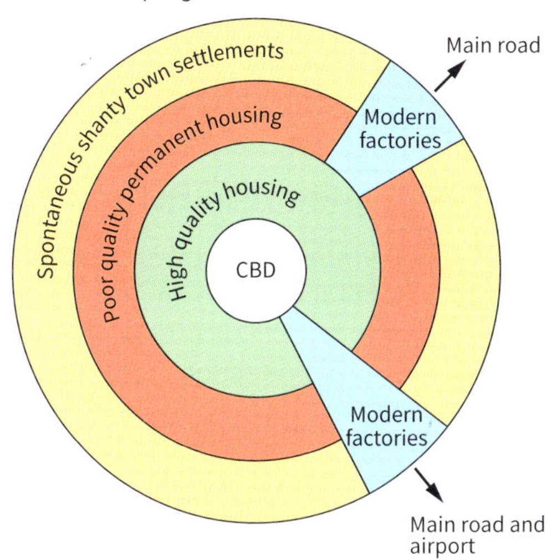

▼ ***Figure 4*** *Land use in Mumbai*

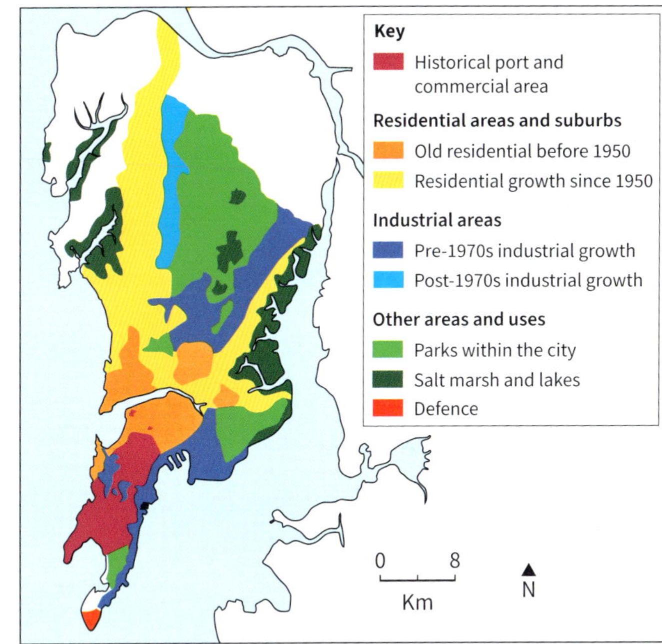

Your questions

1. Write down six things to describe Mumbai's physical location.
2. Draw a sketch map of Mumbai using Figures 2 and 4. Label it to show the following areas: CBD, port, industrial areas, wealthy suburbs, middle-income suburbs, chawls, squatter settlements.
3. Explain two similarities and two differences between Mumbai's structure and a city in a developed country (see Figures 1 and 2 in section 3.6).
4. Make a copy of Figure 3. Explain ways in which Mumbai is **a** similar to, **b** different from this model.

Exam-style questions

5. Define the term 'megacity'. (1 mark)
6. For a named city in a developing or emerging country, explain **two** reasons for its rapid population growth. (4 marks)
7. For a named city in a developing or emerging country, explain why the quality of housing varies. (4 marks)
8. For a named city in a developing or emerging country, assess the influence of its site and location in its growth and development. (8 marks)

3.8 Geographical skills: investigating spatial growth

In this section, you'll identify changes in Mumbai and its spatial growth, using maps and photographs.

The development of Bombay / Mumbai

Before 1995, Mumbai was known by its Anglicised name, Bombay. This dated from the time when India was a British colony (this ended in 1947). In 1995, the Hindu nationalist party, Shiv Sena, won elections in Maharashtra (the state in which Mumbai is located). It announced that Bombay would be renamed after the Hindu goddess Mumbadevi. They argued that the name Bombay was an unwanted leftover from British rule.

In this section, you can trace more gradual **spatial** changes in Mumbai as it grew between 1888 and 2020.

Spatial means 'relating to space' – e.g. the spatial growth of a city means how much extra space it takes up as it grows.

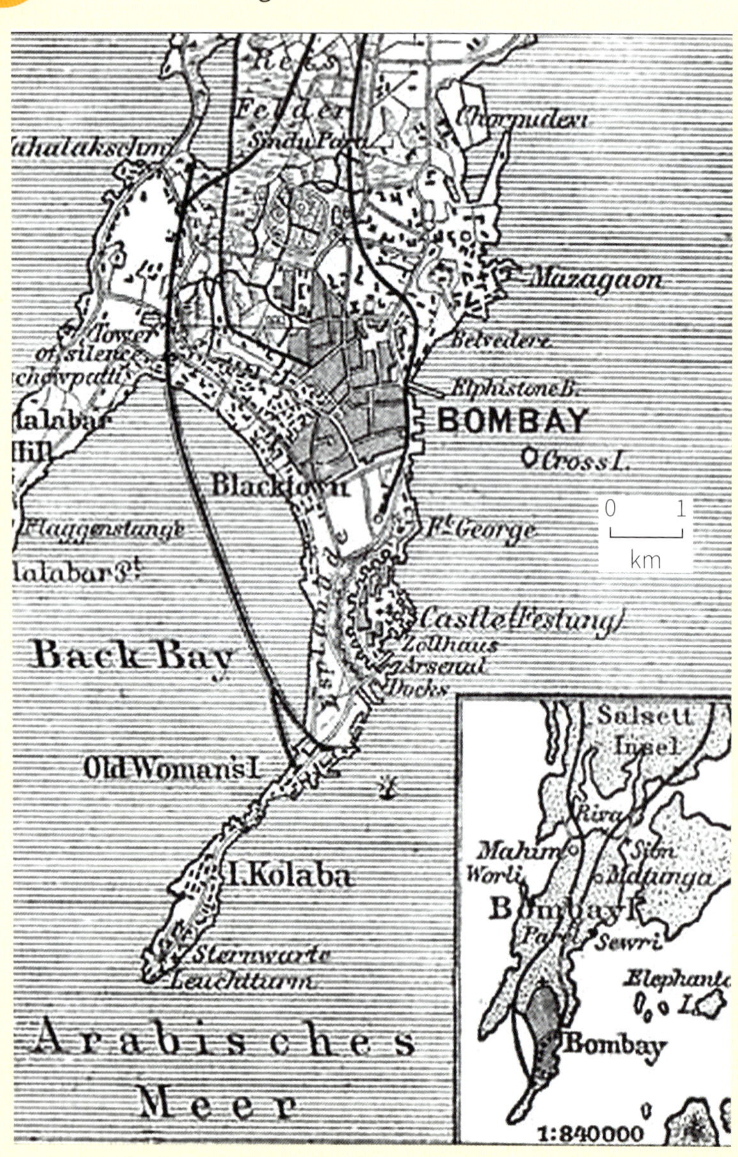

▲ **Figure 1** *Bombay in 1888*

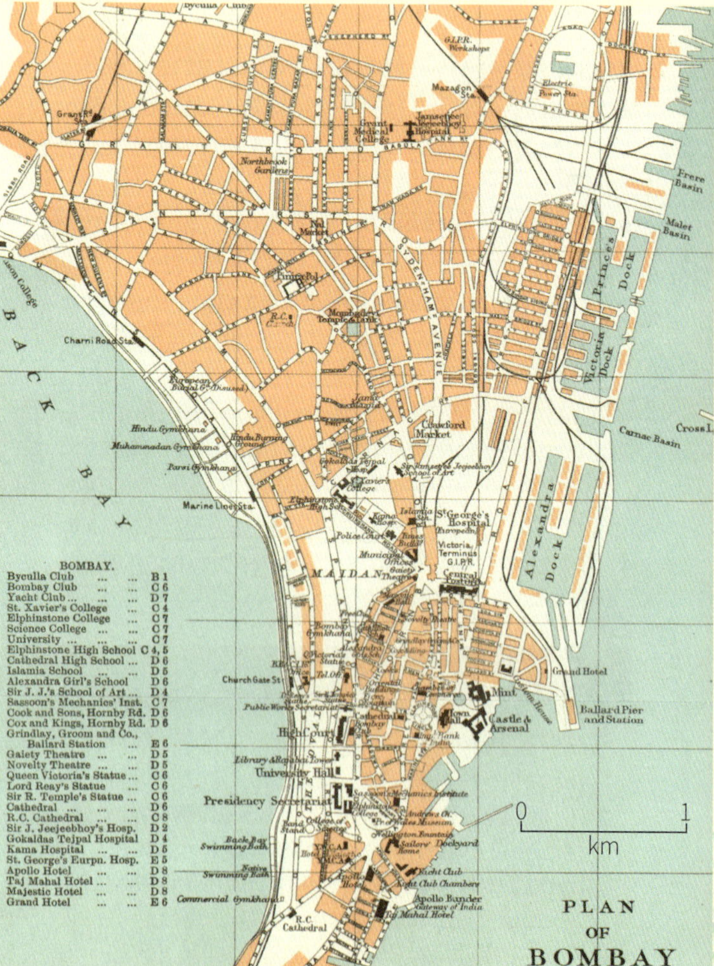

▲ **Figure 2** *Bombay in 1924*

Figure 3 *High density growth in central Mumbai – there is now very little open space.*

Figure 4 *Changing Mumbai – a view of Navi Mumbai (New Mumbai)*

Your questions

1 **a** Describe Bombay as a settlement in 1888 (Figure 1) – where the original settlement was, its location, and the economic activities there.

b Using the scale, measure the size of the settlement N-S and E-W, and then calculate its area in km^2.

c Estimate what percentage of the total map area was built up in 1888.

2 Using the same structure as Question 1 a–c, describe Bombay's spatial growth by 1924 (Figure 2).

3 Describe the environment of central Mumbai shown in Figure 3. Use Google Earth or Google Maps if you need more detail.

Exam-style questions

4 Suggest **two** reasons why middle classes are leaving central Mumbai to live in Navi Mumbai, shown in Figure 4. (4 marks)

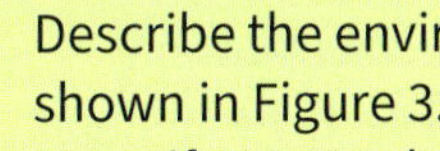

5 Explain the likely effects of people leaving the city on Mumbai's spatial growth. (4 marks)

3.9 Mumbai's changing population

In this section, you'll understand the past and current trends in Mumbai's population and assess how this has affected the city.

A thousand a day

Padmaa lives in Dharavi, a poor suburb near Mumbai's city centre. Her story is typical of many. Her family moved to Mumbai from a rural village when she was young. Migrants to cities in emerging countries often end up living in low income suburbs like this. Her family were lucky, unlike many who became street dwellers; her father became a tour guide and could afford the rent. About 1000 new migrants arrive every day, so that Mumbai's population is growing by 3% a year, doubling every 23 years! Rapid growth like this is called **hyper-urbanisation** – a superfast rate of urbanisation. It's impacting on Mumbai, as shown below.

▶ ***Figure 1*** *Dharavi, one of Mumbai's largest poor suburbs*

Mumbai's population, growth and economy

Population

- The population of the city of Mumbai was about 20 million in 2020 and the United Nations estimates that it will reach 25 million by 2025.
- By 2050 it will probably be the world's largest city.

Pattern of spatial growth

- With increasing population, Mumbai has expanded in size – from 68 km² (the city itself), to 370 km² (including suburban districts), and now to 603 km² (the metropolitan district including areas such as Navi Mumbai).
- New suburbs are developing. Navi Mumbai (New Mumbai) has been built on the mainland. In 2021, 1.1 million people lived there – mostly the middle class moving out of the city.
- Poor housing, in which 60% of Mumbai's inhabitants live, dominates the city landscape.

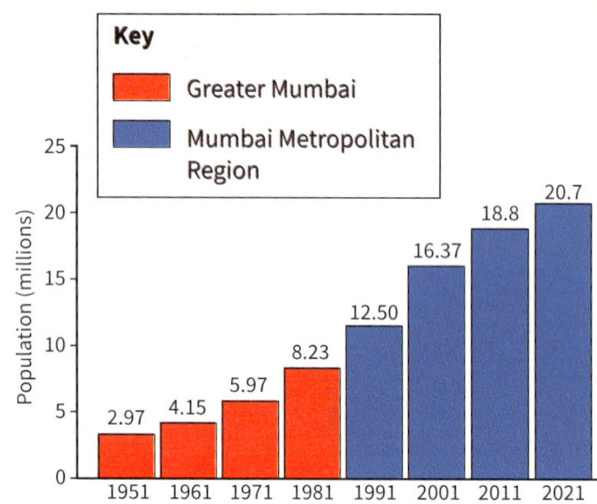

▲ ***Figure 2*** *The population growth of Mumbai*

Changing investment and land use

Mumbai is the commercial capital of India, and investment has grown, increasing the amount of employment rapidly. Investment has been greatest in:

- services (e.g. banking, finance, IT and call centres)
- manufacturing (textiles, food processing and engineering)
- construction (housing, factories and offices)
- entertainment and leisure (Bollywood, hotels and restaurants).

Growth in financial services, entertainment and leisure has put pressure on land in the CBD, making Mumbai one of the world's most expensive cities. Many manufacturers needing large amounts of land are moving out. Audi, Volkswagen, and Skoda car factories are in Aurangabad, 300 km away.

The growth of Mumbai

Mumbai has grown for two main reasons:

- rural-urban migration
- natural increase, i.e. the number of births minus deaths.

1 Rural-urban migration

India's rural areas have few opportunities apart from working on the land. People often live in permanent poverty and have few chances to improve their lives. Factors like this help to 'push' people away from the countryside.

Figure 3 shows that the majority of migrants are fairly short-distance, coming mostly from Maharashtra, the state in which Mumbai is situated. Migrants in India tend to head for states with the three biggest cities – Mumbai, Delhi, and Kolkata. Maharashtra receives the most migrants because it is the wealthiest, and has the most employment. Most migrants are permanent because Mumbai has:

- more jobs
- better education facilities – Mumbai has 12 universities; its literacy rate is 95%, (the national average is 87%)
- entertainment options for those who can afford them
- higher incomes – though the cost of living is also higher.

These factors – and the dream of a better life – 'pull' people to Mumbai.

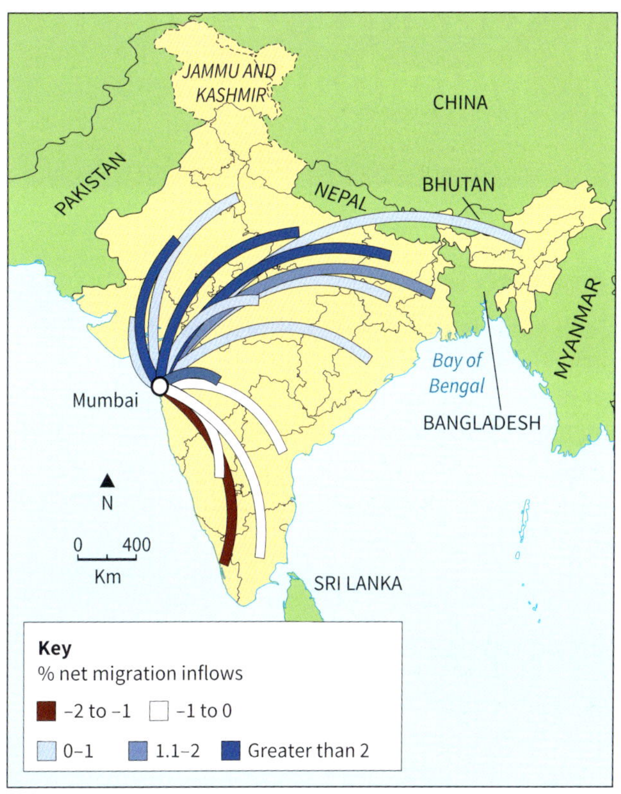

▲ ***Figure 3*** *Origin of migrants to Mumbai*

2 Natural increase

Migrants to Mumbai are like migrants in cities throughout the world – they are young and looking for work. They tend to be in their 20s and 30s, and as long as there is work, they usually stay and settle. Once settled, they start families. Mumbai's natural increase is 1.4% per year – accounting for nearly half of Mumbai's annual growth.

Your questions

1. Using examples, explain the difference between 'push' and 'pull' factors in rural-urban migration.
2. Describe the pattern of increase in Mumbai's population, using Figure 2.
3. Explain why most migrants to Mumbai **a** are young, **b** are likely to start families in Mumbai rather than in rural areas, **c** affect the rate of population increase in cities.
4. Explain how Mumbai's growth has had impacts on **a** its geographical size, **b** the amount of poor quality housing, **c** the location of manufacturing.
5. **a** Suggest two possible reasons to explain why many middle-class people have moved out of Mumbai to Navi Mumbai.

   b Suggest one disadvantage of doing this.

Exam-style questions

6. Explain **two** pull factors which have caused megacities in emerging or developing countries to grow. (2 marks)
7. Study Figure 3. Describe the pattern shown by the flow of migrants moving to Mumbai. (3 marks)

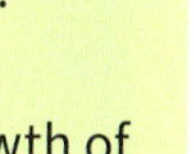

8. Explain **two** reasons for the economic growth of a named megacity in an emerging or developing country. (4 marks)
9. For a named megacity in an emerging or developing country, assess the social and environmental impacts of its rapid population growth. (8 marks)

3.10 Quality of life in Mumbai

In this section, you'll understand the inequalities in lifestyles within Mumbai.

Welcome to India

- **Rajesh and his wife Sevita** live on a Mumbai beach. They support their children by running their home as a makeshift pub. But the council constantly threatens to bulldoze them.
- **Johora** started as a rag-picker, but has built a bottle-recycling business on a railway embankment. She has plans for her seven children.
- **Ashik** buys beef fat from the abattoir and renders it down. It looks disgusting – but his products could be destined for your soap or cosmetics.

(Adapted from a TV series about life in India)

Economic opportunities

Walk through the streets of Mumbai, and you'll see people everywhere working in the **informal economy** (see section 3.4) like the examples above and in Figure 1. There's no regular wage, just people providing goods and services, often 'cash-in-hand'. There's no contract or job security. But the informal economy is essential, worth US$1 billion to Mumbai's GDP each year.

Many who work in Mumbai's informal economy live in areas of low quality housing, of which Dharavi is the largest. It lies between two railway routes. Although many houses are made from brick, wood and steel, quality of life there is poor, as shown in Figures 2, 3 and 4. A BBC report described Dharavi as *'one unending stretch of narrow dirty lanes, open sewers and cramped huts.'* Many homes have electricity, but it's often obtained illegally via dangerous 'hook-ups'.

But Dharavi is affordable. Rents for a small flat are 200 rupees (£2.00) per month and for many it's convenient for work. It's also been described as the 'flywheel of Mumbai's economy'.

▲ ***Figure 1*** *Rag-picking – making money from other people's rubbish – is an important part of the informal economy in Mumbai*

Life on a low income – Dharavi

▲ ***Figure 2*** *Dharavi – a million people live here*

Walking through Dharavi, New York Times journalist Jim Yardley described a journey through 'a dank maze of ever-narrowing pasages'. He noticed that often the houses were so close that daylight would hardly reach the pathways like 'a great urban rainforest, covered by a canopy of smoke and sheet metal'. Jammed into the narrow streets were roadside fruit stalls. Up to 10 families might share a tap – if water was flowing (which it often didn't) and toilets were communal, costing 3 cents to use.

▲ ***Figure 3*** *Adapted from 'A walk in Dharavi', by Jim Yardley, New York Times*

People	
Population of Dharavi	Estimated 800 000–1 million
Area	2.39 km² (the size of London's Hyde Park)
Population density	At least 330 000 people per km²
No of homes in Dharavi	60 000
People per home	Between 13 and 17
Average size of home	10 m² (equivalent to a medium-sized bedroom)

Hygiene and health	
No of individual toilets in Dharavi	1440
People per individual toilet	625
% of women suffering from anaemia*	75%
% of women with malnutrition	50%
% of women with recurrent gastro-enteritis**	50%
Most common causes of death	Malnutrition, diarrhoea, dehydration, typhoid
Education	
Literacy rate in Dharavi	69% (Mumbai average is 91%)

▲ ***Figure 4*** *Dharavi factfile*

**anaemia is a lack of iron leading to tiredness*

*** gastro-enteritis symptoms are diarrhoea and vomiting*

Life on a middle income

[Adapted from news articles]

The Chopra family – two adults, two young children – live in a small flat (living room, kitchen, bedroom and bathroom). The family all sleep in one room. Mr Chopra teaches at a local school. His monthly salary is 23 000 rupees (£230). It is not enough, so he tutors privately, increasing his salary to 75 000 rupees (£750).

India's lower middle class is growing and their incomes rising. Mr Chopra says Mumbai is changing. 'Nearly every family has a TV and mobile phone. There are many who are poor and don't have basic necessities. The extremes are there but it is changing. My children are growing up in a different world. I hope they work in IT or healthcare, or perhaps engineering.'

Life on a high income

Vihaan, 25, works for AccelorMittal, one of India's largest companies, with an annual salary of 1.6 million rupees (£16 000). He has a degree from Mumbai's top engineering college, speaks English, and is part of Mumbai's high income group. He's recently gained promotion to Operations Manager. He works long hours, except evenings when he meets friends at a restaurant, or goes clubbing. He's a shopper, spending a lot on designer clothes and meals.

His company provides him with a luxury 10th floor one-bedroomed apartment in a gated complex overlooking the harbour in upmarket Colaba. He saves a quarter of his salary, but in 2021, high-spec one-bedroomed apartments in Mumbai cost £320 000. He estimates that it will be five years before he can afford a deposit for one.

Your questions

1 List the advantages and disadvantages of the informal economy for **a** people working in it, and **b** Mumbai.

2 **a** Write a news report of about 300 words headed 'Quality of Life in Dharavi'. Use information in Figures 1–4 to help you.

b Which gives a better picture of quality life in Dhavari from Figures 2 to 4 – quantitative data (figures) or qualitative data (descriptions)? Explain your answer.

3 Compare life for the Chopra family with Vihaan using the headings 'Living space', 'Job', 'Salary', 'Future prospects'.

Exam-style question

4 Evaluate the view that living in areas such as Dharavi offers people more benefits than problems. (8 marks)

3.11 Challenges facing Mumbai

In this section, you'll assess the challenges facing Mumbai caused by its population growth.

Problems ahead?

As its population grows, Mumbai has become India's economic giant. Its industries – like the port, factories, and film industry – are its formal economy (see section 3.4). Its informal economy is also booming (see section 3.10). In just one area, Dhavari, there are 15 000 back-street factories.

But **employment conditions** vary. Most of Dharavi's factories are illegal and many are sweatshops. Some families work at home on sewing machines making shirts. One cuts, another sews, and so on. The shirt is sold to a buyer for 15p. It's informal – low pay, no security, and no tax.

Tax is Mumbai's problem. There are few tax collectors, and it's impossible to chase payments from informal workers. The city has set up tax-free zones to attract companies, so these companies also pay no tax. Without tax income Mumbai can't provide services for the population (see Figure 1).

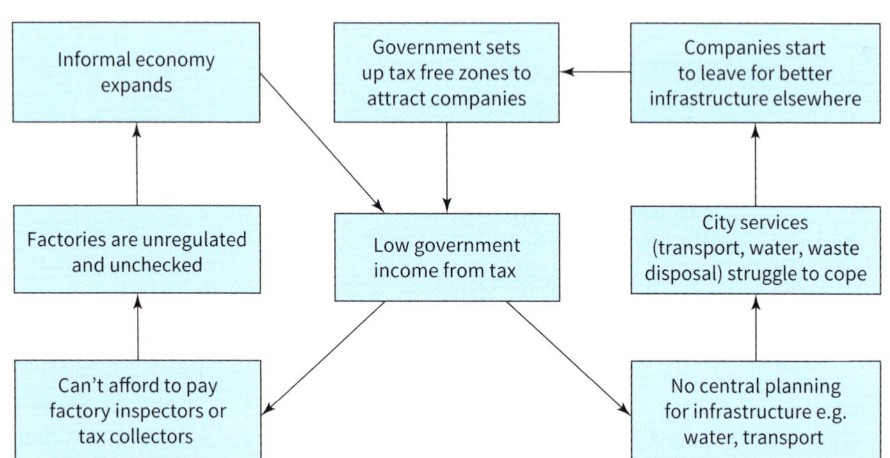

▲ ***Figure 1*** *A low-tax government cycle*

Housing shortages and development

Mumbai's population growth is hard to keep pace with, so there's a housing shortage. City authorities have no money to build housing. Private companies are put off building because the government limits maximum rents, meaning rental income is reduced.

Most people have to put up with poor housing. Many live in cramped, poor quality, expensive rooms, far from work. They are forced into these, while others squat on streets or spare land. Over time, they buy or find materials to build a home. Once there, squatters try to stay – few can afford to move because Mumbai is so expensive. Conditions are unhealthy, as Figure 2 shows.

▲ ***Figure 2*** *Squatter settlements in Mumbai*

Water supply and waste disposal

Only better-off suburbs have private water supplies. Like Dharavi (see section 3.10), 60% of Mumbai's population uses communal taps. But supplies fail when power cuts stop pumps from working. In some low income suburbs, water only runs for 30 minutes a day.

Rapid urbanisation has caused uncontrolled water pollution. Factories use the Mithi River (Figure 3) to dump untreated waste – and the airport dumps untreated oil. 800 million litres of untreated sewage go into the river daily.

However, 80% of Mumbai's waste is recycled – a figure way beyond cities like London. In Dharavi, nothing is rubbish! The recycling industry is worth US$1.5 million a year, and employs 10 000 people. Children collect plastic, glass, cardboard, batteries, computer parts and soap. In workshops, small smelters recycle cans, and vats of waste soap from hotels are melted and remoulded into bars.

▲ ***Figure 3*** *The heavily polluted Mithi River in Mumbai*

Dealing with air pollution and traffic

In 2015, an air quality index was introduced to improve air quality in Mumbai. Suggestions for improvement include:

- using LPG instead of burning coal
- introducing of low benzene petrol
- checking on fuel tampering
- improving public transport
- charging higher road tax on older vehicles.

Traffic congestion is legendary in India. Little is spent on Mumbai's transport infrastructure. There are too few suburban train and bus networks to meet demand, made worse by Mumbai's funnel shape, which restricts transport routes. Commuter trains and buses are overcrowded, and 3500 people die on Mumbai's railway each year. Most deaths are caused by passengers crossing tracks, sitting on train roofs and being electrocuted by overhead cables, or hanging from doors (see Figure 4). One upside of the Covid-19 **pandemic** was that reduced commuting cut the number of deaths by two-thirds during 2020–21.

▲ ***Figure 4*** *Just one of Mumbai's problems – 10 people die on its massively overcrowded rail system every day*

Your questions

1. Explain how the cycles in Figure 1 can lead to poor services in Mumbai.
2. How would you persuade Mumbai's city authorities to invest in piped water and waste treatment?
3. Redraw Figure 1 to show how Mumbai's services could improve with greater tax income.
4. List the advantages and disadvantages of each of the following schemes:
   - **a** Encourage companies to move out of Mumbai to suburbs where there is more land.
   - **b** Build large affordable housing estates on the edge of Mumbai.
   - **c** Introduce a clearance programme.

Exam-style questions

5. For a megacity in a developing or emerging country, explain how quality of life varies in different parts of the city. (4 marks)
6. Explain **two** urban environmental problems in a named megacity in a developing or emerging country. (4 marks)
7. Assess the challenges for managing a named megacity in a developing or emerging country. (8 marks)

3.12 Sustainable Mumbai 1

In this section, you'll evaluate whether top-down strategies can make Mumbai more sustainable.

Sustainability and the future

Think about the problems that Mumbai faces, e.g. low quality housing, and pollution. There are many ideas about how Mumbai could improve. How can we know whether an idea is right or not?

One way is to think about **sustainable development**. This was defined in 1987 by the UN as '*development that meets the needs of the present without compromising the ability of future generations to meet their own needs*'. Any new idea can be tested against this definition.

The sustainability of an idea or proposal can be measured in two ways, shown in Figure 1:

1 as a three-legged **'stool'.** This judges an idea on economic, social and environmental benefits and problems. If benefits outnumber problems, then it's a good idea.

2 as a **'quadrant'.** This uses four questions, shown in Figure 1. Ideally, the answer to all four is 'yes'.

SUSTAINABILITY
ENVIRONMENT
SOCIETY
ECONOMY

EQUALITY
Does it benefit everyone?
FUTURE
Will it last?
PUBLIC PARTICIPATION
Is it bottom-up?
ENVIRONMENT
Is it eco-friendly?

▲ ***Figure 1*** *Two ways of measuring sustainability – the 'three legged stool' and the 'quadrant'. Both are just as useful.*

Top down development – 'Vision Mumbai'

In 2003, a report by American firm McKinsey, called 'Vision Mumbai', suggested investing US$40 billion to improve Mumbai. It involved a partnership between government, property companies and investors – so was typical of **top-down development** (see section 2.13). Completion would be by 2050. The basic problem it tried to solve was Mumbai's worsening quality of life. Poor housing has multiplied, and traffic congestion, pollution and water quality are all worse than in 2000.

The plan was to transform Mumbai with ambitious targets, including:

- building one million low-cost homes and reducing poor quality housing
- improving transport infrastructure by road and rail
- improving air and water quality by reducing pollution.

Top-down development is development imposed by large powerful organisations – such as companies or governments – on local people, involving large investment and usually covering a wide area.

Putting plans into action

Vision Mumbai was based on targets (see Figure 2). Some things were 'quick wins' – they could be done quickly to improve Mumbai, e.g.:

- restore 325 'green' spaces that were polluted and used for dumping waste
- build 300 extra public toilets
- widen and 'beautify' main roads
- improve train capacity and safety.

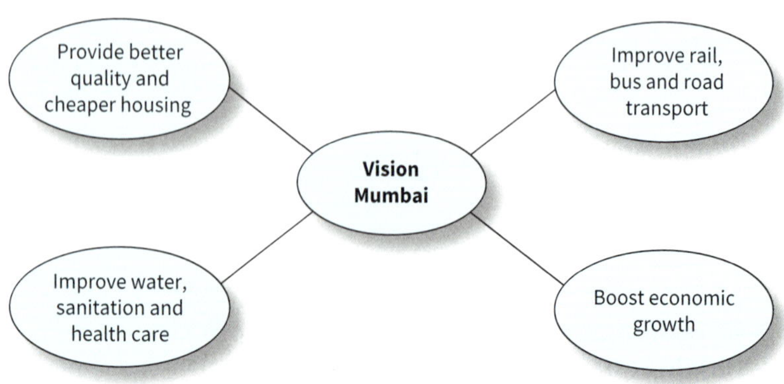

▲ ***Figure 2*** *Vision Mumbai's four core targets*

But the main plan was based on property development.

- Dharavi would be demolished. Its location next to Mumbai's financial district made it worth US$10 billion! Developers would buy land at a discount, and redevelop it. For every square metre of cheap housing built, they could have 30% more for offices, which are highly profitable.
- High-rise blocks for those living in poor quality housing would be built next to shopping malls, offices and luxury apartments.

▲ ***Figure 3*** *New apartment blocks in Dharavi*

Did Vision Mumbai work?

- Since the start of the project, 200 000 people have been moved, and 45 000 homes demolished in Dharavi. New flats replaced low quality housing (Figure 3).
- Piped water and sewerage systems were established for the new flats.
- In 2020, 350 new trains were added to Mumbai's rail network. Platforms were raised to prevent people falling into 'gaps' between trains and platforms – the cause of many deaths.
- In 2015, new laws were introduced to improve air quality (see section 3.11).

But people in Dhavari don't like the changes.

- Many prefer housing improvement (e.g. piped water, sewage treatment) to demolition.
- New 14-storey apartment blocks have split communities.
- Rents cost more than in the housing they replaced.
- Small workshops would have to move or go out of business, affecting Mumbai's recycling industry.

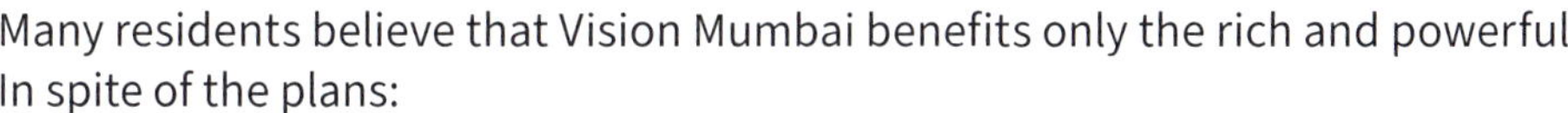

Many residents believe that Vision Mumbai benefits only the rich and powerful. In spite of the plans:

- water quality in Mumbai is worsening because of sewage discharge
- its beaches are unsafe for recreation
- poor quality housing is growing so rapidly that improving sewage treatment and disposal in line with population growth is a long way off.

Your questions

1 Explain what makes Vision Mumbai a 'top-down' development.

2 Why do people living in Dharavi see Vision Mumbai differently compared to other Mumbai residents?

3 Draw a spider diagram with four 'arms' – water supply, waste disposal, transport and air quality (see section 3.11). Explain what Vision Mumbai has done to improve these.

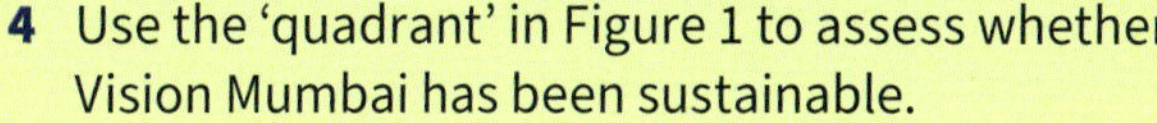

4 Use the 'quadrant' in Figure 1 to assess whether Vision Mumbai has been sustainable.

Exam-style questions

5 Define the term 'top-down development'. (1 mark)

6 For a named megacity in a developing or emerging country, evaluate the view that top-down projects designed to improve quality of life have proved successful. (8 marks)

3.13 Sustainable Mumbai 2

In this section, you'll evaluate whether bottom-up projects have helped to make Mumbai more sustainable.

What happens when you're ill in Mumbai?

Mumbai's poor housing areas are a challenge because of poor water quality, waste disposal and lack of sanitation. As the Covid-19 pandemic showed, disease is rife and spreads easily. India has no national health service. If you need health care you have to pay for it or use insurance. But better health would improve the sustainability of communities.

Some charities focus on health issues, often on specific diseases and conditions. They are Non-Government Organisations (NGOs), and normally work in communities on **bottom-up** development schemes (see section 2.14).

Bottom-up development is where experts work with communities to bring about change by identifying their needs, offering assistance, and letting people have more control over their lives.

Leprosy is a slow-developing, contagious, bacterial disease, which affects the skin, mucous membranes, and nerves. It causes skin discoloration and lumps, and at worst can cause body deformities. It is treatable and curable.

LSS – a case study of a health charity

Lok Seva Sangam (LSS) is a health charity in Mumbai. It raises funds to employs volunteers. It was set up in 1976 to control leprosy in Chunabhatti, on the edge of Dharavi. Leprosy patients often experience prejudice because it disfigures their body (Figure 1). It is commonly found in areas of poor housing quality, and is contagious, increasing risks of infection. LSS set up surveys to detect leprosy, with dermatology (skin) clinics, helped by pharmacies (chemists) dispensing treatments.

LSS expanded its work to Baiganwadi, a nearby community (Figure 2). By 2021, it had under 300 leprosy patients, having treated 28 000 people since the 1980s; most were cured. It expanded its work to treating tuberculosis (TB), employing physiotherapists, nurses, paramedics, and volunteers.

LSS and the Covid-19 pandemic

Since the Covid-19 pandemic, LSS has:

- run education programmes in Baiganwadi for 20 000 people on Covid-19 awareness, including how the virus spreads, using masks, hand sanitising, social distancing, and promoting vaccination
- provided 200 leprosy patients with masks, sanitisers and soaps
- provided food to those families whose have lost their jobs
- screened 2300 patients at its mobile health clinic for symptoms of Covid-19; in May 2021 (the height of India's pandemic), no patients were found.

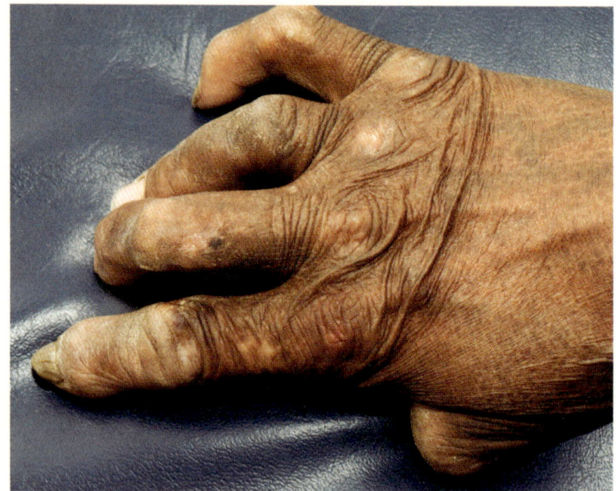

▲ ***Figure 1*** *The effects of leprosy on a person's hand. Many people are affected like this over their entire body.*

▲ ***Figure 2*** *Baiganwadi, where LSS carries out its work*

Education

LSS now works mainly in Baiganwadi, focusing upon education. It calls its work 'SET' – **S**urvey, **E**ducation and **T**reatment. It employs five teachers. Their work:

- **surveys** and detects cases of TB, leprosy and Covid-19
- **educates** people about symptoms, so that they can recognise these. They also teach about care and treatment.
- **treats** people by persuading them first that treatment is easy. They follow up patients to complete treatments, and carry out check-ups.

▲ ***Figure 3*** *Women's craft group run by LSS*

Community work

LSS also works with communities to teach about health, using:

- exhibitions and group talks about diseases such as typhoid, and personal care
- street theatre, using drama to shows disease symptoms, and accessibility of health clinics
- DVDs played to community groups about health care.

Much of its work is with women (see Figure 3), focusing on:

- **sanitation** – like boiling drinking water, washing hands before food preparation, and waste disposal.
- education about **vermiculture** (worms composting waste). Worms reduce the bacteria in household waste by 'eating' it, producing compost as a result – which can then be sold!
- **activities** e.g. paper bag making, or sewing, to aid discussion as well as making items to raise money.

Of course, this is just one project in one community. It focuses on three diseases and education, and tries to extend community understanding of health. It is not city-wide, nor country-wide.

Your questions

1 Read the definition of 'bottom-up development'. Describe three things about work done by LSS that fit this definition.

2 Look back at the 'three legged stool' of sustainability (section 3.12). List the economic, social and environmental benefits of work done by LSS.

3 What difficulties would there be in trying to expand the work of LSS outside a single community?

Exam-style questions

4 Define the term 'bottom-up development'. (1 mark)

5 Compare the difference in meaning between 'top-down' and 'bottom-up' development. (3 marks)

6 For a named megacity in a developing or emerging country, evaluate the view that bottom-up projects designed to improve quality of life have proved successful. (8 marks)

COMPONENT TWO
UK Geographical Issues

The Isle of Skye in the Hebrides

What is Component Two?

- Pearson Edexcel's GCSE Geography specification B consists of three Components.
- Each Component consists of three Topics, making nine in all.
- Each Component is assessed by its own exam paper – so Component Two is assessed by Paper 2.

What Topics will I study in Component Two?

- **Topic 4 The UK's evolving physical landscape** is about UK landscapes, Coastal change and conflict, and River processes and pressures.
- **Topic 5 The UK's evolving human landscape** is about the changing and varied human landscape of the UK, including the socio-economic and political processes that influence it, and a detailed case study of a major UK city – London is included in this book.
- **Topic 6 Geographical investigations** is about fieldwork. You'll have two days on fieldwork:
  - one day will be on **physical** fieldwork investigating **either** Coastal change and conflict **or** River processes and pressures
  - one day will be on **human** fieldwork investigating **either** Dynamic urban areas **or** Changing rural areas.

You'll also learn several **geographical skills** (interpreting maps, satellite images, diagrams, statistics, and photos) as well as fieldwork skills (including data collection, presentation and analysis).

What is Paper 2 like?

- **Time:** 1 hour 30 minutes
- **It has four Sections:** A and B assess Topics 4 and 5, and C1 and C2 assess Topic 6 fieldwork.
- **It's worth 94 marks:** 90 on the four sections and another 4 for Spelling, Punctuation, Grammar and use of specialist geographical terminology (SPaG) which is assessed on the 8-mark question in Section B.
- **It counts for:** 37.5% of your final grade.

Where can I get help in preparing for Paper 2?

Chapter 11 gives you all the guidance that you need on all three exam papers. It includes advice relevant to Paper 2 about:

- the exam format, handling different sections, and how exam papers will be marked (sections 11.1 and 11.2)
- how to answer shorter questions worth 1–4 marks (sections 11.3 and 11.4)
- how to answer longer 8-mark questions in Paper 2 (section 11.6)
- how to answer questions on fieldwork in Paper 2 (sections 11.7 and 11.8).

4.1 Landscapes from the past

In this section, you'll assess the role of geology and past processes in creating the UK's upland landscapes.

Coral reefs – in the Pennines?

If you love the UK's uplands, Malham Cove in the Yorkshire Pennines is a spectacular sight. It was once a huge waterfall, like Niagara Falls. At over 80 metres high, it was 30 metres higher than Niagara!

But Malham Cove poses a mystery for **geologists** (people who study rocks). The pale grey rock from which it's formed is **limestone**. It consists of crushed shells of corals that lived in tropical seas 300 million years ago. So how did this limestone get to 300 metres above sea level, where it is now?

▲ ***Figure 1*** *Malham Cove in the Yorkshire Pennines*

Explaining the past

The landscape around Malham Cove results from three factors:

- its geology (rock type)
- past tectonic processes (like plate tectonics – see section 1.16)
- past processes caused by glaciation.

Geology

To see living corals now you could go to Australia's Great Barrier Reef. Geologists know that fossils at Malham Cove are just like the coral species living in the Great Barrier Reef. By testing fossils from Malham Cove using **carbon dating**, they know that they lived during a geological period called the **Carboniferous** (250 to 350 million years ago).

At the time, the UK was covered by tropical seas, just like the Barrier Reef. Tropical fish and corals thrived. As they died, skeletons fell to the sea floor, forming horizontal layers (or **strata**). Two processes turned them into solid rock:

- as skeletons fell, they crushed those beneath, eventually squeezing out water and compacting them into rock
- calcium carbonate (which occurs naturally in sea water) crystallized around the fragments. This cemented them together and even preserved some fossils intact.

Later, other rock strata were deposited on top of the limestone, e.g. sandstone and shale. All the rocks in Figure 2 were formed in Carboniferous times. They vary in hardness (see section 4.3). The **most resistant** rock in Figure 2 is millstone grit. It resists **erosion** so well that it forms the highest peaks of the Pennines, and protects weaker sands and shales beneath.

Carbon dating uses radioactive testing to find the age of rocks which contained living material.

Erosion means wearing away the landscape.

Upland and lowland landscapes

Upland areas of the UK consist of resistant igneous, metamorphic and some sedimentary rocks. You can see how geology affects the distribution of uplands on pages 110–111.

Lowland areas of the UK generally consist of younger, and less resistant sedimentary rocks. There is more about lowland landscapes on page 115.

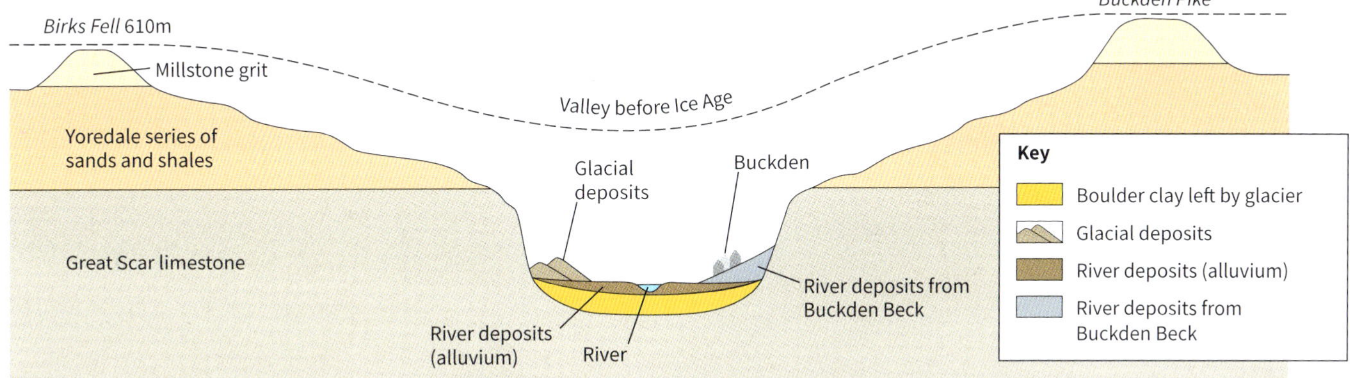

▲ ***Figure 2*** *Geological cross section of Wharfedale, a valley in the Pennines*

▶ ***Figure 3*** *Giggleswick Scar, a fault scarp in the Pennines*

Tectonic processes

Over 300 million years, three tectonic processes (see section 1.16) affected rocks in the Pennines:

- The plate on which the UK sits shifted, away from the tropics!
- Tectonic plate movement **uplifted** rocks from below the sea – becoming land.
- During uplift, some rocks snapped and moved along **faults** in a series of earthquakes over thousands of years. Each movement disturbed the strata so that they tilted. Sometimes, the faults form a steep edge (called a **fault scarp**) where uplift has raised some parts more than others. You can see a fault scarp at Giggleswick Scar in Figure 3, and where uplift has left the rocks on the left higher than those on the right.

Glaciation

As the Pennines were uplifted, rivers like the Wharfe eroded into them, creating V-shaped valleys (see section 4.16). But the most recent Ice Age, over 10 000 years ago, brought huge glaciers to the Pennines. They had two effects:

- Glaciers don't create their own valleys, but instead follow existing river valleys, making them deeper and widening them into U-shaped troughs, like the valley in Figure 2.
- As the glaciers melted, their meltwater created fantastic features like Malham Cove – a spectacular waterfall! That's now dry – because the glacier has disappeared.

Your questions

1. Draw a sketch of Malham Cove from Figure 1. Label its features and evidence to show it was once a waterfall. 
2. Using Figure 2, identify and explain which rock is most resistant and which is least resistant. 
3. Study Figure 3. Explain three things that uplift has done to the landscape. 
4. Describe three changes that glaciers brought to the valley of the Wharfe.
5. Draw a spider diagram with three arms – geology, tectonic processes, and past climates. Add details to explain how each has created the Pennine landscape.

Exam-style questions

6. Define the term 'erosion'. (1 mark)
7. Explain **one** way past tectonic processes influenced the physical landscape of the UK. (2 marks)
8. Explain how tectonic movements have created uplands such as the Pennines. (4 marks)

4.2 The UK's relief and geology

By using these maps, you'll assess the relationship between landscape and geology. Use these maps in connection with Section 4.3.

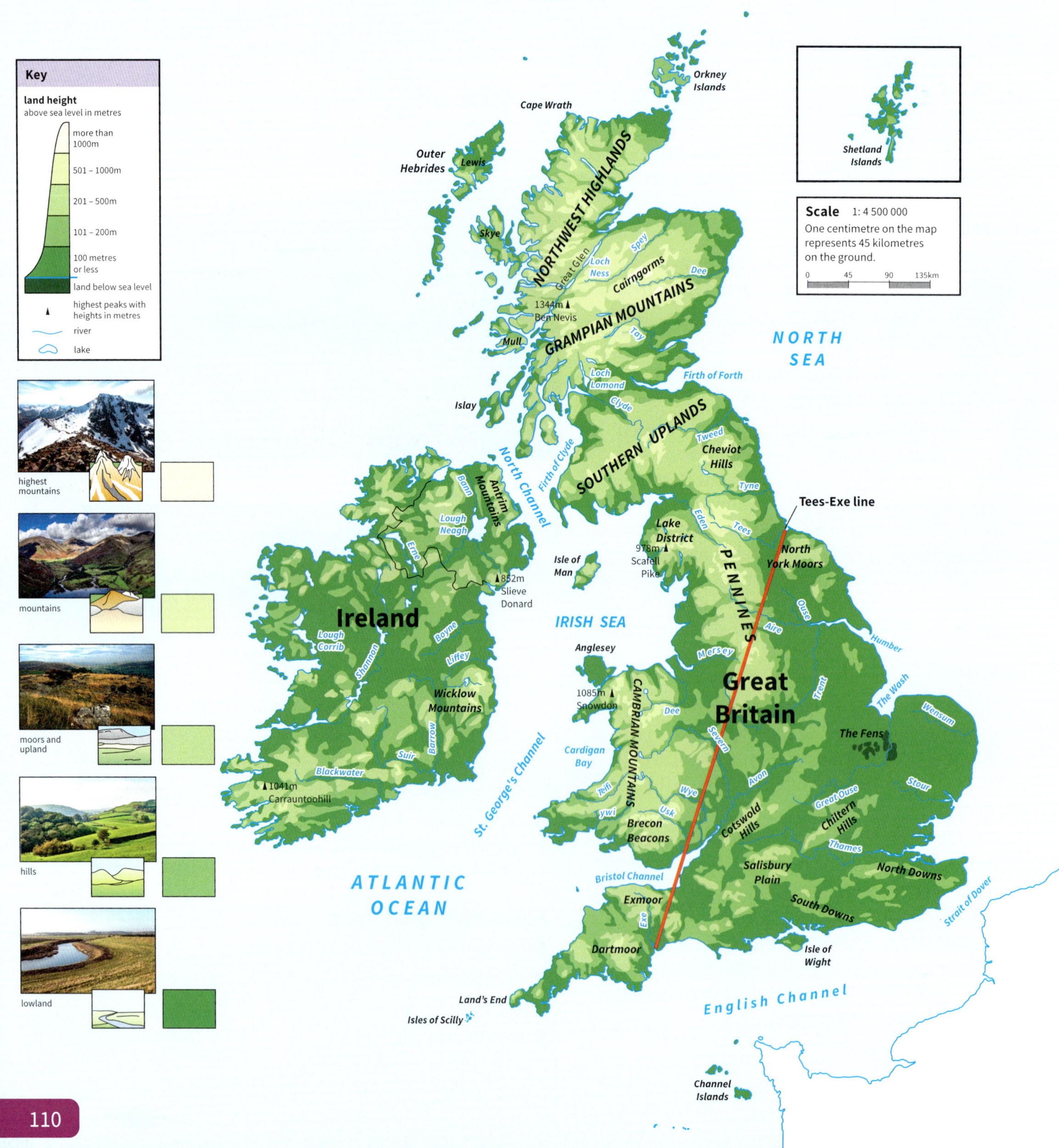

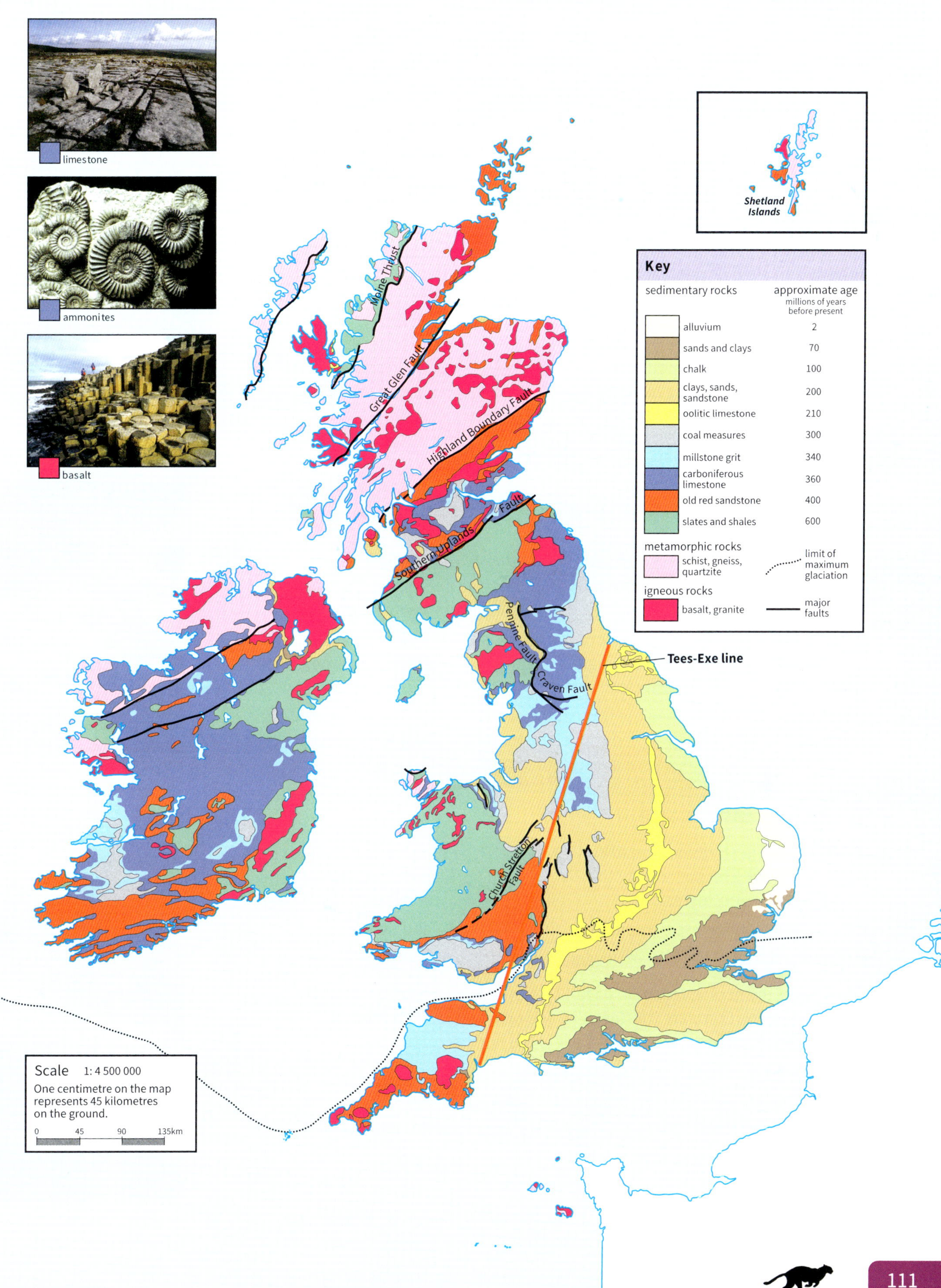
limestone
ammonites
basalt
Shetland Islands
Key
sedimentary rocks
approximate age
millions of years before present
alluvium 2
sands and clays 70
chalk 100
clays, sands, sandstone 200
oolitic limestone 210
coal measures 300
millstone grit 340
carboniferous limestone 360
old red sandstone 400
slates and shales 600
metamorphic rocks
schist, gneiss, quartzite
limit of maximum glaciation
igneous rocks
basalt, granite
major faults
Moine Thrust
Great Glen Fault
Highland Boundary Fault
Southern Uplands Fault
Pennine Fault
Craven Fault
Church Stretton Fault
Tees-Exe line
Scale 1: 4 500 000
One centimetre on the map represents 45 kilometres on the ground.
0
45
90
135km

4.3 It's all about rocks

In this section, you'll get to know the characteristics and distribution of the UK's main rock types.

The shape we're in

For a small group of islands, unravelling the geology of the UK can be a headache! The height and shape of its highlands and lowlands all depend on the rocks from which they are formed. Parts of Cornwall and Scotland have still not been properly surveyed by geologists because rocks there are so complex!

Britain's geology is valuable too.

- Cornwall has large amounts of tin and copper which made some people wealthy.
- Huge strata of coal helped to make Britain the world's first industrial nation. Other resources include building stone (e.g. marble) and raw materials (e.g. iron ore).

▲ ***Figure 1*** *19th century copper mines in west Cornwall*

How do rocks differ?

Rocks that make up the UK were formed in different ways. There are three main types of rock:

- **Igneous** rocks – the Earth's oldest rocks, formed from lavas and deep magmas. They were once molten, then cooled and crystallised. Most igneous rocks are resistant to erosion (e.g. Giant's Causeway shown in Figure 2).
- **Sedimentary** rocks – formed from sediments eroded and deposited by rivers, the sea, or on the sea bed (see Figure 3). Some are resistant (e.g. limestone) while others crumble easily (e.g. shale).
- **Metamorphic** rocks – sedimentary and igneous rocks that were heated and compressed during later igneous activity. Heating and compression harden them and make them resistant – e.g. limestone becomes marble.

▲ ***Figure 2*** *Giant's Causeway in Northern Ireland. It formed from basalt that cooled to form natural columns from a lava flow about 50 million years ago.*

Rocks and landscape

Relief (landscape) depends greatly on rock type. Page 110 shows a UK relief map. Notice the 'Tees-Exe line' joining the River Tees in north-east England with the River Exe in the south-west.

- To the north and west are uplands of England, Wales, and Scotland.
- South and east of the line are lowlands of central and southern England.

Now look at the geology map on page 111. Notice that the key is arranged by age and type of rock. Compare the Tees-Exe line on this map with the relief map. You should see the following north and west of the line:

- Most rocks are older.
- Most resistant igneous and metamorphic rocks are found here.
- There are more faults, where upland areas were uplifted by tectonic activity.

To the south and east of the line, most rocks are:

- younger
- weaker sedimentary rocks. Limestones are found there too, but they are younger and less resistant than Carboniferous limestone (see section 4.1).

▲ ***Figure 3*** *Chalk cliffs in southern England*

Name of rock	How it was formed	Characteristics
Igneous		
Granite (Figure 5)	Formed from magma cooling deep underground.	Contains crystals of quartz (glassy), feldspar (white) and mica (shiny black). Very resistant.
Basalt	Formed from lavas mainly on the surface.	Almost black, and dense. Very resistant.
Sedimentary		
Chalk (Figure 3)	A purer, younger form of limestone.	Very porous. Medium resistance but stronger than clays and younger sands.
Carboniferous limestone	For formation, see section 4.1.	Permeable, with underground rivers, passages and caves. Generally resistant.
Clay	Formed from muds deposited by rivers or at sea.	Soft and crumbly. When compacted it becomes shale. Generally weak.
Sandstone	Formed from sand grains compacted together.	Slightly porous. Those less than 100 million years old are weak; those more than 300 million years old are resistant.
Millstone grit	Sandstone which has been firmly cemented and compacted.	Very resistant.
Metamorphic		
Slate	Formed from heated muds or shale.	Very resistant.
Schist (Figure 6)	Formed by further metamorphosis of slate, where it partly melted and solidified.	Very resistant.
Marble	Formed from heated limestone.	Very resistant.

▲ ***Figure 4*** *Ten rocks you need to know!*

▲ ***Figure 5*** *Granite sample*

▲ ***Figure 6*** *A schist sample. You can see the original layers from when it was a sedimentary mud. They've been altered by heating. The white bands are injections of quartz.*

Your questions

1 Using the maps in section 4.2, mark and label these on a blank map of the UK: **a** ten upland areas, **b** the highest points in England, Wales, Scotland and Northern Ireland, **c** the rivers Thames, Severn, Trent, Ouse, Clyde and Forth, **d** lowland areas – Thames Basin, Fens, Cheshire Plain, Vale of York, and Central Lowlands.

2 Identify and mark on your map the geology of each of the upland and lowland areas.

3 **a** Research the economic uses of any five of the ten rocks listed in Figure 4.

b Copy and complete this table for all five. Use the relief and geology maps on pages 110–111.

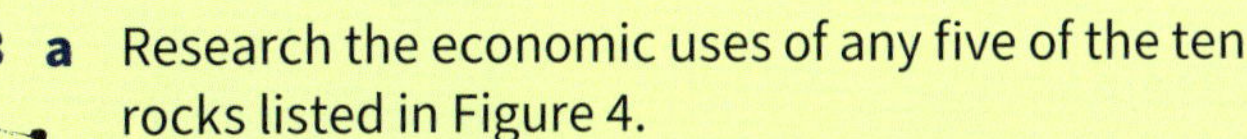

Rock	Where found in UK	Economic uses

Exam-style questions

4 Define the term 'sedimentary rock'. (1 mark)

5 Explain **one** way rock type influences the relief of the land in the UK. (2 marks)

6 Analyse Figures 1 and 2 in section 4.2. Assess the influence of geology on the relief of the UK. (8 marks)

4.4 Physical processes in the landscape

In this section, you'll understand how upland and lowland landscapes result from different physical factors.

1: The Lake District – an upland landscape

Every year, the Cumberland Fellrunners Association in the Lake District holds an annual race above Wastwater, one of the lakes. It's not for the faint-hearted – the race is long (34 km) and runners have to climb a total of 2750 metres over several peaks of hard igneous rock! The ground is rough, and the race includes England's highest mountain, Scafell Pike. Reaching its summit involves climbing a near-vertical rock face!

Weathering and slope processes

What makes the ground rough are rock fragments, known as **scree**, shown in Figure 1. Scree consists of angular rock pieces created by freeze-thaw **weathering** (see section 4.16). Each winter, temperatures are often below freezing at night and warmer during the day. Rainwater gets into cracks in the rock, freezes, and expands by 10%. Expansion widens the crack, and eventually the rock breaks into pieces. Every winter adds more scree.

In addition, rapid **slope processes** affect valley sides.

- Scree fragments are unstable and move easily during **rockfalls**, increasing dangers for walkers.
- **Landslides** are common. The Lake District is the UK's wettest region (over 2000 mm of rain a year). Rain adds to the weight of weathered rock so it slides easily.

> **Weathering** is the physical, chemical or biological breakdown of solid rock by the action of weather (e.g. frost, rain) or plants. See section 4.16.

Post-glacial river processes

Like the Pennines, the Lake District was once **glaciated**. Glaciers created deep U-shaped valleys and hollows filled now by lakes. Today, rivers flow in the valley bottom instead of glaciers. These rivers are small compared to their valleys, and are known as **misfits**. They deposit silt and mud (known as **alluvium**) in the valley bottoms, making them fertile for farming.

▲ ***Figure 1*** *Screes on the edge of Wasdale*

▲ ***Figure 2*** *A misfit river in Lakeland Valley, the Lake District*

2: The Weald – a lowland landscape

Although southern England is much lower than the Lake District, some parts consist of **undulating** (gently rolling) hills. One such area is the Weald, in Kent and Sussex. The landscape is also affected by geology (see Figure 3). The Weald was once a dome of folded rocks, forming an arch called an **anticline**. Study Figure 3 – look how strata across the anticline used to be continuous, linking what is now the North and South Downs. Erosion has left alternate strata of more and less resistant rock to form a landscape known as **scarp and vale topography** (see Figure 4).

- Resistant rocks, like chalk, form steep **escarpments**.
- Behind the escarpment, gentle slopes follow the angle at which the rocks were tilted (called the **dip**), known as a **dip slope**.
- Softer clays are lower and flatter, forming the vales.

▼ ***Figure 3*** *A cross section of the geology of the Weald*

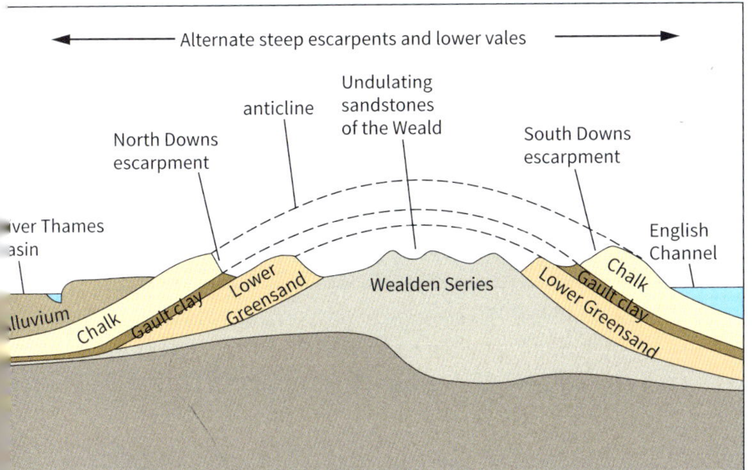

▲ ***Figure 4*** *Scarp and vale topography*

Weathering

Because southern England is warmer than the Lake District, different types of weathering occur.

- Chalk is calcium carbonate, an alkali, so is affected by solution (**chemical weathering**, see section 4.16).
- Tree and shrub roots break up solid rock (**biological weathering**, see section 4.16).

Post-glacial river and slope processes

Chalk is porous, so it is unusual to find rivers in chalk areas, except when it's saturated after wet weather. However, clay is impermeable so rivers are common in vales. During and after the last Ice Age, water in the chalk froze, making it impermeable. Then, fresh water formed rivers and valleys. As the climate warmed, water seeped through the chalk once again leaving **dry valleys** where rivers had once flowed.

Slope processes are slower than in the Lake District; the most common is **soil creep**. It's caused by rain dislodging soil particles (see section 4.16).

Your questions

1 Copy and complete the following table to compare features of the Lake District and Weald landscapes. 

Feature	Lake District	Weald
Geology (see section 4.2)		
Highest points (see section 4.2)		
General shape of landscape		
Main weathering processes		
Main slope processes		
Effects of last Ice Age		

2 Research three pictures online of **each** area to illustrate **a** its highest points, **b** valleys, **c** slopes.

Exam-style questions

3 Define the term 'weathering'. (1 mark)

4 Explain **two** weathering processes that affect the UK's landscapes. (4 marks)

5 Explain **two** factors that make the landscapes of southern England distinctive. (4 marks)

4.5 People in the landscape

In this section, you'll understand how distinctive landscapes result from human activities.

Living among the trees

The village of Ae, near Dumfries in southern Scotland, has Britain's shortest place name! It's a remote place in the Southern Uplands. It's different from most villages in the UK because it was specially created for forestry workers in the early 20th century. The government set up the Forestry Commission in 1919 after using so much wood in the trenches of World War I. New woodlands were needed, and rural areas needed jobs as they were suffering economic depression. Forests were planted and the commission built houses to attract workers. That's how Ae came about, surrounded by thousands of hectares of forest (see Figure 1).

▲ ***Figure 1*** *The village of Ae, with forestry around it planted by the Forestry Commission. The deliberate planting of forests in some upland areas of the UK has changed the landscape.*

Ae is recent, but most settlements in the UK are not! Many were here in Roman times 2000 years ago. Almost every UK settlement we see now was recorded in the Domesday Book in 1086! Every landscape has the stamp of people who settled there many centuries ago – their economic activities and buildings made from whatever materials they had around them.

The Yorkshire Dales

The Yorkshire Dales landscape tells the story of Norse farmers and settlers of the 8th and 9th centuries, who left their mark. The limestone on the valley sides shown in Figure 2 made excellent building stone. So did the boulders and rocks left by the rivers of melting glaciers in the valley bottoms. By clearing stones from the valley bottom, the settlers improved the land for farming and built dry stone walls as field boundaries, as shown in the photo.

▶ ***Figure 2*** *Dry stone walls and barns make the Yorkshire Dales landscapes in the Pennines unique*

The Norse settlers and their farming were each influenced by the Pennine climate.

- Winters are cold and the growing season short, so sheep farming is best. In winter, sheep were kept in fields in the valley bottom. In summer, they grazed the upland fells (or moors) shown in the background of Figure 2. The fields in the valley bottom produced hay for winter feed. The practice still survives now.
- Winter hay was stored in the stone barns shown in Figure 2. It took time to look after animals if you lived far from the fields, so early farmers lived near their animals. They built longhouses like the one shown in Figure 3, consisting of a house and barn together, so animals could be kept inside during bad weather. This led to a dispersed settlement pattern of isolated farms, instead of villages.

▲ ***Figure 3*** *A Yorkshire Dales longhouse, which combined homes (on the right) and barns (on the left) in one building*

East Anglia

The East Anglian landscape in eastern England is very different from that of the Pennines! It's low-lying, almost all under 100 metres above sea level, and much of it is flat. Its coastline faces Europe, so it was settled by waves of European Angles (hence its name) and Vikings. They settled in communal villages.

On the surface, East Anglia's geology is mainly sands and clays, known as **till**, which were deposited by glaciers in the last Ice Age. Till produces fertile soil for arable (crop) farming, but nothing solid for building. Hedges are used as field boundaries (shown in Figure 4) instead of dry stone walls. Below the surface the solid geology is chalk. Chalk is too crumbly for building, but within it are pieces (known as **nodules**) of flint, a hard crystalline form of quartz. Many older buildings were built from this.

▲ ***Figure 4*** *Field boundaries in East Anglia's flat landscape – sometimes hedges, or ditches on low-lying and wetter ground*

Your questions

1 Using the UK relief and geology maps on pages 110–111, identify and mark the location, relief and geology of the Yorkshire Dales and East Anglia on a blank UK map.

2 **a** Draw a table to compare the following landscape features in the Yorkshire Dales and East Anglia: geology, relief, settlement, building materials, field boundaries, farming.

b Explain how these factors have combined to produce distinctive landscapes in each area.

3 In pairs, research one upland and one lowland UK region. For each region, produce six PowerPoint slides to show **a** landscape photos, **b** how geology, relief, settlement, building materials, and economic activity have created the landscape.

Exam-style questions

4 Explain **one** way in which human activity has influenced the UK's physical landscape. (2 marks)

5 Explain **two** ways in which distinctive landscapes can result from human activity over time. (4 marks)

6 Analyse Figures 2 and 4. Assess the role of human activities in creating distinctive landscapes. (8 marks)

4.6 Contrasting coasts

In this section, you'll understand how geological structure and rock type can influence coastal erosional landforms.

The coastal zone

Geographers study coasts because they are dynamic places, always changing. The **coastal zone** is the changing boundary between land and sea. It's a popular zone economically because it gives access to the sea for fishing, trade and resources such as oil and gas. It's also popular as a place to live, and for tourism.

Geology and rock type

The most important feature of a coast is its geology, and rock resistance. Coasts vary depending upon the resistance of rocks to erosion:

- **Hard rock coasts** consist of resistant rocks, such as igneous granite, and resistant sedimentary rocks, e.g. sandstone, limestone or chalk. Examples include Flamborough Head (East Yorkshire) and Lulworth Cove (Dorset).
- **Soft rock coasts** consist of less resistant rocks such as clays and shales, which are more easily eroded. Examples include the Holderness Coast (East Yorkshire), Christchurch Bay (Dorset and Hampshire) and the North Norfolk Coast.

This resistant sandstone layer forms a very large overhang

Coal, a very weak layer. A wave cut notch has formed.

Shale, a weak layer

Ironstone, a resistant lay

▲ ***Figure 1*** *A cliff with different rock types – all of which have different resistances to erosion*

Rock structure

Rock structure means the way different rock strata are arranged. There are often several rock types in one cliff, as Figure 1 shows. The cliff is only as resistant as its weakest strata. Rock strata can be arranged in two ways along coastlines:

- If strata are at right angles to the coast, the coast is known as **discordant**. These have different rock types.
- If strata are parallel to the coastline, the coast is known as **concordant**. Concordant coasts have the same type of rock parallel to the coastline.

As these two types of coast erode, different landforms are produced.

Discordant coasts: headlands and bays

South-west Ireland is an example of a discordant coast (Figure 2). It's an unusual coastline, with long headlands and bays. Resistant sandstones and softer limestones are found alternately along the coast. Waves have eroded limestone to form bays, leaving harder sandstone as headlands.

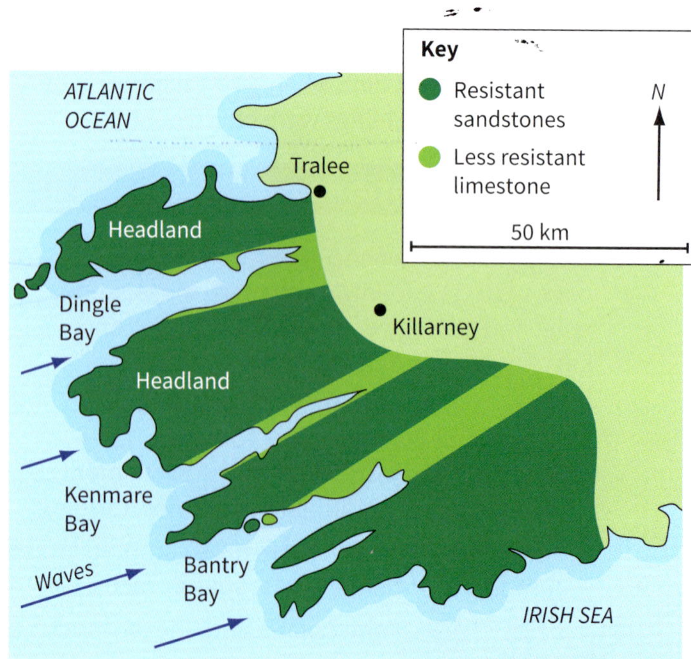

▲ ***Figure 2*** *The discordant coastline of south-west Ireland*

Concordant coasts: coves and cliffs

One of the UK's best examples of a concordant coast is Lulworth Cove in Dorset, shown in Figure 3. It's a World Heritage Site because the geology is so unique. Follow this sequence to show how Lulworth Cove has developed:

- A resistant layer of Portland limestone runs along the coast at Lulworth.
- Erosion processes (see section 4.7) have created a gap through the limestone, exposing less resistant sands and clays behind.
- As waves reach the sands and clays, the cove has quickly widened.
- At the back of the cove, waves have reached the more resistant chalk, which forms a steep cliff.
- Nearby is Stair Hole, a cove which is beginning to form. The sea has eroded a gap in the limestone, and is already widening the cove behind.

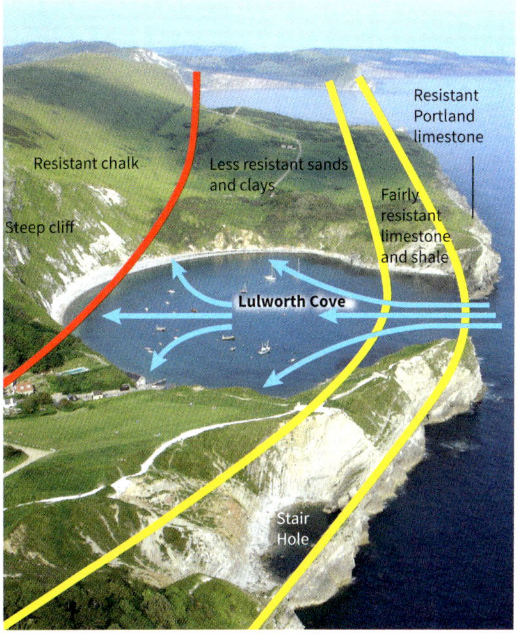

Resistant chalk
Less resistant sands and clays
Resistant Portland limestone
Steep cliff
Fairly resistant limestone and shale
Lulworth Cove
Stair Hole

▲ ***Figure 3*** *Lulworth Cove – an example of a concordant coastline*

Weaknesses in rock structure

Natural weaknesses in rocks influence coastal erosion. There are two types of weakness:

- **Joints** are small, usually vertical, cracks found in many rocks.
- **Faults** are larger cracks caused by past tectonic movements, where rocks have moved.

The more joints and faults there are in rocks, the weaker it is, because waves can erode easily. Figure 4 shows how joints in a cliff can widen to form a **cave**. These in turn erode into an **arch** which then collapses to form a **stack**. Eventually, this erodes down to become a **stump**.

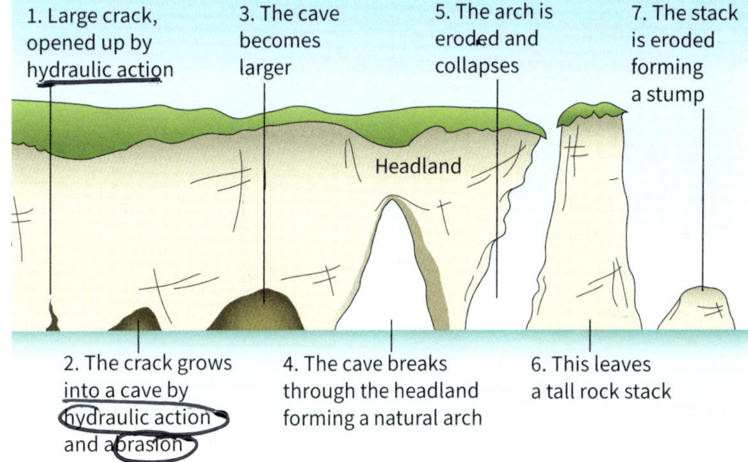

▶ ***Figure 4*** *Hard rock cliffs at Flamborough Head in East Yorkshire have been eroded to form caves, arches, stacks and stumps*

Your questions

1 **a** Draw a sketch of the cliff shown in Figure 1. Label it to show the most and least resistant rocks.

b Explain using diagrams how it got to look like this.

2 Draw a flow diagram to show the stages of development of Lulworth Cove.

3 Make a copy of Figure 4, titled 'Stage 1'. Then explain in a series of labels how the following might change over time: **a** the large crack (point 1), **b** the headland arch (4), **c** the stump (7).

Exam-style questions

4 Define the term 'coastal zone'. (1 mark)

5 Explain the difference between a concordant and discordant coastline. (3 marks)

6 Analyse Figures 1, 2 and 3. Assess the influence of geological structure on the formation of coastlines. (8 marks)

4.7 The UK – climate and coastline

In this section, you'll assess how climate and coastal processes affect the UK's coastal landscapes.

What causes waves?

As a rule, the UK climate is pretty windy! Winds are seasonal, and are strongest in winter. It's wind that is responsible for forming waves in the sea. When wind blows across the sea, friction between the wind and the water surface causes waves (see Figure 1). Wave size depends on:

- wind strength, and how long it blows for
- the length of water the wind blows over – called the **fetch**. Some waves have a huge fetch. Waves reaching Cornwall begin near Florida and travel across the Atlantic Ocean, a fetch of 6000 km.

Winds are strongest in autumn and winter, and it's then that the UK coastline is affected most. The **prevailing winds** (i.e. those which blow most often) are from the south-west, so coastlines facing this way feel the full force of storms.

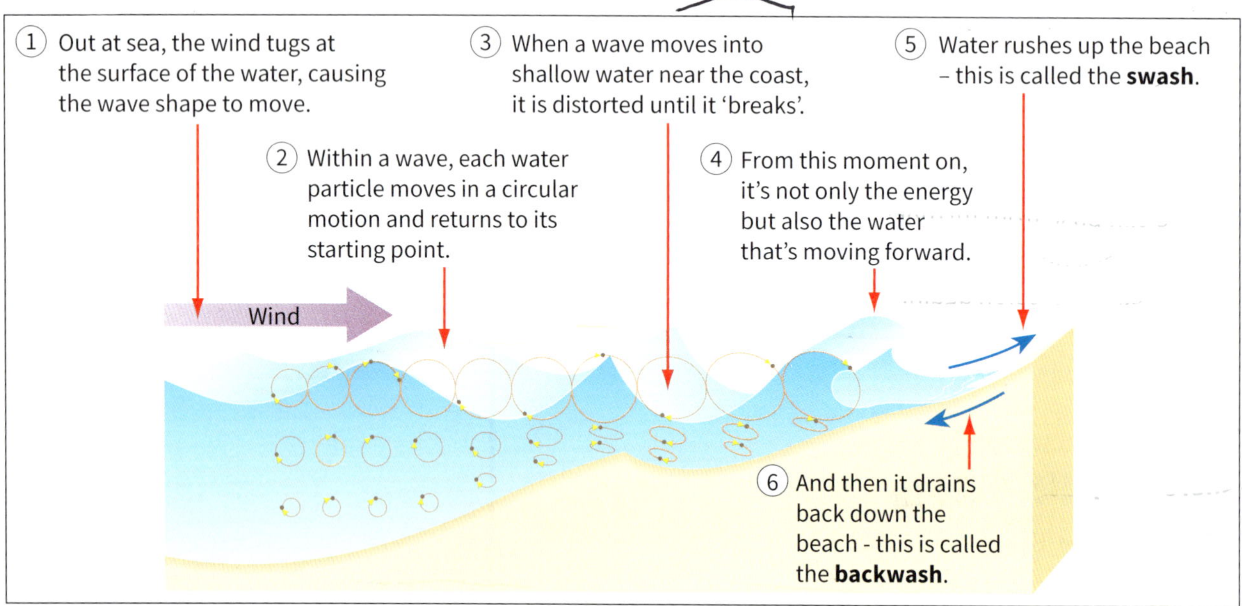

▲ ***Figure 1*** *How waves form and develop*

Summer and winter waves

Waves vary. The shape of a beach (or **beach profile**) results from how waves break.

Summer waves tend to be small. They are called spilling waves – as they break, they spill up the beach. They arrive slowly, with long wavelengths and low amplitudes.

- They have a strong **swash**, which transports sand up the beach.
- The gentle slope means that the backwash is slow, so sand is deposited.
- Deposited sand forms a bank or **berm**.

Because these waves build up a beach, they are called **constructive waves** (see Figure 2a).

In winter, strong winds are common. Waves are taller (larger amplitude) and closer together (shorter wavelength). These are called **plunging waves**. They arrive much more quickly.

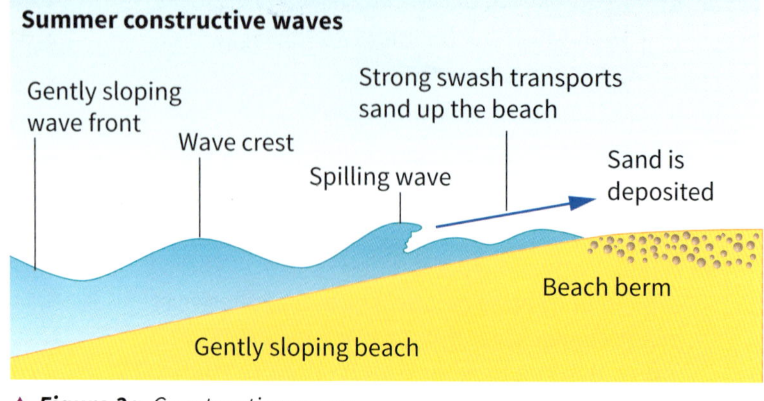

▲ ***Figure 2a*** *Constructive waves*

- They have a strong **backwash**, eroding sand from the beach.
- The backwash flows under the next incoming wave, forming a **rip current** (see Figure 2b). These can be strong and drag swimmers out to sea.
- Waves like this create a steep beach profile.
- The sand is carried away and deposited offshore, forming a bar.

Waves which erode beaches are called **destructive waves** (Figure 2b).

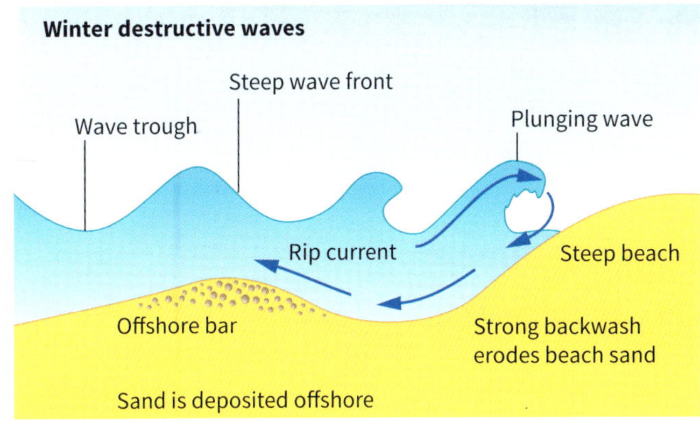

▲ ***Figure 2b*** *Destructive waves*

Coastal erosion

On coasts with resistant rocks, erosion is slow. Most erosion happens during winter storms. (See page 129 to find out about erosion in Christchurch Bay.)

- Wave power is concentrated at the cliff base where **abrasion** forms a wave-cut notch.
- As the notch grows, a cliff overhang develops.
- The overhang becomes unstable and eventually collapses, forming a pile of rock debris. The debris protects the cliff base from further erosion.
- Over time, the rock debris is eroded by attrition, exposing the cliff to erosion again.

Over long periods of time, a succession of wave-cut notches form, and the cliff collapses again and again, retreating inland. A level area of smooth rock is left where the cliff line once was, stretching out to sea, called a **shore-line platform**.

Your questions

1 Explain the difference between **a** prevailing wind and fetch, **b** abrasion and attrition.

2 **a** Using an atlas, explain why the largest waves experienced by the British Isles are to be found in Cornwall, south-west Ireland, and north-west Scotland.

b Identify those parts of the UK which would have the smallest waves, and explain why this is.

3 Explain how and why **a** the rate of cliff erosion, and **b** beach angle vary seasonally.

Exam-style questions

4 Define the term 'fetch'. (1 mark)

5 Describe **one** process of coastal erosion. (3 marks)

6 Explain how beach formation can be influenced by different types of waves. (4 marks)

Know your coastal erosion processes!

Erosion means wearing away and breaking down rocks. There are three main types of coastal erosion, shown below – abrasion, hydraulic action and attrition.

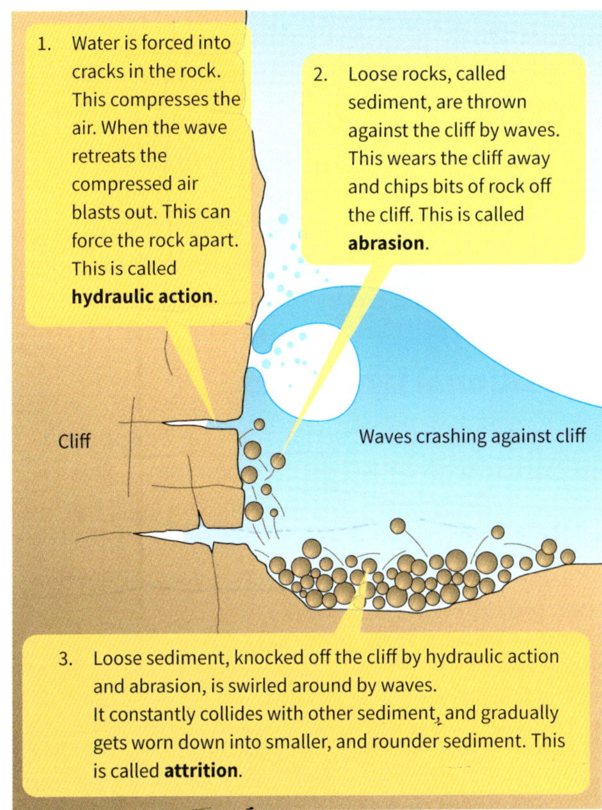

▲ ***Figure 3*** *The three main types of coastal erosion*

Thinking beyond

Coastal change is often a seasonal thing – with deposition occurring in the calm of summer, and erosion and cliff collapse occurring during winter storms.

4.8 Coastal deposition

In this section, you'll understand how coastal processes create depositional landforms.

Beach sediment

The material eroded from cliffs by hydraulic action and abrasion is called **sediment**. Sediment comes in many sizes, from tiny clay particles to larger sand and silt, right up to pebbles, cobbles and boulders. Sand is material that is 0.06–2 mm in size. Over time, two things happen:

- attrition makes sediment smaller and rounder
- sediment is transported from where it was eroded to new locations.

Get the drift?

The main way that sediment is transported is by **longshore drift**, shown in Figure 1. This happens when waves break at an angle to the coast, rather than parallel to it.

Because prevailing winds are mostly from one direction, longshore drift is usually in one direction too. Longshore drift transports sediment along coastlines, as Figure 1 shows, sometimes for hundreds of kilometres before it is eventually deposited.

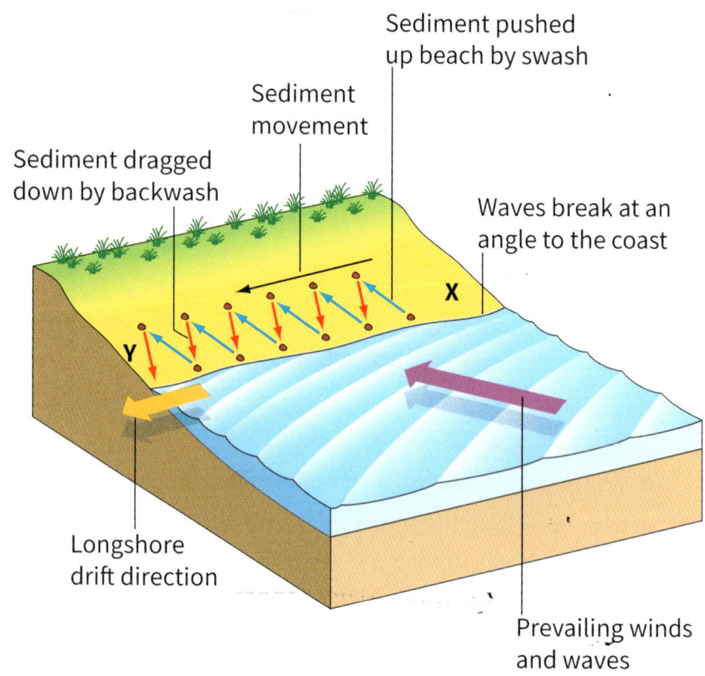

▲ ***Figure 1*** *How longshore drift works*

Depositional landforms

When rocks are eroded, sediment is first deposited very close to where it eroded. In a sheltered area such as a cove or bay, a beach forms because the sediment is trapped in the bay. Sediment transported by longshore drift creates new landforms where it is deposited, as Figure 2 shows.

Many **beaches** are simply rivers of sand and shingle (pebbles) slowly moving along the coast as sediment is transported by longshore drift.

- Strong onshore winds can blow sand inland, forming **sand dunes** parallel to the shoreline.
- Small bays on the coast can sometimes be blocked by a **bar** of sand which grows across the mouth of a bay due to longshore drift. A good example is shown in Figure 3, from south Devon. Behind the bar, a shallow **lagoon** forms. These are often important habitats for birds.

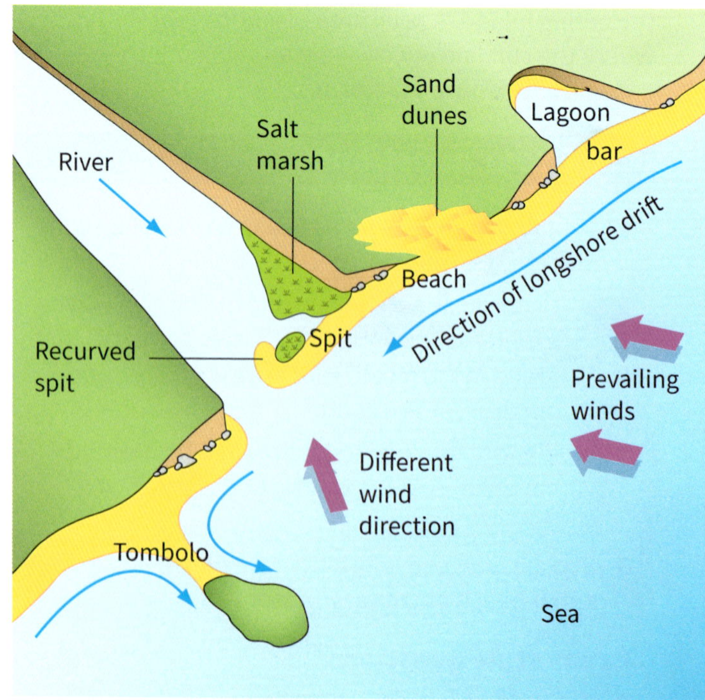

▲ ***Figure 2*** *Different landforms created by coastal deposition*

- Longshore drift carries sand along the shore until it reaches a river estuary where it gets pushed out into the river channel. The river flow halts the drift, so sand is deposited, forming a long sandy neck, called a **spit** (see Figure 4). The spit stops growing when deposition of sand, by longshore drift, is balanced by erosion from the river.
- The river moves out to sea at low tide, whilst at high tide, the sea flows inland. Each tide erodes the spit and causes it to curve back on itself. Many spits have a hooked or **recurved** end. The water behind a spit is protected from storms and tides so remains calm, allowing **salt marshes** to form, as shown in Figure 4.

Depositional landforms are made of loose sediment, so they are not very stable. Sand dunes can be stabilised by plants growing on the sand. Plants that grow on beach sand need to be tough:

- They have long roots to hold them in place in the strong winds, e.g. marram grass.
- They have tough, waxy leaves to stop them getting sandblasted.
- They can survive being sprayed by salt water.

▲ *Figure 3* *Slapton Ley bar in south Devon*

▶ *Figure 4* *Hurst Castle spit at the end of Christchurch Bay, in Hampshire*

Your questions

1. **a** Rank sand, boulders and pebbles in order of size.
   b Explain how and why sediment size would change between X and Y in Figure 1.
2. Draw a labelled copy of Figure 1 to show how longshore drift occurs. Add the following labels: **a** large sediment at this end of the beach, **b** fine sediment at this end of the beach.
3. Explain the differences between a beach, a bar, a spit, a tombolo, a salt marsh, and a sand dune.
4. Explain why spits **a** do not grow across the mouths of rivers (unlike bars), **b** have a curved end.

Exam-style questions

5. Define the term 'longshore drift'. (1 mark)
6. Describe **two** features of a coastal spit. (2 marks)
7. Using only a labelled diagram, explain the processes that lead to the formation of a sand bar. (4 marks)
8. Explain why salt marshes often form behind coastal spits. You may use a diagram to help you. (4 marks)

4.9 Human activities and the coast

In this section, you'll understand how human activities affect coastal landscapes.

We love the coast!

On a summer's day, it's easy to see why south-west England is the UK's largest tourist destination. Its lengthy coast provides stretches of sand such as Woolacombe Bay in north Devon (see Figure 1) which provides space for sunbathing, walking and surfing. It's a popular coast, but one which is surprisingly unaffected by people. Take people out of the photo, and you're left with a landscape of the beach and an undeveloped coastline. Few signs of pressure here – and it's that which makes it unusual in a crowded country like the UK.

Four impacts of human activity affect the coast – development, agriculture, industry and coastal management.

The effects of development

It's common in parts of the UK to speak of a 'crowded coast'. It's not just hotels or campsites for tourists that are the cause; two other types of development also put pressure on coasts.

- **Housing.** Many people who work in London can no longer afford housing there, so some coastal towns and cities offer good alternatives for people who commute each day. The coast is also important for those seeking a place in which to retire. Resorts such as Bournemouth, Blackpool, and Scarborough are all popular retirement destinations.
- **Office development.** The high cost of London's property also affects companies, so some of those too are moving out, such as J P Morgan, an investment bank, which moved to new office buildings in Bournemouth in 2015 (see Figure 2). Both Brighton and Bournemouth are popular locations with younger populations as universities and companies expand there.

Each of these demands puts pressure on already crowded coasts.

The effects of agriculture

Romney Marsh is a wild place on the Kent coast. It consists of a shingle bank behind which a wetland and marsh has developed, which is a habitat for birds.

▲ ***Figure 1*** *Woolacombe Bay in north Devon – its wide stretches of sand make it popular with tourists*

▲ ***Figure 2*** *Bournemouth – relocation destination for many financial companies*

Sarah Cullen, Director of Digital Butter, explains why there's a strong sense of community in Bournemouth. "Being able to get out at lunchtime and experience the sea air fuels creativity and perspective, enhancing productivity.
This, along with a strong support network and a commitment to invest in the creative industry, is what attracts a steady stream of talent who now call Bournemouth their home. The Covid-19 pandemic proved it's no longer necessary to be in cities to attain a dream job. That's what makes Bournemouth attractive to live in – it's possible to achieve a work/life balance."

The marsh also provides summer grazing pasture for cattle. It faces two pressures:

- The price of good farmland has risen sharply, from £2400 per hectare in 1995 to £30 000 in 2015. Farmers have to maximise their income by using whatever land they can. The need for extra grazing is putting pressure on wildlife habitats.
- Climate change and rising sea levels are likely to lead to flooding by salt water during winter high tides, which could threaten the pastures.

The effects of industry

Bacton, a village on the Norfolk coast, is critical to the UK economy. North Sea gas is piped onshore at the terminal shown in Figure 3. It's an example of how an essential development brings conflicts with tourists. Figure 3 shows this stretch of coast is sandy, meaning it's popular with tourists. But a holiday next to a gas terminal is not necessarily what tourists want!

▲ ***Figure 3*** *Bacton gas terminal on the Norfolk coast*

Similar industrial developments have taken place at:

- the Solent (Southampton), the Severn estuary (Bristol), the Mersey and Dee estuaries (Liverpool), and the Tees estuary (Middlesbrough) – all important oil and chemical refining locations with huge industrial installations
- the Thames estuary east of London – important for shipping and power stations to supply London.

The effects of coastal management

Figure 4 shows Milford-on-Sea in Christchurch Bay in Hampshire. It's a location that has suffered from erosion and is now influenced by coastal management to prevent further problems. You can find out more about methods of managing coasts in section 4.12.

▲ ***Figure 4*** *Milford-on-Sea – can you identify the recurved sea wall, stone and wooden groynes? See 4.12 for further reading.*

Your questions

1. On a blank outline map:
   - **a** locate each of the places mentioned in this section
   - **b** annotate each location showing the pressure it faces.
2. In pairs, design a spider diagram to show how housing, offices, agriculture and industry affect the coastal landscape.
3. Produce a 400 word news report titled 'The impacts of human activity on coastal landscapes'. Summarise the impacts using material in this section.

Exam-style questions

4. Describe **one** pressure placed on coastal landscapes by agriculture. (2 marks)
5. Explain **two** ways in which human activities affect coastal landscapes. (4 marks)
6. Explain how human activities can lead to environmental pressures on coastal landscapes. (4 marks)

4.10 The risks from coastal flooding

In this section, you'll examine how climate change might increase the risk of erosion and flooding on coastlines.

Rising sea levels

Many climate scientists fear that global warming will cause sea levels to rise. The amount by which they will rise is not known, but there are estimates of between 30 cm and 1 metre by the year 2100 (see section 1.8). Sea levels are rising today, as the sea is warming up and expanding. Melting ice sheets are likely to speed this up.

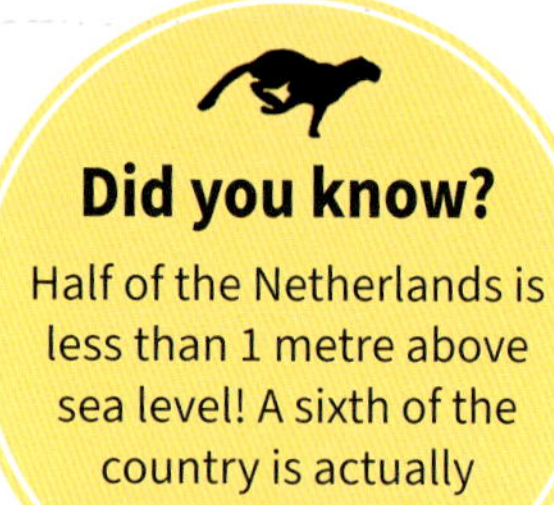

Did you know?

Half of the Netherlands is less than 1 metre above sea level! A sixth of the country is actually below sea level!

For people who live on very low-lying land next to the sea, this could spell trouble. There are many areas around the world at risk:

- In Bangladesh, if sea levels rose by 1 metre, up to 15% of the country might be flooded (see section 1.12).
- In the UK, London and Essex are at risk, because they are low lying.
- Many small coral islands in the Pacific and Indian Oceans, like the Maldives and Tuvalu, could disappear underwater.

Flood risks and the future

Sea levels are constantly changing.

- Twice a day, due to the gravity of the moon, high tides cause raised sea levels.
- Twice a month there are exceptionally high tides, called spring tides. During spring tides the flood risk rises. If spring tides coincide with large waves, the sea rises even more.
- If air pressure falls to very low levels then a storm surge occurs (see section 1.9). During a storm surge, the sea level rises by 10 mm for every 1 millibar drop in air pressure. The worst situations occur when spring tides and large waves coincide with low air pressure, forming a severe storm surge. They are most serious when caused by very low-pressure weather systems such as cyclones and depressions.

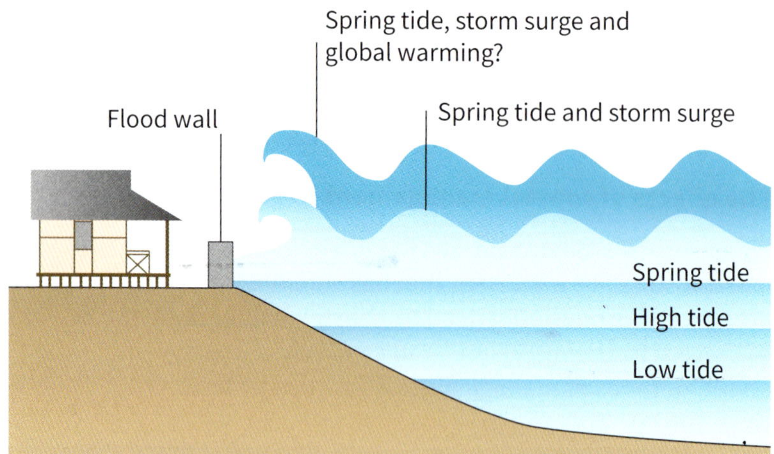

▲ ***Figure 1*** *People will become increasingly vulnerable to spring tides and storm surges as sea levels rise due to global warming*

It is possible that global warming could make hurricanes and depressions more powerful. They might also become more frequent (see section 1.8). This would mean storm surges would happen more often. If melting ice sheets raise sea levels as well, the combined results could be very serious indeed.

The storm surge of December 2013

Coastal flooding is a risk. In December 2013, a storm surge struck coastal areas of eastern and south-east England, as shown in Figure 2. High winds and a 7 m surge caused the worst flooding since 1953.

There were several impacts:

- two people died
- 1400 homes were flooded (300 in Lincolnshire, 500 in the Humber region, and 500 in Kent)
- thousands of residents were evacuated in Norfolk, Suffolk, Essex and Lincolnshire
- insurers calculated the cost of damage at £100 million.

Improved flood barriers and tidal walls prevented the surge from causing the same damage as in 1953, a previous surge that killed 326 people. The Environment Agency claims that 800 000 homes are now protected.

▲ ***Figure 2*** *Flooded homes in Great Yarmouth, Norfolk, 2013*

What of the future?

With higher sea levels and increased storminess in future, there are several challenges:

- Climate scientists estimate that '1 in 50-year' events like that of 2013 could be '1 in 20-year' events by 2050.
- Beaches, spits, and river deltas may be eroded faster, and become submerged.
- For coastal southern and eastern parts of the UK, future flooding is a problem. A sea level rise of just 50 cm would make existing sea defences useless. The only choice would be to build higher defences or abandon some areas to the sea.

East coast towns flooded as tidal surge hits

There have been mass evacuations overnight along the east coast of the UK. Flood defences and flood warnings, built to avoid a repeat of 1953, have given time for mass evacuation and saved many lives.

Adapted from a Guardian newspaper extract, 6th December 2013

Norfolk: the tidal surge and its impact on wildlife

Volunteers from groups such as the RSPB have said that nature reserves were affected by floods, with damage at the Snettisham reserve in Norfolk. Freshwater habitats have been inundated by salt water.

Adapted from a BBC news extract, 10th December 2013

Your questions

1. Use an atlas to identify and mark on a world map regions and cities in the world where rising sea level could put people and property at risk.

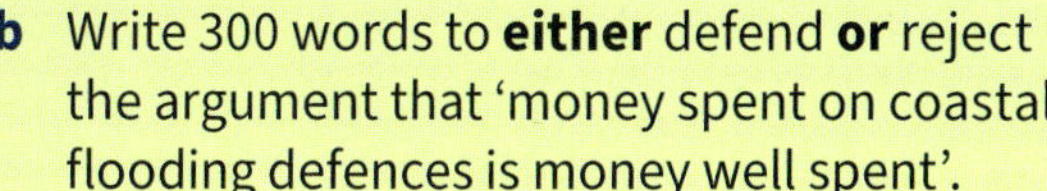

2. Explain how a storm surge forms.
3. List the impacts of the 2013 storm surge, then classify them into economic, social and environmental. Which impacts were most significant?
4. **a** In pairs, list the arguments for and against improving sea defences.

   b Write 300 words to **either** defend **or** reject the argument that 'money spent on coastal flooding defences is money well spent'.

Exam-style questions

5. Define the term 'storm surge'. (1 mark)
6. Explain why future sea level rise could threaten people and property. (4 marks)

4.11 Falling into the sea

In this section, you'll understand why coasts vary in their rate of erosion, and the impacts of erosion on people.

A complex problem

Some coastlines rarely change. Others are eroding quickly, by over 2 metres a year. Three factors explain this:

- **Geology.** Cornwall's coast has many resistant cliffs of granite and slate. They withstand the energy of even the biggest Atlantic waves for long periods. The Dodman Point in south Cornwall has an Iron Age fort that has lasted 3000 years – and suffered little erosion!
- **Cliff processes.** Meanwhile, the Holderness coast in East Yorkshire, the North Norfolk coast, and some coastal areas of Hampshire and Dorset have high erosion rates due to weak geology. They all suffer from **cliff foot erosion** (hydraulic action and abrasion). However, weathering and **mass movement** make the problem worse. Together, these are known as **sub-aerial processes**, and are the cause of **cliff face erosion**. Water movement on, and within, the cliff causes weak cliffs in these three areas to collapse.
- **Waves.** As well as geology, much depends on wave energy, which in turn depends on the fetch over which the wave has travelled (see section 4.7).

Mass movement is the movement of materials downslope, such as rock falls, landslides or cliff collapse.

Why cliffs collapse

Follow these stages in Figure 1 below.

Marine cliff foot processes

1. The base of the cliff is eroded by hydraulic action and abrasion, making the cliff face steeper.

Sub-aerial cliff face processes

2. Weathering weakens the cliff face. This can be mechanical weathering, (e.g. freeze-thaw), chemical (e.g. solution) or biological (e.g. plant roots).
3. Heavy rain saturates the permeable rock at the cliff top. Rainwater may also erode the cliff as it runs down the face or emerges from a spring.
4. The water flows through permeable rock, adding weight to the cliff geology which is too weak to support itself.

Human actions

5. Building on the cliff top adds a load, which can weigh down on the weakened cliff.

During a big storm, heavy rain saturates the permeable rock, and erosion by the sea undermines it. Eventually a large chunk of cliff gives way and slides down the cliff – a rotational slip.

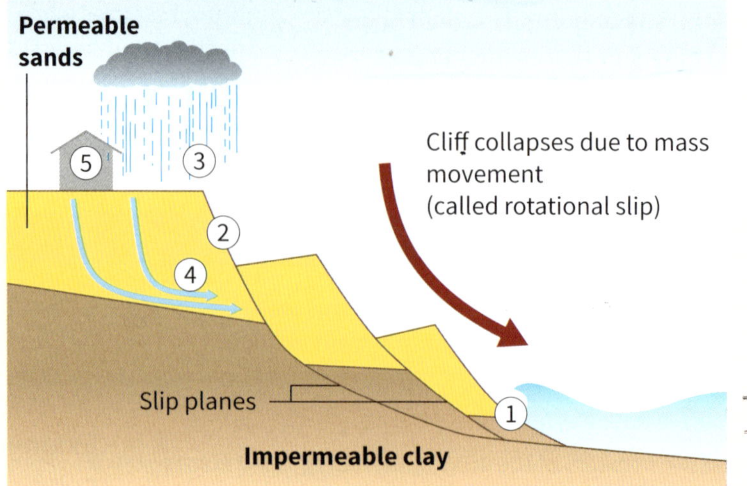

▲ ***Figure 1*** *The causes of cliff slumping and rotational slip*

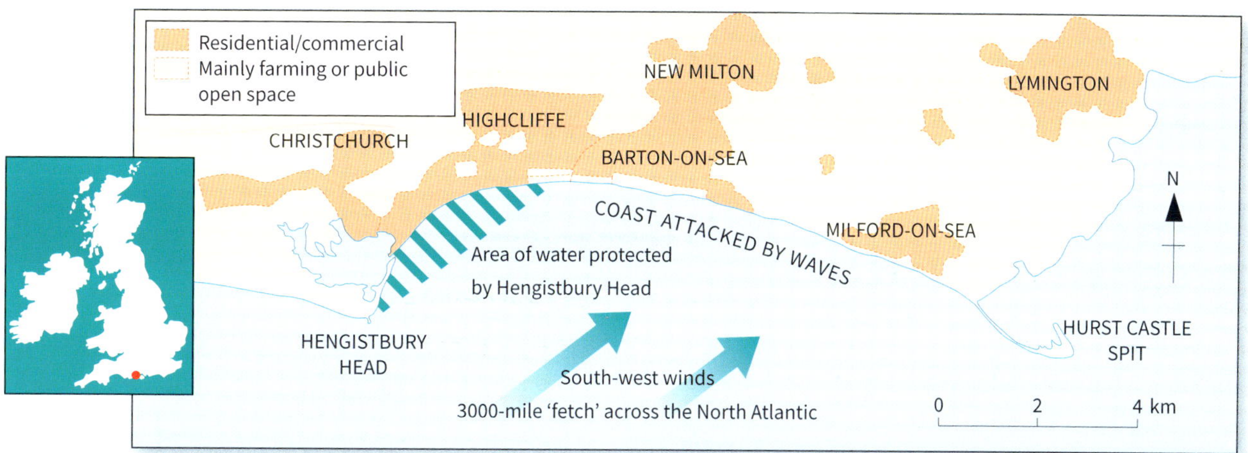

▲ ***Figure 2*** *The causes of erosion in Christchurch Bay*

▲ ***Figure 3*** *Erosion at Christchurch Bay*

Christchurch Bay

Christchurch Bay on the UK's south coast has a severe problem. Without management, the cliffs erode by over 2 metres a year, threatening residential areas along the coast, such as Barton-on-Sea (see Figure 2). At Barton-on-Sea, mass movement is the major problem, caused by weathering and water movement. But cliff foot erosion plays a part, as the Atlantic fetch brings big waves. This makes managing erosion there very difficult.

Impacts of erosion

Erosion in Christchurch Bay affects many people:

- Homeowners lose their homes to the sea. House values fall, and insurance is impossible to get.
- Rapid cliff collapses are dangerous for people on the cliff top, and on the beach.
- Roads and other infrastructure are destroyed.
- Some people think that erosion makes the coastal environment look less attractive. This could reduce the number of tourists who visit, and therefore affect local businesses.

People living in Christchurch Bay argue that they need sea defences to protect their coast. These are expensive, and there is no agreement about which defences work best. Some people prefer sea walls, which create a promenade for tourists, but which are expensive. Others might prefer less attractive, but cheaper, ways of protecting the coast.

Your questions

1. Explain why **a** abrasion and hydraulic action are found at the cliff foot, whereas **b** sub-aerial processes are found on the cliff face.
2. Draw a flow diagram to show the stages leading to the collapse of a cliff.
3. Why should cliff foot erosion be more important in areas such as Cornwall, whereas in areas of weaker geology, cliff face erosion is more important?
4. Explain the factors that make erosion rapid in Christchurch Bay.
5. Make a list of the impacts of erosion in Christchurch Bay. Classify these into economic, social and environmental.

Exam-style questions

6. Define the term 'sub-aerial processes'. (1 mark)
7. For a stretch of coast you have studied, explain **two** reasons why erosion is rapid. (4 marks)
8. Suggest why some people might benefit more from coastal protection than others. (4 marks)

4.12 Managing the coast

In this section, you'll understand the reasons for coastal management, and assess traditional ways of protecting a coast.

Managing the coast

Engineers who build sea defences to protect a coast from erosion or flooding have two choices:

- 'hard' engineering – using concrete and steel structures, such as sea walls, to stop waves
- 'soft' engineering – using smaller structures, often built from natural materials, to reduce wave energy.

The 'hard' method is the traditional method of coastal management. However, it has two major problems:

- It is very costly.
- It often makes a coast look unnatural, or ugly.

Sometimes, hard sea defences are the only way to stop erosion. There are many types, as Figure 1 shows. Often engineers use several types of defence together. The decision to use a method often depends on **cost-benefit analysis** – that is, its **costs** (environmental as well as economic) versus the **benefits** of what is saved.

Type of defence	Cost	Benefits and problems
Sea wall	Up to £5000 per metre	• Reflects waves back out to sea. • Can prevent easy access to the beach. • Suffers from wave scour, where plunging waves erode the beach and attack the wall's foundations.
Sea wall with steps and bullnose	Up to £5400 per metre	• Steps help to dissipate wave energy; the bullnose throws waves up and back out to sea.
Revetments	Up to £3000 per metre	• Breaks up incoming waves. • Restricts beach access and looks ugly. • Can be destroyed by big storms.
Gabions	Up to £500 per metre	• A cheap type of sea wall. • Permeable, so absorbs wave energy. • Not very strong.
Rock armour (rip-rap)	Up to £6000 per metre	• Cost depends on location. • More expensive if built in the sea. • Dissipates wave energy; looks 'natural'.
Groynes	£10 000 to £100 000 each, with five needed per km	• Prevents longshore drift, trapping sand and shingle. • Larger beach **dissipates** wave energy, reducing erosion. • May increase erosion downdrift.

▲ ***Figure 1*** *Different hard defences and their costs*

Dissipate means to reduce wave energy, which is absorbed as waves pass through, or over, sea defences.

▲ ***Figure 2*** *Coastal management at Hornsea, East Riding of Yorkshire*

Conflict in Christchurch Bay

People who live in Barton-on-Sea in Christchurch Bay know their coast is eroding rapidly (see section 4.11). Most favour hard engineering to protect their coast. However, not everyone agrees, as the table in Figure 4 shows.

'Terminal groyne syndrome'

Beaches are natural sea defences. Wide beaches absorb wave energy before it reaches the cliff. Coasts with rapid erosion often have narrow beaches or no beach at all.

The traditional way of solving this is to allow a beach to grow. Wooden or stone groynes are built at right angles to the coast (see Figures 2 and 3). They're expensive; each stone groyne costs £250 000. Groynes trap sediment from longshore drift, allowing a beach to build up. They're popular in seaside resorts like Hornsea, shown in Figure 2.

But trapping sand in one place stops it reaching another. One location, Naish Holiday Village, shown in Figure 3, suffers rapid erosion because groynes at Highcliffe further west have starved it of sand (see section 4.11). Further east, Barton-on-Sea's beach has almost disappeared for the same reason. This is called 'terminal groyne syndrome'. Without a beach, cliff erosion at Naish Holiday Park has increased, and the local council has had to spend millions of pounds to protect Barton.

▲ ***Figure 3*** *Erosion near Naish Holiday Village in Christchurch Bay, caused by the stone groynes at Highcliffe*

Stakeholders	Views on costs/benefits of coastal protection
Coastal residents and business owners in Christchurch, Barton and Milford	They prefer a 'hold the line' policy to protect their homes, businesses and tourist sites. Most favour hard engineering, but it's expensive. Ugly sea defences can be an eyesore and deter visitors.
Local politicians and the council	Need the support of all local people, so have to be careful not to favour one group more than another; they want coastal protection, but not at any price.
Local people living further inland	Unaffected by coastal erosion, but fear that local taxes will rise to pay for coastal protection; they prefer low cost options. They don't want low value land, e.g. farmland, protected.
Environmentalists	Fear that habitats and sensitive ecosystems will be affected by the construction of sea defences; they prefer a 'do nothing' approach and soft engineering.
Residents and businesses downdrift	They worry that defences built updrift will reduce beach size and protection; they want management to benefit everyone affected.
Fisherman and boat users	Their priority is access to the sea, so they want to protect harbours, marinas and some beaches.

▶ ***Figure 4*** *How opinions about coastal protection vary between different stakeholders in Christchurch Bay*

Your questions

1. In pairs, draw a table to show **a** different methods of coastal protection, **b** their benefits, **c** their costs.
2. Explain why cost-benefit analysis means that **a** resorts like Hornsea (see Figure 2) usually succeed, **b** farmers rarely succeed in getting money for coastal protection.
3. Explain why places like Hornsea need more than one method of coastal protection.
4. Explain the arguments **a** for and **b** against protecting Naish Holiday Village (see Figure 3).
5. Why might not everyone in Christchurch Bay agree with protecting Naish Holiday Village?

Exam-style question

6. Analyse Figures 2, 3 and 4. Assess the costs and benefits of hard engineering in managing coastal erosion. (8 marks)

4.13 Managing the modern way

In this section, you'll understand why coasts are now increasingly managed in a more holistic, sustainable way.

Managing the whole coast

Coastal management is expensive, and has to be worth the money. Now, **holistic** coastal management is adopted; it means looking at the coastline as a whole instead of just, for example, Barton-on-Sea. Some places are protected from flooding or erosion if they are worth it, whereas others aren't. This is because:

- the value of land and buildings might not justify the cost
- building defences might cause more erosion elsewhere (see section 4.12)
- climate change is likely to bring rising sea levels
- it might be better for the environment, e.g. creating new areas of marsh.

This involves making different choices about a whole stretch of coast, such as Christchurch Bay — and not just about one place, such as Barton-on-Sea.

Holistic management takes into account the:

- needs of different groups of people
- economic costs and benefits of different strategies today, and in the future
- environment, both on land and in the sea.

The name for this approach to managing coasts is **Integrated Coastal Zone Management (ICZM)**. For long stretches of coast, a plan is drawn up, called a **Shoreline Management Plan (SMP)**, which sets out how the coast will be managed. In theory, this prevents one place building groynes if they then cause more erosion downdrift.

The choices

In the UK, local councils pay for sea defences. They may get some money from the Government or the Environment Agency if there is a flood risk. There are four possible choices that councils can make about how to manage the coast:

1. **Hold the line** – use sea defences to stop erosion, and so the coast stays where it is today. This is expensive.
2. **Advance the line** – use sea defences to move the coast further into the sea. This is very expensive.
3. **Strategic realignment** (also known as **strategic retreat**) – gradually let the coast erode, and move people and businesses away from areas at risk. This may involve financial compensation for people when their homes are lost.
4. **Do nothing** – take no action at all, and let nature take its course.

These choices can cause conflict. Choices 3 and 4 may mean some people lose land, businesses or homes. It depends on costs and benefits (see section 4.12). The diagram below shows the choices that have been made along part of the North Norfolk coast.

▼ ***Figure 1*** *Choices about coastal management along the Norfolk coast*

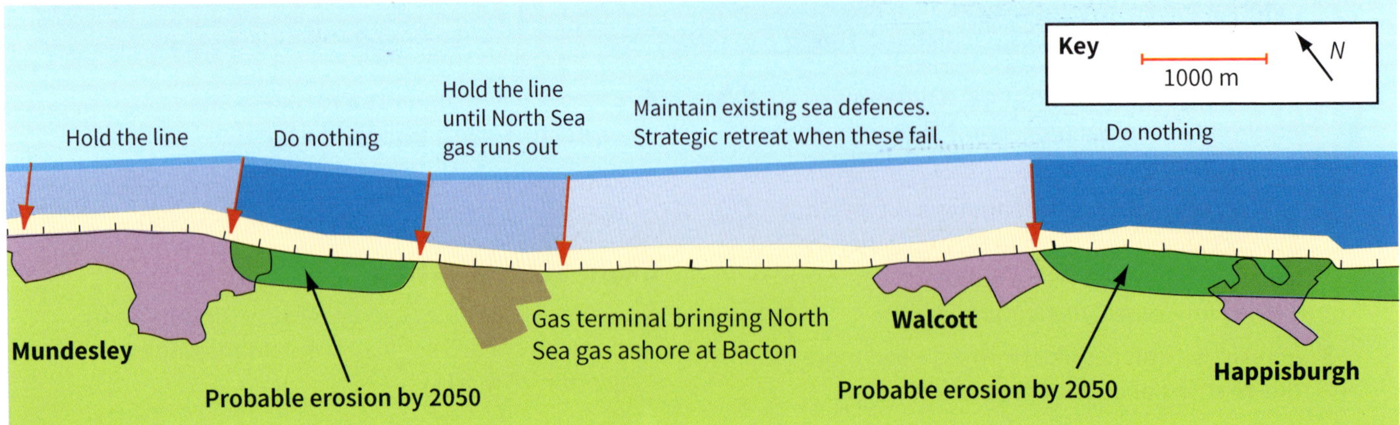

Soft engineering

On many coasts, **soft engineering** is replacing hard engineering. Soft engineering means using natural processes to protect coasts. Figure 2 shows some soft engineering solutions. These try to limit erosion by stabilising beaches and cliffs, and reducing wave energy. Soft engineering solutions are often much cheaper than hard, and are less intrusive.

- Planting vegetation: £20 – £50 per square metre.
- Beach nourishment: up to £2000 per square metre (see Figure 3).
- Offshore breakwaters: £3000 per metre.

Soft engineering may not always work. If rocks are very weak, hard engineering or 'do nothing' are the only choices.

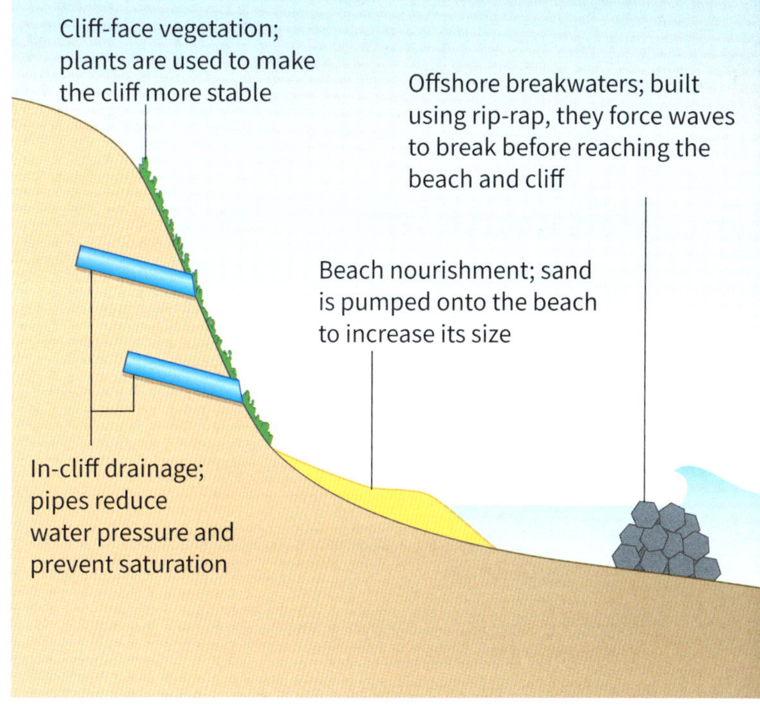

▲ ***Figure 2*** *Soft engineering solutions*

Here comes the sea

Holistic management and soft engineering solutions are seen as more sustainable. In some places in the UK, hard defences have been abandoned allowing nature to take its course. In the next few decades, the UK faces many difficult decisions about how best to protect the coast.

- At present, the Government thinks it is too expensive to protect farmland and isolated houses.
- Residents, councils and businesses often disagree.
- It is very hard to persuade people who have lived by the coast all their lives that protecting their property is not sustainable.
- Planning defences is difficult in an era of planning for climate change. Network Rail in the UK has recently spent £80 million on new protection for the railway linking Dawlish in Devon with London and the North.

▲ ***Figure 3*** *Beach nourishment in Poole, Dorset*

Your questions

1. Explain the differences between 'holistic management', 'hard engineering', and 'soft engineering'.
2. In pairs, suggest the advantages and disadvantages of **a** local councils, **b** national government paying for coastal protection. Which do you think should pay?
3. **a** Study Figure 1 and list the land uses and features that will be lost to the sea by 2050 in Norfolk.
   b Explain how people's attitudes towards coastal management are likely to vary between Happisburgh, Walcott and Mundesley.
4. Draw a table to show the costs and benefits of the soft engineering solutions shown in Figure 2. 

Exam-style questions

5. Define the term 'soft engineering'. (1 mark)
6. Explain **two** benefits of soft engineering. (4 marks)
7. Analyse Figures 1 and 2. Assess the costs and benefits of soft engineering in managing coastal erosion. (8 marks)

4.14 Geographical skills: investigating coasts

In this section, you'll use geographical skills to investigate coastal erosion and its impacts in Christchurch Bay.

Christchurch Bay

Figure 2, on the opposite page, shows part of Christchurch Bay in Dorset and Hampshire where erosion has been a problem for a century. Compare this map with Figure 1.

- Hengistbury Head (in square 1890) protects Christchurch and Highcliffe from strong prevailing winds and waves from the south-west. Erosion is limited here.
- Barton-on-Sea is not protected, so those same waves hit it full on! Erosion has been a problem there and the coast now is not where it was a century ago!
- Millions of pounds have been spent on hard engineering to protect Barton-on-Sea.

You can find the key used for all OS maps on page 327. To help you with skills such as grid references, see page 326.

Figure 1 shows the coastline at different dates from 1908 until 2012. By studying these, we can work out whether:

- erosion has changed as a result of coastal management
- money spent on managing the coast is justified.

In this way, we can judge whether the **stakeholders** in Christchurch Bay – i.e. those people who live and work there and who are affected by the coast – are getting what they want.

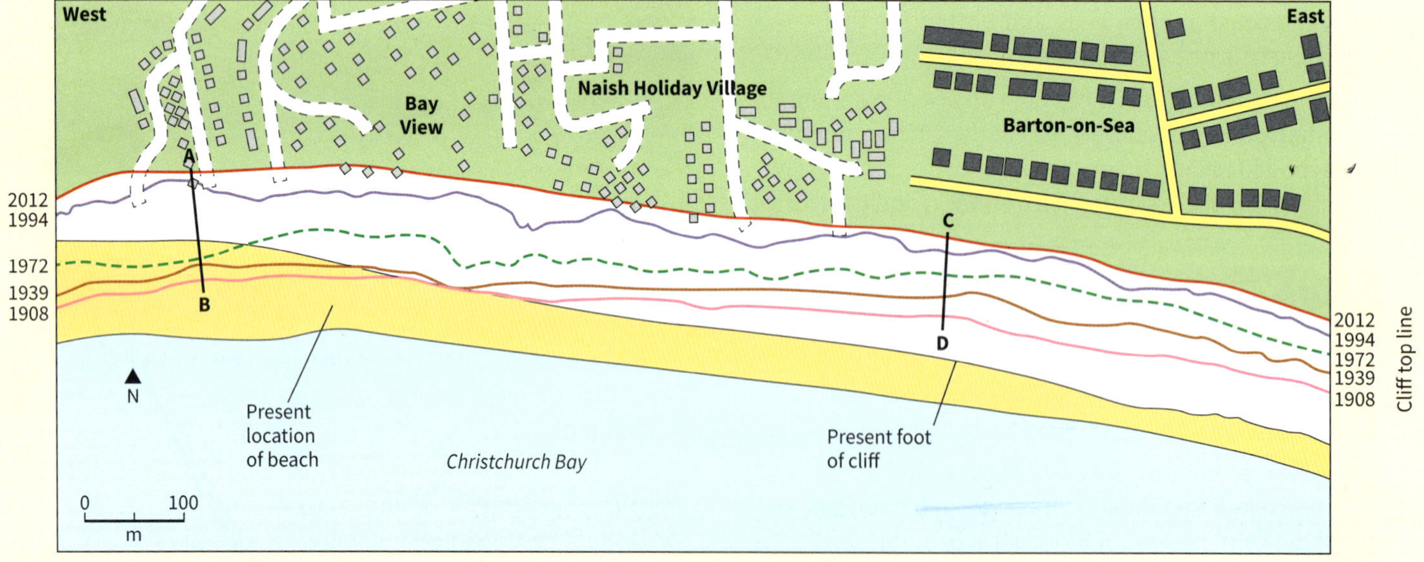

▲ ***Figure 1*** *Rates of erosion at two locations in Christchurch Bay, 1908–2012*

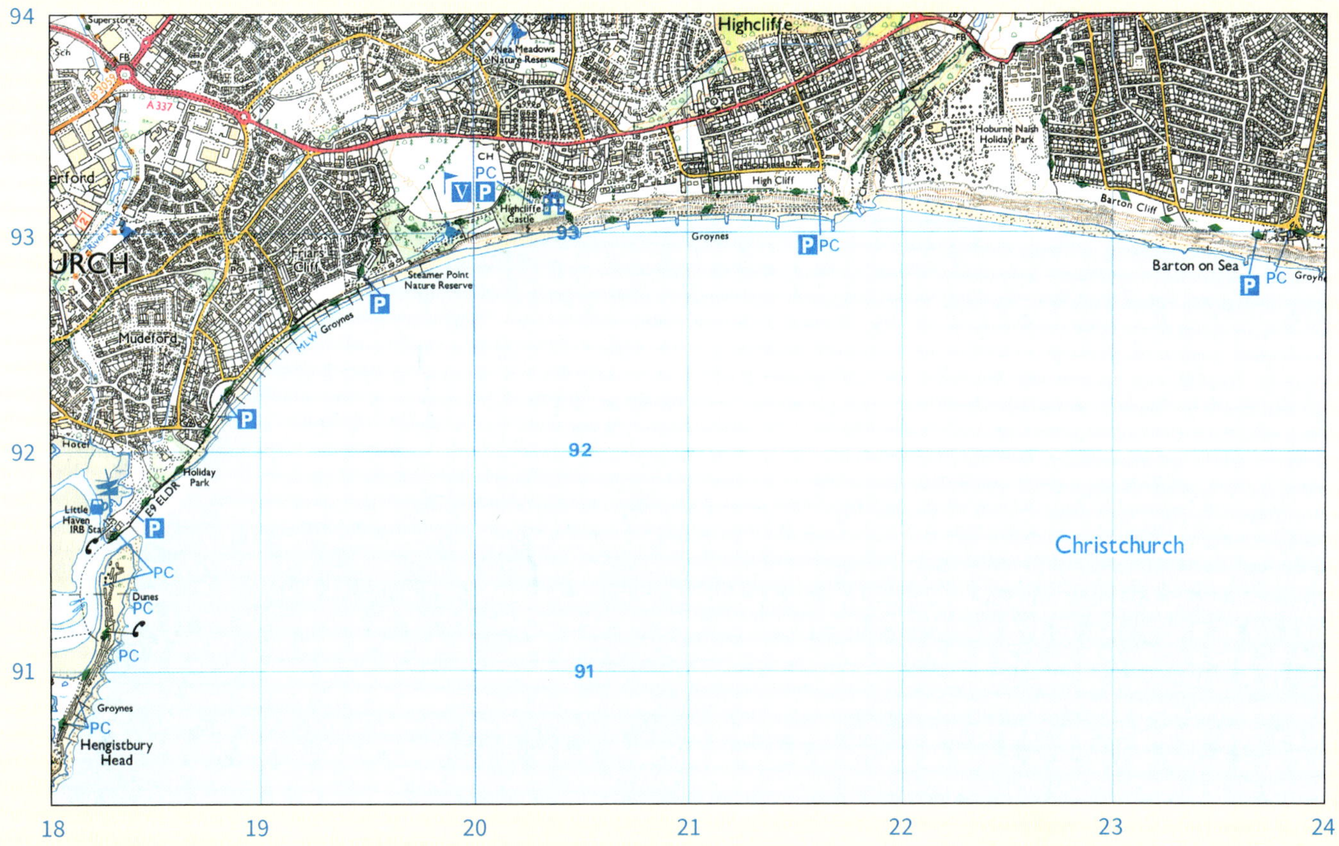

▲ ***Figure 2** 1:25 000 Ordnance Survey (OS) map extract of part of Christchurch Bay*

Your questions

1 Using Figure 2, **a** identify the coastal landform in grid square 1891. **b** Give four-figure references for **i)** a beach, **ii)** a cliff.

2 Identify four pieces of evidence to show that this area is used for tourism and leisure.

3 Using Figure 2, **a** identify and give a six-figure reference for one method of coastal management used. **b** Compare the use of this method at Barton-on-Sea and Highcliffe with Naish Holiday Village/Park.

4 Copy the table below. Calculate and complete the blank cells to show the amount of erosion at A–B and C–D on Figure 1.

From A–B (Naish Holiday Village East)

Period	Number of years	Amount of erosion in metres	Amount of erosion per year
1908–39	31	10	0.32 metres
1939–72	33	5	0.15 metres
1972–94	22		
1994–2012	18	6	0.33 metres
Total	**104**		

From C–D (Barton-on-Sea)

Period	Number of years	Amount of erosion in metres	Amount of erosion per year
1908–39	31	17	0.54 metres
1939–72	33		
1972–94	22		
1994–2012	18	11	0.61 metres
Total	**104**		

5 Use the completed tables.

a Identify in which period erosion was greatest at each location A–B and C–D.

b The decision to build hard and soft engineering at Barton-on-Sea (C–D) was taken in the 1970s. Using the data, explain whether these have been successful.

6 Using all the evidence in sections 4.10–4.13 inclusive, explain the potential **costs** and **benefits** of investing in hard coastal engineering to protect Naish Holiday Village (square 2293).

4.15 River processes in the upper course

In this section, you'll understand river processes in upland areas.

Buckden Beck

Figure 1 shows Buckden Beck, a small stream – or tributary – which flows into the River Wharfe in northern England. The Wharfe joins the River Ouse, and eventually flows into the Humber Estuary, where it reaches the sea.

Look at the small rapids and waterfalls in the photo. This is typical of a river's upper course. The river loses height rapidly, making the stream look very fast. In fact, the stream is flowing slowly because so much of the river's energy is lost through friction with the stream bed.

However, rainstorms increase the river's energy, and allow it to carry large angular boulders and stones which wear away – or erode – the channel. Material carried by the river is known as the **load** (see Figure 2). Except for dissolved material, the load becomes the tool with which running water can erode rock. Therefore, most erosion is done during periods of wet weather when most load is carried (see Figure 3).

Channel refers to the bed and banks of the river.

▶ ***Figure 1*** *Buckden Beck – the* ***gradient*** *(slope) of the river course is steep*

▼ ***Figure 2*** *How a river carries its load*

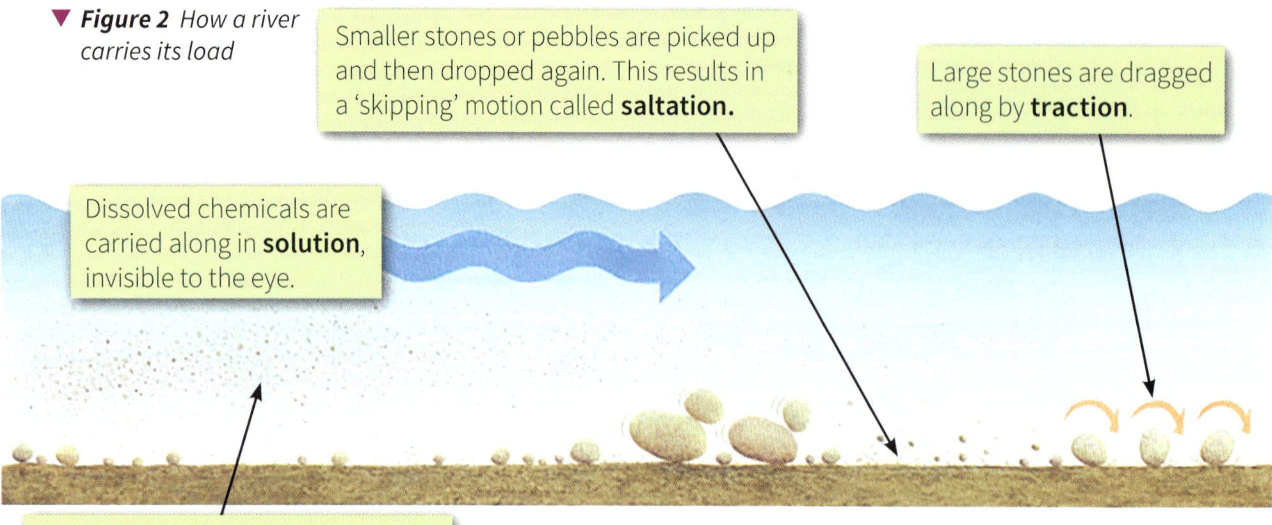

▲ ***Figure 3*** *River erosion*

In the upper course, most erosion is vertical (downward). Whenever the river meets a hard – or resistant – rock, a step is formed. This can eventually become a waterfall. The river gains a great deal of energy where it falls over the lip of the waterfall, allowing it to erode rapidly, as shown in Figure 4.

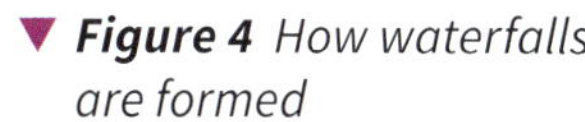

▼ ***Figure 4*** *How waterfalls are formed*

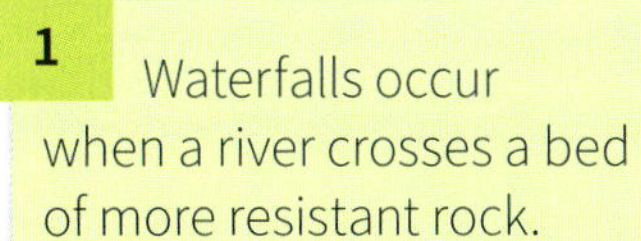

1 Waterfalls occur when a river crosses a bed of more resistant rock.

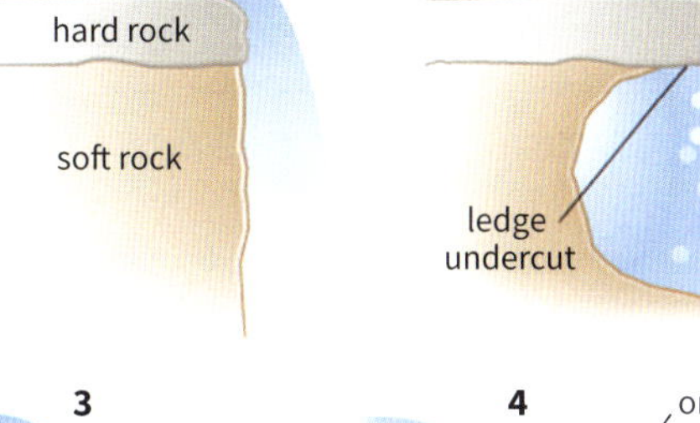

2 Erosion of the less resistant rock underneath continues – undercutting the hard rock above it. The river's energy creates a hollow at the foot of the waterfall, known as a **plunge pool**.

3 The less resistant rock beneath is eroded more rapidly by abrasion and hydraulic action. This creates a ledge, which overhangs and collapses.

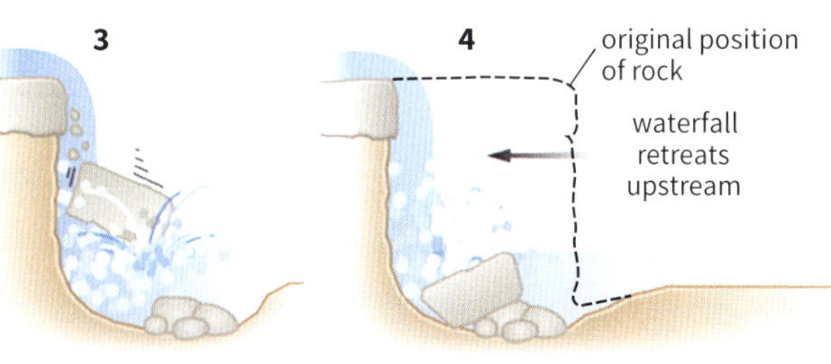

4 The waterfall takes up a new position, leaving a steep valley or **gorge**.

Your questions

1 For each set of words below, **a** say which is the odd one out, **b** explain your choice.
- stream, tributary, river, waterfall
- solution, attrition, plunge pool, abrasion
- suspension, gorge, traction, load

2 Explain how the following processes in a stream might change during wet weather:
- **a** The amount of water, and its energy.
- **b** The load carried by the stream.
- **c** The amount of erosion that a stream can do.

3 Type 'British Geological Survey Geology of Britain' into Google. Use the locator to find Buckden in North Yorkshire.
- **a** Identify Buckden Beck, and the solid geology there.
- **b** How resistant are these rocks? Check the 'Ten rocks you should know' on page 113.
- **c** Explain how the rock type has helped to make the river and valley features so steep.

Exam-style questions

4 Define the term 'load' of a stream. (1 mark)

5 Explain **two** reasons why the energy of a river can vary. (4 marks)

6 Explain the processes that lead to the formation of a waterfall. (4 marks)

4.16 River valleys in the upper course

In this section, you'll understand how and why river valleys develop in upland areas.

The valley of Buckden Beck

Look at the steep valley sides in Figure 1. This is typical of a river's upper course in upland areas. It has three main features:

- The rock type is carboniferous limestone (see section 4.1) which is resistant.
- The valley sides are steep and the bottom narrow – that's why valleys like this are called **V-shaped**.
- As the river cuts vertically into the limestone, it produces steep sides and winds around areas of more resistant rock. This produces **interlocking spurs** – or ridges of land – which jut into the river valley, looking as though they're interlocked. They are steeper on one side – where the river cuts into the side – and gentler on the other.

Although the stream is vital in eroding the valley, what happens on the valley sides is also important. Two things happen: **weathering** and **mass movement**.

▲ ***Figure 1*** *Buckden Beck is typical of a valley in its upper course – it forms a V-shape, with interlocking spurs*

Weathering on the valley sides

The pale grey limestone on the valley sides in Figure 2 has been attacked and broken down by weathering (see section 4.4). The cliffs of rock are known as **rock outcrops**, where rock comes to the surface. Below, the valley is covered in limestone scree (see section 4.4), which has broken away from the cliffs above because of weathering. There are different types of weathering – physical, chemical and biological, shown in Figure 2.

Biological weathering

Although rocks look solid, small cracks allow plant roots to penetrate in search of water and nutrients. As they grow, root cells force the cracks apart, widening them and breaking the rock into pieces.

Physical weathering

Physical weathering occurs when physical force breaks rock into pieces. In winter, cracks in the limestone rock fill with rain. This freezes, expanding in volume by 10% and widening cracks so that more water gets in. This process is known as **freeze-thaw**. If repeated often enough, pieces of rock break away, becoming **scree** at the base of the cliff.

Chemical weathering

Chemical weathering is any chemical change or decay of solid rock. Rainwater mixes with atmospheric gases, e.g. CO_2, to form weak acids which dissolve alkaline rocks such as limestone.

▶ ***Figure 2*** *Weathering processes on the valley sides of Buckden Beck*

Mass movement means all processes that cause rock material to move downslope under gravity.

Mass movement

Once rock is weathered, the fragments move downslope towards the stream. Some move quickly, others slowly. This is known as mass movement, but there are different processes:

- **Rapid** – such as landslides and mudflows. They tend to occur on railway cuttings and along cliff coastlines.
- **Slow** – the most common of which is **soil creep**. It's caused by rain dislodging tiny soil particles each time it rains. It's slow – perhaps 2 cm a year – but over decades has many effects, like those in Figure 3.

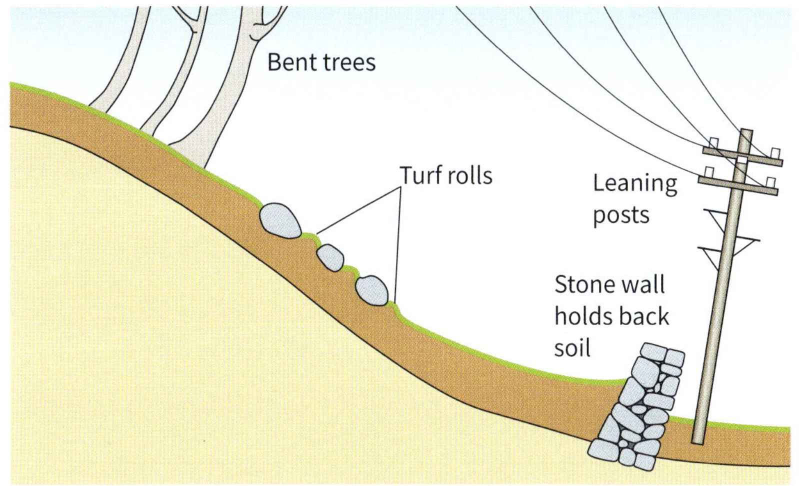

▲ ***Figure 3*** *Different ways in which soil creep can be recognised in the landscape. Although it is slow, soil creep can cause walls or telegraph poles to lean.*

The shape of the valley

Valley shape is affected by three things, as shown in Figure 4:

- The rate of weathering. If scree piles up, weathering is taking place rapidly.
- The rate of mass movement.
- How quickly the river can remove the material brought by mass movement.

If the river has plenty of energy, it takes the material away and uses it to help erode the valley, making it steeper. However, if it is slow and cannot cope with all the material, weathered rock collects at the bottom of the slope, making the valley gentler and flatter.

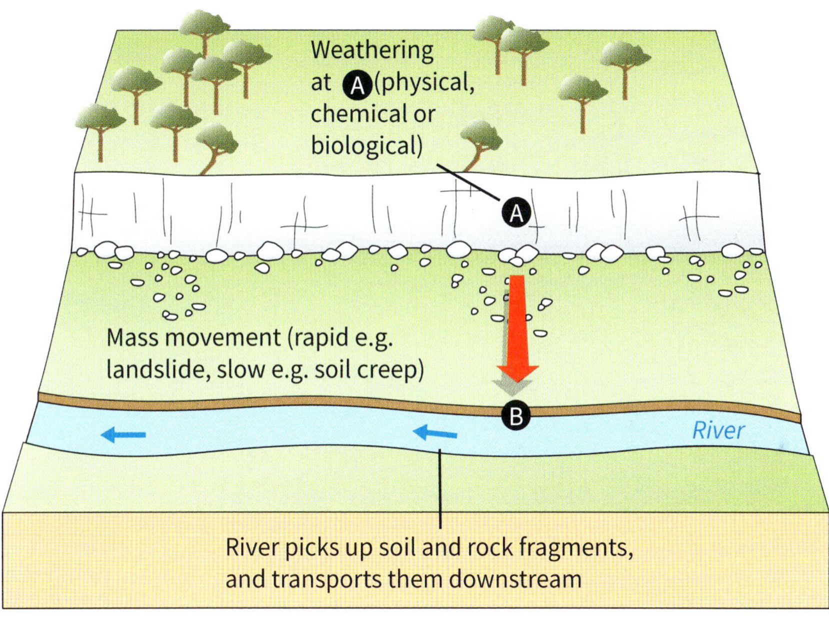

▲ ***Figure 4*** *Weathering takes place at A on the valley sides to break down solid rock into fragments. These move down the slope to the river at B, where they are removed. The river then uses these to wear away – or erode – the river bed.*

Your questions

1 Draw labelled diagrams to explain these features of a valley in the upper course:
  - a How scree is formed.
  - b Why valley slopes get covered in scree.
  - c Why hedges and stone walls can fall over.
  - d How trees can grow out of solid rock cliffs.
  - e Why sediment is large and sharp-edged.

Exam-style questions

2 Define the term 'mass movement'. (1 mark)

3 Explain the relationship between volume of water in a stream and the amount of erosion. (2 marks)

4 Explain how weathering and mass movement can affect the shape of river valleys. (4 marks)

4.17 Rivers and valleys in the middle course

In this section, you'll understand how both the river and its valley change in the middle course.

The River Wharfe in its middle course

▲ ***Figure 1*** *The valley of the Wharfe in its middle course. This is a meander and flood plain near Kettlewell in Wharfedale.*

Figure 1 shows the River Wharfe between Kettlewell and Starbotton, downstream from Buckden Beck. By now, the river is in its **middle course**. Several streams like Buckden Beck have joined the Wharfe, making it wider and deeper.

By the time the River Wharfe reaches its middle course, the gradient is gentler but the river's **discharge** has increased as more streams and tributaries add to it. The river channel is also smoother because smaller, rounder pebbles, muds and sands have replaced large stones and boulders creating less friction to slow down the river. As a result, **velocity** increases. This provides more energy for the river to erode laterally (i.e. side to side, not downwards), creating meanders, point bars and flood plains (shown in Figure 1). The valley shape changes from a V-shape to a U-shape, as the flood plain widens the valley floor.

During wet weather, the volume of water can be so great that the river sometimes floods over the **flood plain**. During this time, sands and clays brought down by the river from upstream settle over the flood plain and form layers of **alluvium**. Flood plains are at risk of flooding, but alluvium is fertile and attracts farmers.

Velocity is the speed of a river, measured in metres per second. It's obtained by timing how long it takes a floating object (a dog biscuit works well!) to cover a distance (e.g. five metres).

Discharge is the volume of water flowing in a river, measured in cubic metres per second. It's measured by multiplying area (width times average depth across the channel) by velocity.

How meanders change the valley

1

The river bends in its middle course; each sharp bend is called a meander. Meanders are natural; rivers almost never flow in a straight line. Water flows naturally in a corkscrew pattern. This is called **helicoidal flow**.

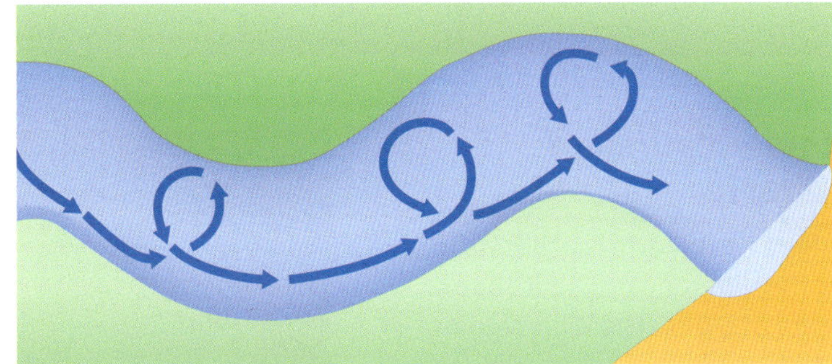

2

Helicoidal flow sends the river's energy laterally – i.e. to the sides. The fastest current (called the **thalweg**) is forced to the outer bend (A), where it undercuts the bank. This produces a steep edge, or **river cliff**, which eventually collapses. Gradually, the river undercuts more and more, so that the channel moves to a new position. Helicoidal flow shifts sediment across the channel to the inner bank (B). Here the sediment is deposited by the slower moving water, to form a point bar. As the river channel shifts, the point bar gets bigger and bigger, forming a flat area known as a **flood plain** (C).

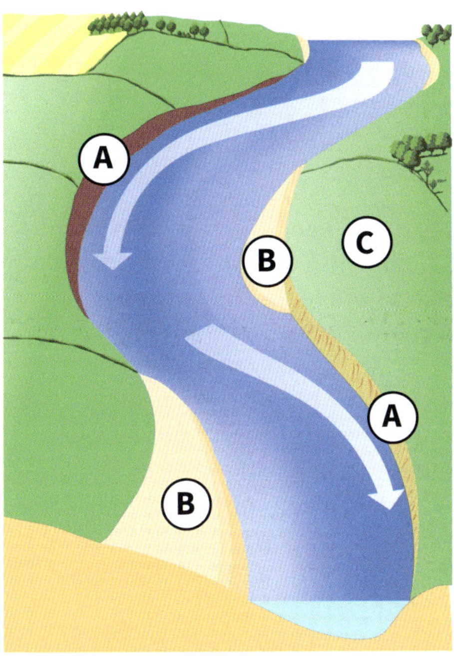

3

Continued erosion can create a narrow neck between two meanders (X). Eventually, the neck will be breached at (Y), cutting off the meander to create an **ox-bow lake** (Z).

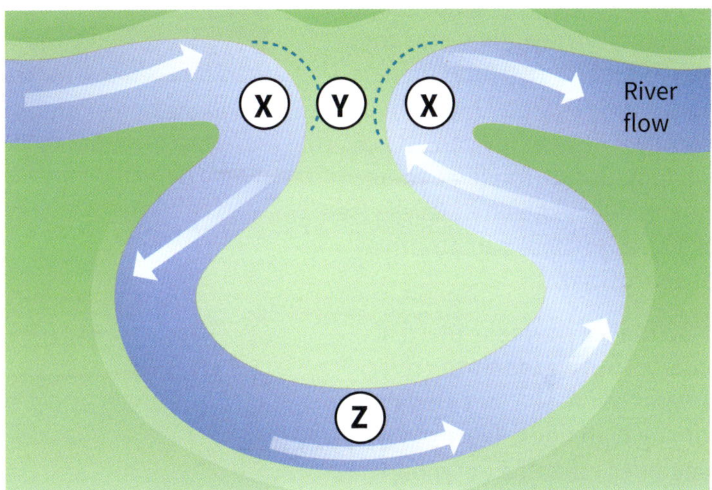

Your questions

1 Classify each of these features by whether they are part of **a** the river in its middle course, **b** its valley in its middle course: flood plain; meander; point bar; alluvium; helicoidal flow; meander neck; U-shape; thalweg; ox-bow lake; lateral erosion; river cliff.

2 Draw a sketch of the photo in Figure 1, adding the following labels: flood plain, alluvium, lateral erosion, meander, thalweg, undercutting, point bar.

Exam-style questions

3 Define the term 'river discharge'. (1 mark)

4 Name **one** feature of a river channel in the middle course. (1 mark)

5 Explain the processes that lead to the formation of an ox-bow lake. You may use diagram(s) to help with your answer. (4 marks)

4.18 Rivers and valleys in the lower course

In this section, you'll understand how both the river and its valley change in the lower course.

The lower course of the Wharfe

By the time the Wharfe reaches its lower course, the differences with its upper course are obvious. The Wharfe joins the Ouse River, which in turn joins the Humber to form an **estuary** – where the river meets the sea. The river is tidal, affected by incoming and outgoing sea tides – hence the muddy bed of the river, shown in Figure 1. The river – once narrow, with waterfalls – is now wide and deep, flowing over a gentle (almost flat!) gradient. Its meanders are large, discharge is much larger, and there is a wide, flat, low-lying flood plain which floods easily. The geology consists of fine alluvium. Beside the river, there are embankments known as levées (see Figures 1 and 2). These can be either natural or artificial.

Two directions of flow take place at the estuary – **outwards** by the river, taking water out to sea, and **inwards** by incoming high tides. Twice a day, incoming tides meet the outgoing river and the flow stops, forcing the river to deposit sediment. This forms a broad wide area of mud – hence the name **mudflats** (shown in Figure 1). A few rivers deposit so much sediment that they extend beyond the coastline, forming a **delta**. The only example in the UK is The Wash in Norfolk and Lincolnshire.

The tidal estuary is submerged by the sea twice a day, so **salt marshes** form where plants have to be able to stand both salt and fresh water. Salt marshes are valuable for wildlife; migrating birds use them to shelter during stormy weather, and the mud is rich in shellfish and worms. But salt marshes are under threat from ports and industry.

▲ ***Figure 1*** *Mudflats and a tidal river in the lower course of the River Humber. Notice the artificial levée (embankment) used to protect nearby housing and farmland from flooding.*

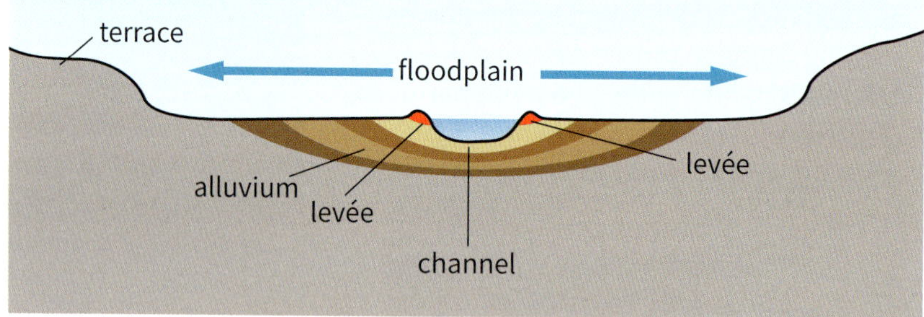

▲ ***Figure 2*** *Features of a flood plain in a river's lower course*

Natural levées form beside the river's bank where it first floods. As a river reaches **bankful** – i.e. before it spills on to the flood plain – it deposits sand and clay particles where the flow is slower. These build up beside the river as a bank.

Artificial levées are built by engineers to protect farms or towns from flooding. These are common in the UK along rivers like the Severn and the Ouse, where flooding is common, and along some of the world's giant rivers, e.g. the Mississippi in the USA.

Conclusion – spot the changes!

Four main changes occur during the upper, middle and lower courses of a river.

- Changes in the channel (items 1–4 in Figure 3)
- Changes in river sediment (items 5–7 in Figure 3)
- Changes in its gradient (item 8 in Figure 3) – what is known as its **long profile**. Figure 4a shows the long profile and how it changes. Put simply, it's steep in upland areas, and gentle in the lowlands.
- Changes in its **cross profile**, or valley shape (see Figure 4b). Put simply, the valley is V-shaped in uplands and almost flat by the lower course.

Upper course
Mid course
Lower course
1 River discharge
2 Channel width
3 Channel depth
4 Velocity
5 Sediment load volume
6 Sediment particle size
7 Channel bed roughness
8 Slope angle

▲ ***Figure 3*** *The Bradshaw model summarises the changes to river characteristics from source to mouth down the long profile. Five features increase downstream, and three decrease.*

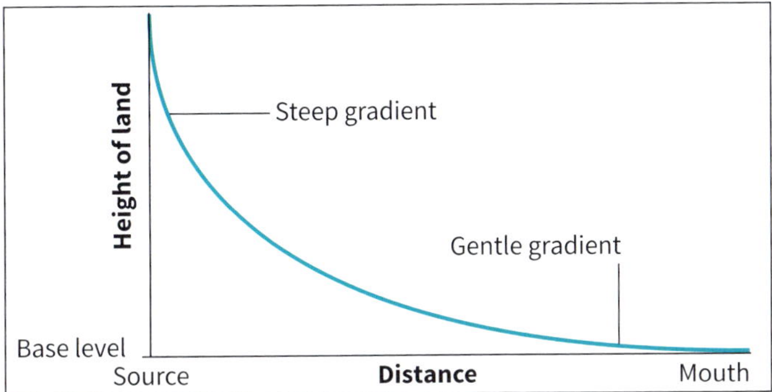

▲ ***Figure 4a*** *The long profile of a river – showing how river gradient changes between the upper and lower course*

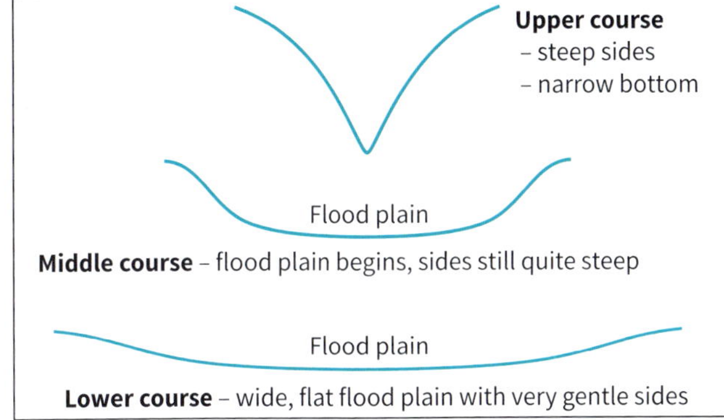

▶ ***Figure 4b*** *The cross profile of a river valley – showing how valley shape changes as the river moves from its upper to its lower course*

Your questions

1. Explain the differences between **a** long profile and cross profile, **b** an estuary and a delta.
2. Study Figure 3. Copy and complete the table opposite to show how each feature of a river changes between upper and lower courses, and why these occur.

Feature	Increase or decrease downstream?	Reasons for the change
1 River discharge		
2 Channel width		
3 Channel depth		
4 Velocity		
5 Sediment load volume		
6 Sediment particle size		
7 Channel bed roughness		
8 Slope angle (gradient)		

Exam-style questions

3. Define the term 'estuary'. (1 mark)
4. Describe the features of a river channel in its lower course. (3 marks)
5. Describe the features of a river valley in its lower course. (3 marks)
6. Explain how river channel characteristics change along a river's long profile. (4 marks)

4.19 Geographical skills: investigating rivers and their valleys

In this section, you'll use geographical skills to investigate river features and valleys.

Identifying rivers and valley landforms

You will find it helpful to use the following:

- page 327 which contains the key for 1:25 000 scale maps
- page 326 which will help you to develop your skills in interpreting maps.

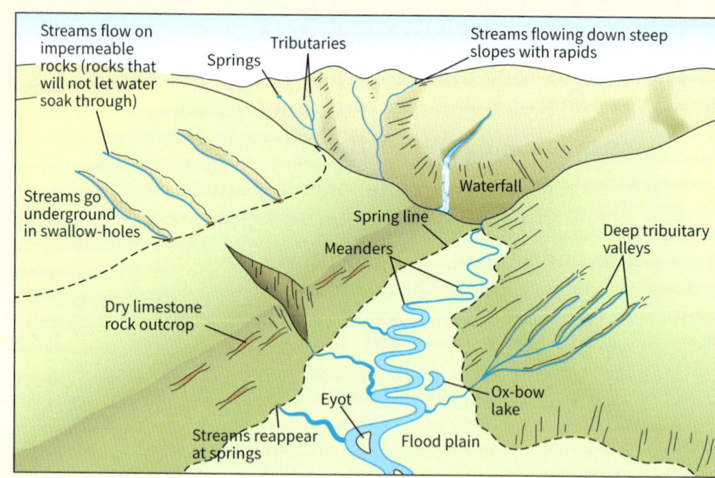

▶ ***Figure 1*** *Landforms of a typical upland region*

How to draw a map cross section

a Place a piece of paper wide enough to cover the length of the cross section on the map.

b Mark the boundaries on the west and on the east on the piece of paper.

c Each time your piece of paper crosses a contour line, mark it with a pencil AND write down the height in metres (see Stage 1).

d When you have done these, draw a graph axis as wide as the section, and as high as the height of land needs.

e Plot points on your graph as in Stage 2.

f Join the x points with a line – you should now have a clear picture of the landscape along the length of the cross section!

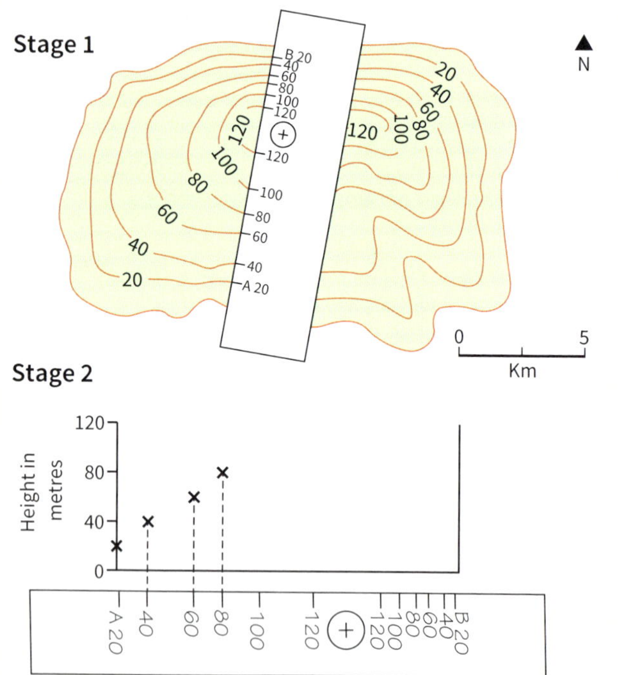

▶ ***Figure 2*** *How to draw a map cross section*

Your questions

1 Is the River Wharfe flowing from south to north or north to south across the map in Figure 3? Explain your answer. 

2 Use 6-figure references to locate the following features on Figure 3: a meander, an ox-bow lake, an eyot, a flood plain, a deep tributary valley, a spring.

3 a Draw a cross section along grid line 74 between eastings 94 and 97. Use Figure 2 to help you. 

b Annotate your drawing with key features. Use Figure 1 to help. 

4 Using Figure 3, compare the valley of Cam Gill Beck (square 9575) with that of the River Wharfe (square 9573). 

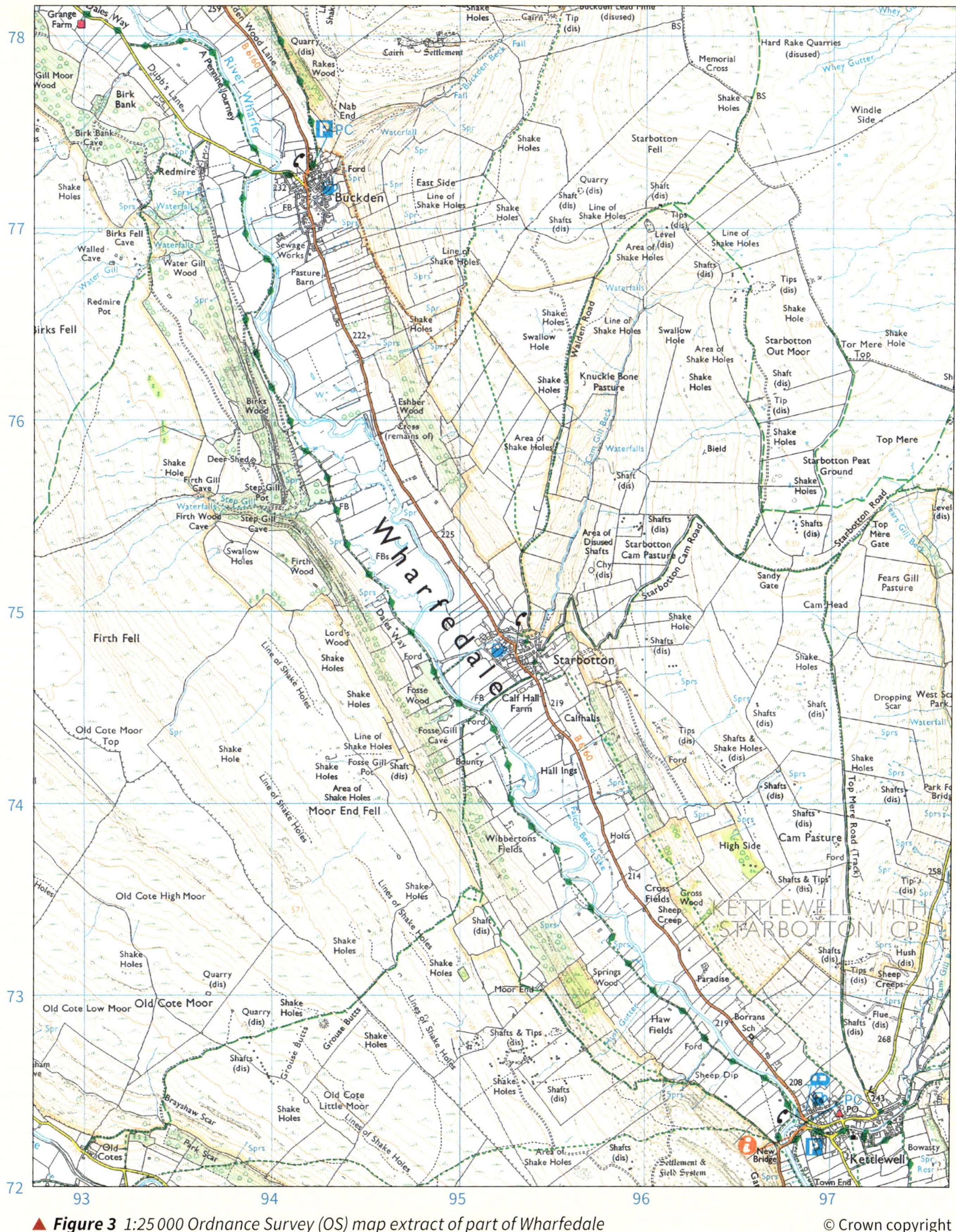

▲ ***Figure 3*** *1:25 000 Ordnance Survey (OS) map extract of part of Wharfedale*

4.20 Understanding storm hydrographs

In this section, you'll understand how storm hydrographs can be affected by physical and human factors.

Stage 1 – Precipitation and runoff

When it rains, very little falls directly into rivers; most falls elsewhere. As Figure 2 shows, leaves and branches of plants trap a lot of the rain that falls. This is called the **interception zone**. The amount intercepted depends on the vegetation, and also the season. For example, deciduous plants (those which lose their leaves in winter) intercept more rain in summer, when they're in leaf.

Some intercepted water is **evaporated** into the atmosphere. The rest drips from leaves to the soil and soaks in – this is called **infiltration**.

Eventually, the soil becomes saturated and cannot take any more. Any extra rain flows overground – called **surface runoff** (see Figure 1). How quickly this happens depends on:

1. How much rain has fallen recently – known as **antecedent rainfall**. If the weather has been wet, the soil may already be **saturated**, unable to absorb any more, making flooding more likely.
2. How permeable the geology and soil type are. Sandstone and chalk are permeable and absorb water easily, so surface runoff is rare. Clay soils are impermeable, so runoff occurs quickly.
3. How heavily rain falls. Heavy storms cause low infiltration and rapid runoff.
4. The shape of a river basin. Circular shaped basins increase flood risk whereas longer, thinner-shaped basins have low flood risk.

▲ ***Figure 1*** *Saturated soil here is leading to surface run-off*

Stage 2 – Throughflow and groundwater flow

Once water enters the soil:

- some is taken up by plants and **transpired** through leaves into the atmosphere
- some seeps into the river through soil air spaces – known as **throughflow**
- some continues into solid rock, and saturates it. The upper limit of saturated rock is known as the **water table**. From here, water seeps slowly towards the river as **groundwater flow**, which keeps a river flowing even when there is no rain.

▼ ***Figure 2*** *A cross section of a river valley to show how water moves down valley sides to a river*

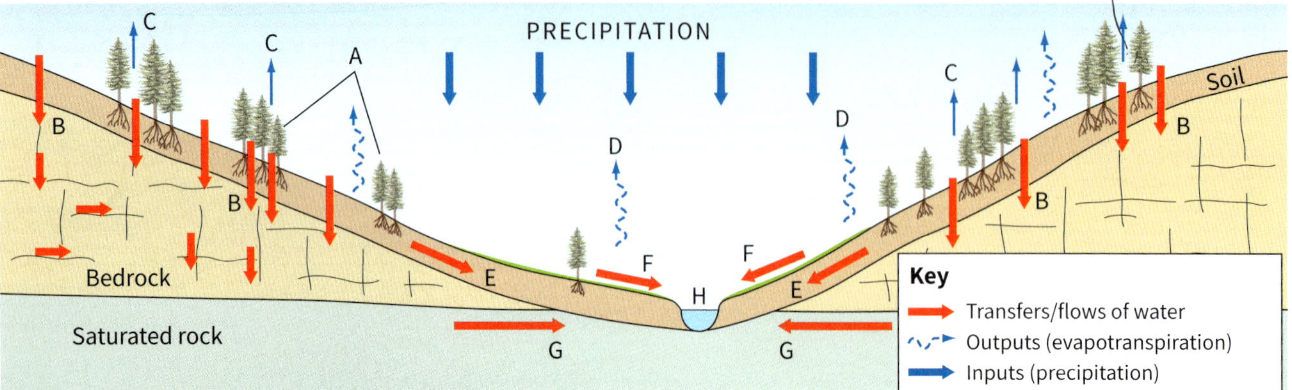

Understanding storm hydrographs

A **storm hydrograph** (Figure 3) shows how a river reacts to a rainfall event. Rain takes time to reach a river because most falls on the valley sides and reaches a river as throughflow or surface runoff. Discharge rises from baseflow to peak discharge (the rising limb) and then falls (the falling limb) as the water flows away.

A **storm hydrograph** is a graph which shows how a river changes as a result of rainfall. It shows two things: rainfall and discharge.

Hydrograph shape

- Hydrograph A in Figure 3 reacts to rainfall quickly. This happens during heavy rainfall, and in urban areas where impermeable surfaces like roofs allow surface runoff and little infiltration. Drains force water into rivers rapidly so river levels rise quickly.
- Hydrograph B in Figure 3 is how a river behaves in rural areas, woodland, on permeable soils (where water is absorbed) or in summer (where water is evaporated). Each increases the time for water to reach a river, so lag time is longer and peak discharge lower.

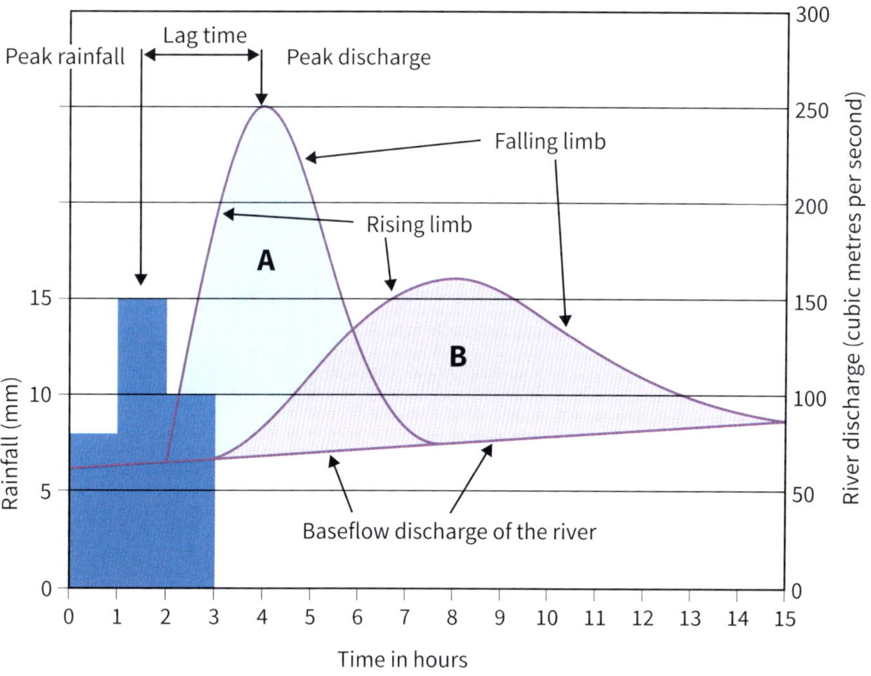

▲ ***Figure 3*** *Two hydrographs, 'A' (a river with a quick response to a rain storm) and 'B' (a slow response)*

Natural factors influence a hydrograph's shape. However, human activity also affects the shape of a hydrograph:

- **Land use change** can replace fields or woodland with buildings and roads, or pasture with arable (crops). Both changes increase runoff, so rain reaches a river quickly, reducing lag time. The same is true of **urbanisation**.
- **Deforestation** can also change the hydrograph shape from 'B' to more like 'A', by reducing interception and infiltration. Afforestation (planting trees) has the opposite effect and is often used in upland areas to reduce flooding.

Your questions

1. Read the text, then match the letters in Figure 2 with these terms: groundwater flow, evaporation, infiltration, interception zone, river channel, surface runoff, throughflow, transpiration.
2. Compare lag times and peak discharge levels of hydrographs A and B in Figure 3.
3. Explain how antecedent rainfall, vegetation and geology can affect storm hydrograph shape.
4. Sketch and explain two storm hydrographs (like Figure 3) for **a** double the rainfall in Figure 3 occurring in winter, **b** half the amount of rain falling on to a chalk area in summer.

Exam-style questions

5. Define the term 'infiltration'. (1 mark)
6. Explain **two** factors that can affect infiltration of moisture into the soil. (4 marks)
7. Explain **two** ways in which moisture can be lost into the atmosphere in the hydrological cycle. (4 marks)
8. Analyse Figures 2 and 3. Assess the impact of changes of land use upon the hydrological cycle. (8 marks)

4.21 Flood threats and the future?

In this section, you'll assess the increasing risks from river flooding and the threats to people and the environment.

Moorland, Somerset, January 2014

Some parts of the UK are much more vulnerable to flooding than others. One area at serious risk is the Somerset Levels (see Figure 1), a broad stretch of flat, low-lying land between Taunton and Glastonbury. Figure 2 shows Moorland, a small village in the Levels. In January 2014, serious flooding occurred, when the only way into the village was in an amphibious vehicle or tractor.

▲ ***Figure 1*** *The Somerset Levels, and the village of Moorland*

What are the Somerset Levels like?

The Somerset Levels are one of the UK's lowest-lying areas – no part is higher than 8 m above sea level. During the last Ice Age, they were covered by the sea. Now it's a landscape of wetlands, which were artificially drained in the 17th century, using a system of ditches. They were drained for farming; the soils are fertile (though too wet for crops), and wetlands make good summer grazing.

During drainage, the land shrunk, so much of it is below high tide water level. Seven rivers drain into the Levels (see Figure 1). River water pouring into the Levels during a spring tide in the Bristol Channel has nowhere to go except to flood the land. Floods are therefore common – but none so serious as those of winter 2014.

▲ ***Figure 2*** *Aerial photo showing the village of Moorland under floodwater in winter 2014*

What caused the floods?

Like most floods, flooding in the Somerset Levels had both physical and human causes.

Physical causes – rainfall

The floods resulted from a unique set of coincidences in the weather. **Jet streams** – high level winds at about 6–10 km altitude – blow across the Atlantic towards the UK. They drive low pressure weather systems with wind and rain. Normally, their track takes them to northern Scotland, so that's the stormiest part of the UK. In winter 2014, they moved 800 km south over southern England. As a result, winter 2014 brought:

- 12 major storms – the stormiest winter in the UK for over 20 years
- more severe gales than any winter since 1871 (see Figure 3)
- the most rainfall since 1766 – 235% of average winter amounts!
- the highest number of rainy days since 1961
- more days when high winds combined with very high tides and tidal surges.

▲ ***Figure 3*** *Coastal gales along the south-west England coast in February 2014*

Human causes

For years, drainage ditches and rivers have been **dredged** (or dug out) to make them artificially deeper. Dredging machinery scoops mud from river beds and deposits it on the banks to create levées to prevent flooding. The theory is that deeper channels hold more water and can stop flooding. Somerset farmers claimed that public spending cuts had resulted in less dredging, causing the flooding.

Although the argument is logical, it is not accurate.

- Deepening river channels (as shown in Figure 4) does increase capacity – but only for a while. During high rainfall, sediment is deposited on the channel bed, raising it to where it was before dredging! It's the opposite of beach nourishment (see section 4.13) – no sooner is sand added than a storm removes it! But like all hard engineering, dredging wins public support because 'something' is visibly being done.
- Building levées can make the problem worse by raising the river bed further, as the river naturally adjusts to its new channel banks!

▲ ***Figure 4*** *Dredging the River Parrett in 2014*

What of the future?

Climate scientists are now clear that climate change will bring the following for the UK:

- greater storminess with damaging winds
- higher, longer-lasting floods
- higher spring tides
- more storm surges.

Severe flooding in the UK is now twice as likely as it was in 2000, because of climate change. Given this, should more money therefore be spent on flood protection?

Your questions

1. Describe the extent of the flooding of the Somerset Levels in Figure 1.
2. **a** Using Figure 2, list the likely impacts of flooding on the residents of Moorland.
   b Classify these into economic, social and environmental.
3. Explain the costs and benefits of dredging river channels to reduce the flood risk.
4. Explain how and why climate change is likely to mean increased risk of flooding for the Somerset Levels in future.

Exam-style questions

5. Explain **two** reasons why flood risks in the UK are rising. (4 marks)
6. Analyse Figures 1 to 4. Assess the potential impacts of climate change on low-lying areas such as the Somerset Levels. (8 marks)

4.22 Managing the flood risk

In this section, you'll assess the costs and benefits of managing flood risk by hard and soft engineering.

The options ahead

Just as there is a debate about managing coasts, residents of Sheffield (where major flooding occurred in 2007) and Moorland (section 4.21) question what can be done to manage flood risks. Politicians are involved too, because flood protection is done by a government-funded agency known as the Environment Agency.

Engineers whose job is flood management have to choose between 'hard' and 'soft' solutions to managing rivers as well.

- 'Hard' solutions are structures built to defend areas from floodwater.
- 'Soft' solutions adapt to flood risks, and allow natural processes to deal with rainwater.

Each has advantages and disadvantages and each varies significantly in cost.

▲ ***Figure 1*** *Diverting the River Don in Rotherham to prevent flooding in the centres of Rotherham and Sheffield, where major floods occurred in 2007*

Hard engineering methods

In many places, hard engineering methods are used to try to combat the flood risk. The methods used are shown below.

▼ ***Figure 2*** *Summary of hard engineering methods used in flood protection*

Method	How it works	How effective is it?	Cost per km
Flood walls	Build a high wall alongside a river to increase its capacity to prevent flooding.	• These are fairly cheap, and are 'one-off' costs – once a wall is built, it's done. • Useful for city centres where space is limited. • However, they disperse water quickly and can increase the flood risk downstream.	• Depends on type of wall and material used.
Construct levées	Like flood walls, building levées increases the capacity of the river – and are normally built at a distance from the river which increases capacity even more.	• Expensive, but they allow people to live beside rivers or farm with reduced fear of flooding. • They can increase the flood risk downstream. • Levées can fail by overtopping (water rises over the levée), slumping, or by erosion.	• Up to £1m depending on materials used.
Dredging	Dredge the river to increase channel capacity, or line it with concrete to speed up river flow to get flood water away quickly.	• Dredging needs to be done every year as the channel fills with sediment each time it rains heavily. • Concrete lining is expensive to build but cheap to maintain. • Speeding up flow increases the flood risk downstream.	• £50 000
Flood relief channel	Create extra channels to divert excess water from city centre (e.g. in Exeter and Rotherham).	• It protects built up areas but could cause flooding elsewhere.	• A 1 km channel in Rotherham (see Figure 1) cost £14 m in 2008. • A 7 km scheme in Exeter in 2015 cost £30 m.

Soft engineering

Method	How it works	How effective is it?	Cost per km
Flood plain retention	The level of flood plains is lowered, and their surfaces restored to shrubs or grassland, so they retain water over a period and release it slowly into the river.	• Increased ability to store floodwater. • Flooding was reduced in 2007 in spite of heavy rains. • The only flooding in Darlington that year was caused by backlogged water in drains, not by the river.	• £1.2 million in total for a 2 km stretch.
River channel restoration	Some meanders were rebuilt, lengthening the river and slowing water down. Banks were lowered to make the river flood the park instead of Darlington. Concrete and other hard engineering materials were stripped away and replaced with sediment, and planted with trees (see Figure 4).	• Improved ecology with a 30% increase in birds and insects, within one year. • People like the more natural look – in a survey, 82% of people liked it 'mostly' or 'strongly'.	

▲ ***Figure 3*** *Summary of soft engineering methods used along the River Skerne in Darlington*

The Environment Agency (EA) believes that hard defences cost a great deal and can rarely cope with the largest floods. Instead they:

- plant upland areas with trees to reduce surface runoff
- restore river channels to their natural state
- no longer approve planning permission to build near rivers.

In Darlington, the EA restored 2 km of the River Skerne, using methods shown in Figure 3. The work created a riverside park (see Figure 4). The EA believes that these methods make flood management more sustainable, and make the River Skerne's response to rainfall more like hydrograph B on page 147 when it rains.

▲ ***Figure 4*** *Skerne Park along the river Skerne in Darlington, created by soft engineering flood management*

Your questions

1 Compare the impacts of flood management on the landscape in Rotherham (Figure 1) with Darlington (Figure 4). 

2 a Copy the table below to show the costs and benefits of all the flood protection methods mentioned in this section. Award up to 5 points for each benefit (where 5 is best), and up to 5 minus points for each cost.

Method of protection	Costs	Benefits

b Add up the totals for each method. Which is best?

3 a Make a copy of hydrograph B on page 147. Label it to show how managing the River Skerne should produce a hydrograph like this in future.

b Why is hydrograph B more desirable than hydrograph A? 

Exam-style questions

4 Explain the difference between 'hard engineering' and 'soft engineering'. (3 marks)

5 Explain why soft engineering is often preferred to hard engineering when managing flood risk. (4 marks)

6 Analyse Figures 1 to 4. Assess the merits of soft engineering in managing areas of flood risk. (8 marks)

5.1 Where we live 1

In this section, you'll understand the population density of the UK and its urban core regions.

Is the UK Europe's most overcrowded country?

The UK is becoming more crowded. Of the four UK nations, England is four times more densely populated than France. Only Malta – a tiny island – has a higher density per square kilometre.

▲ ***Figure 1*** *Adapted news article*

Our overcrowded islands?

Read news headlines like Figure 1, and you get the impression that the UK is crowded, and can't take any more people.

But Figure 2 shows a different view. It's a **population density** map of the UK, and shows how varied the UK is in its distribution of people. Some areas are very densely populated – for example, London has 5500 people for every km^2! But in 2019, England's population density overall was 432 people per km^2 – Wales was lower (151), as were Northern Ireland (133) and Scotland (65). Towns and cities – the urban areas – account for only 7% of the area of the UK. That means 93% of the UK is not urban, and not crowded!

But urban areas are really important – it's the UK's towns and cities that drive the economy. London alone produces 22% of the UK's annual GDP with just 13% of its population. Urban areas like this are known as the **core regions** of the UK, like those in India (section 2.12).

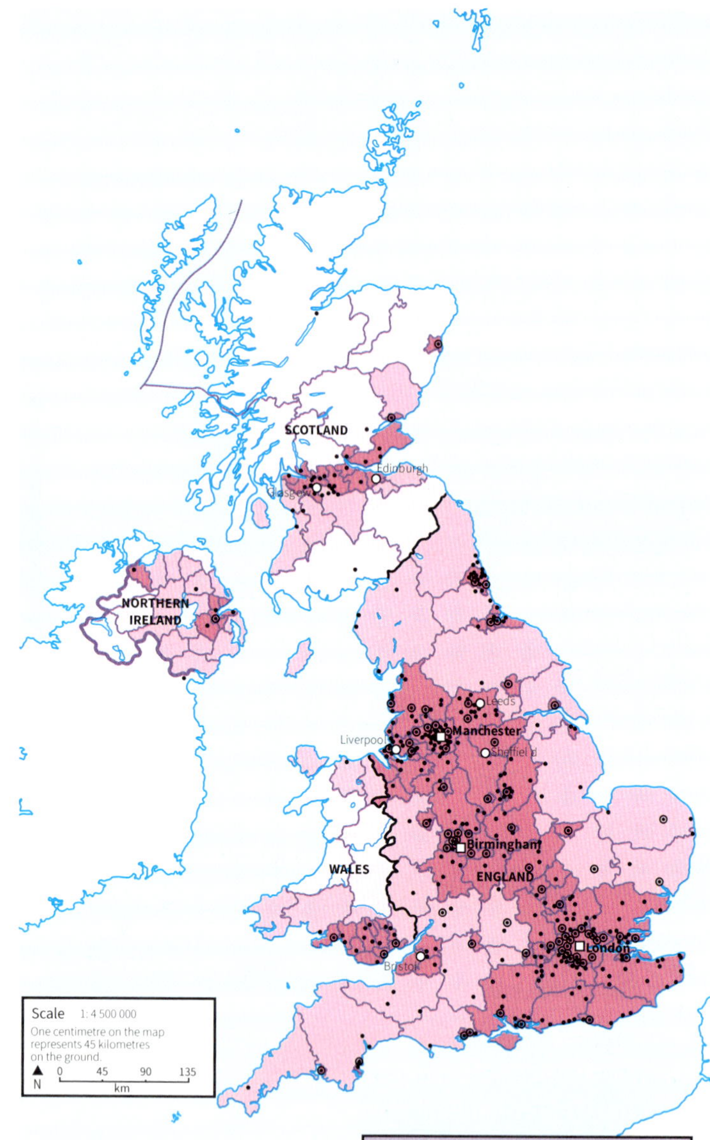

▲ ***Figure 2*** *Population density of the UK*

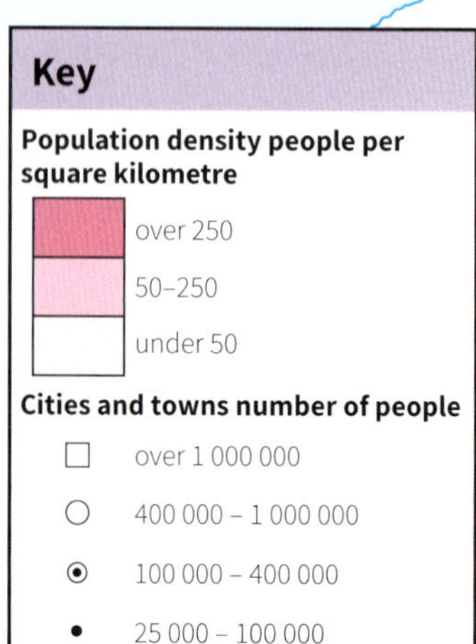

Understanding population density

Population density is the average number of people in a given area, expressed as people per km^2. It's obtained by dividing population by area.

- Dividing the UK's 66.7 million people (in 2020) by its area (243 610 km^2) gives an average density of 274 people per km^2.
- In world rankings, that makes it the 52nd most densely populated country – compare India (32nd with 411 per km^2), and Bangladesh (11th with 1087).

Urban core regions

The most densely populated areas of the UK are its major cities, such as London, Birmingham, Manchester, and Glasgow.

- People migrate to these cities and regions for work and to spend money earned there on housing, goods and services. This creates more jobs, called the **multiplier effect** (section 2.12).
- As the multiplier effect develops it spreads beyond the city, which becomes a centre of a core region. Cities merge with smaller towns into **conurbations** (see section 3.2). These then begin to influence a wider area, for example, the region from which people commute to work. People earn money in the city, then take it home and spend it locally, boosting a wider region. Figure 3 shows the UK's major cities, conurbations, and wider core regions.

Major city	Population of city 2017	Conurbation	Population of conurbation 2017	Wider core region
1. London	8.9 million	Greater London	11 million	London region
2. Manchester	554000	Greater Manchester	2.8 million	South Lancashire and Cheshire
3. Birmingham	1.15 million	Greater Birmingham	2.45 million	West Midlands
4. Leeds	503000	Leeds–Bradford	2.3 million	West Yorkshire
5. Newcastle	282000	Tyneside	1.7 million	Tyne and Wear
6. Sheffield	544000	Sheffield–Rotherham	1.4 million	South Yorkshire
7. Southampton	269000	Southampton–Portsmouth	1.55 million	South Hampshire
8. Nottingham	312000	Nottingham–Derby	1.54 million	East Midlands
9. Liverpool	579000	Merseyside	1.4 million	South Lancashire
10. Glasgow	612000	Glasgow and Clydeside	1.2 million	Central Lowlands of Scotland

▲ ***Figure 3*** *Core cities, conurbations and regions in the UK, ranked by size of conurbation*

Economic activity

Outside London, the conurbations in Figure 3 began as manufacturing cities using energy from Britain's coalfields (Figure 4). Many (e.g. Birmingham) began as mining, metalworking and engineering cities. Newcastle and Glasgow were shipbuilding centres. Others (e.g. Manchester) began as textile centres. Now, many manufacturing industries have moved overseas. Services such as finance and property development have replaced them. But the northern cities of Liverpool, Manchester, Leeds, Bradford and Sheffield form a major core region that rivals London's population for size. Politicians use the term '**northern powerhouse**' for this region because of its potential to drive the economy of northern England.

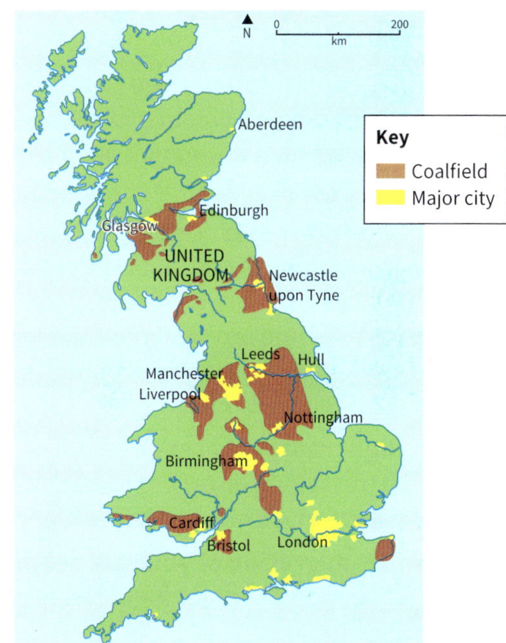

▶ ***Figure 4*** *The location of coalfields in Great Britain*

Your questions

1. Using Figure 2, describe the population density of the four nations of the UK.
2. Using an atlas and Figure 2, identify and mark on a map the cities and conurbations in Figure 3.
3. In pairs, research one conurbation listed in Figure 3. Produce a four-slide PowerPoint that shows **a** key cities on a map, **b** its original industries, **c** its main employment now, and **d** key features e.g. housing, buildings, transport.

Exam-style questions

4. Define the term 'population density'. (1 mark)
5. Compare the population density of England and Wales. (3 marks)
6. Explain **two** reasons why population density varies across the UK in Figure 2. (4 marks)
7. Explain how conurbations develop over time. (4 marks)

5.2 Where we live 2

In this section, you'll understand the UK's rural periphery.

Happy or not?

Ever heard of Allerdale? It's a local authority in the Lake District National Park on the Solway Coast, with just under 100 000 people. According to Hampton's estate agents in 2015, it had the UK's cheapest homes and happiest residents! Hampton's map is shown in Figure 1 – look for light-shaded areas to find Great Britain's happiest and most affordable places. The map uses a combination of where people feel happiest (according to the government's Life Satisfaction Index) and where houses are cheapest.

Allerdale is part of the UK's **rural periphery** – areas away from the urban core (section 5.1). The rural periphery tends to have:

- **Low population density.** Rural densities of people gradually become lower still the further you move from urban core regions. No cities, just smaller towns and villages, and isolated farms.
- **Older populations** (see Figure 2)**.** They are popular places for retirement, creating inward migration; balanced by outmigration of younger people to urban areas for jobs and lifestyles.
- **Lower incomes.** This is partly due to older populations living on pensions, but many jobs in farming and tourism in rural areas are low-wage. There is little industry, or high-income professional employment. Most tourist jobs are seasonal, rarely pay more than minimum wage, and are often part-time.
- **High transport costs.** Many rural households have to travel long distances to work without public transport, so must have a car.
- **Out-migration of younger people,** faced with fewer opportunities for university or employment.

But as Hampton's survey shows, these peripheral areas are happy and property is cheaper. In 2020, London households spent more than double the amount spent per week on housing compared to north-east England.

Reducing the gap

One problem faced by peripheral areas is that they don't receive the same level of investment which **affluent** (higher-income) urban core regions receive. It's not surprising – if you are setting up a business (e.g. a restaurant), then it makes sense to do so where

Did you know?

'Life satisfaction' means how people assess their life (their career, where they live, their work-life balance) as a whole, rather than their feelings on a particular day.

'Life in the country, a good view and not too many neighbours seem to be the formula for happiness,' according to Johnny Morris, Hamptons estate agents, 2015.

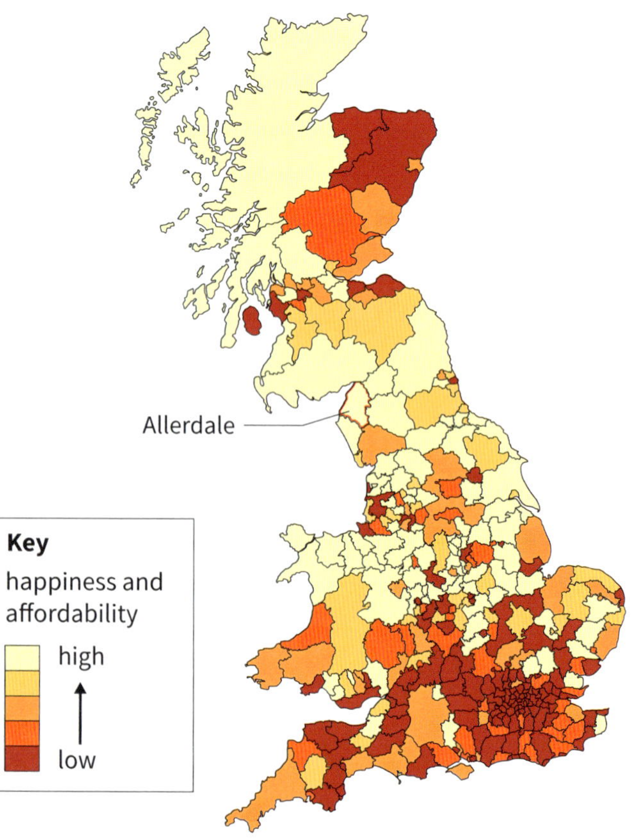

▲ ***Figure 1*** *Who's happy and where*

there are plenty of people with money to spend! To make rural peripheral areas more attractive to companies, the UK government uses four incentives for investors:

- **Enterprise Zones.** These are places where the UK government offers companies help with start-up costs, reduced taxes on profits, and access to superfast broadband. In 2021, there were 48 Enterprise Zones (see Figure 3) – but most are in urban locations, and all are in England!
- **Regional development grants.** These are similar, but available over more of the UK, with most in Scotland, Northern Ireland and Wales. They include grants to help businesses start up. Most are targeted at peripheral areas, but funds are small – investors first have to raise several times the amount of any government grant.
- **Former EU grants.** These were EU funds to invest in the poorest parts of the UK – Cornwall, and west and north Wales. The UK government has said it will replace these funds after leaving the EU in 2021.
- **Improvements to transport.** Transport improvements are vital to rural regions. Neither Cornwall, north Wales nor the Scottish Highlands have any motorways. Most transport investment is taking place in England's urban core, e.g. the HS2 railway linking London and Birmingham, and then Manchester and Leeds. Cuts in government budgets have reduced spending on transport elsewhere. However, Scotland's government has invested in a new Borders Railway between Edinburgh and Tweedbank, the A9 is to be dual-carriageway between Perth and Inverness, and a second road bridge over the River Forth has been completed.

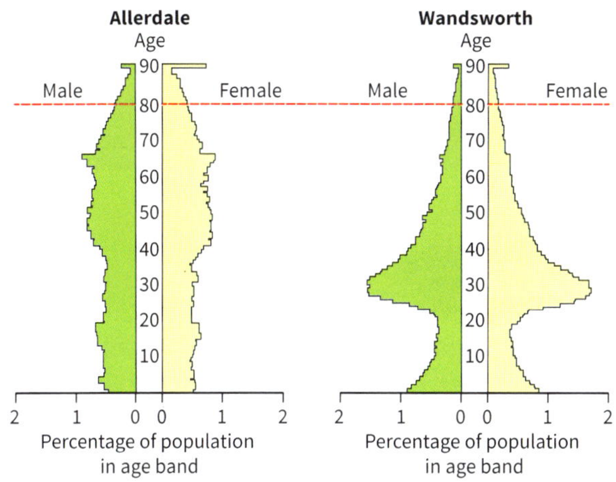

▲ ***Figure 2*** *Comparing Allerdale's age-sex pyramid with that of Wandsworth in inner south-west London*

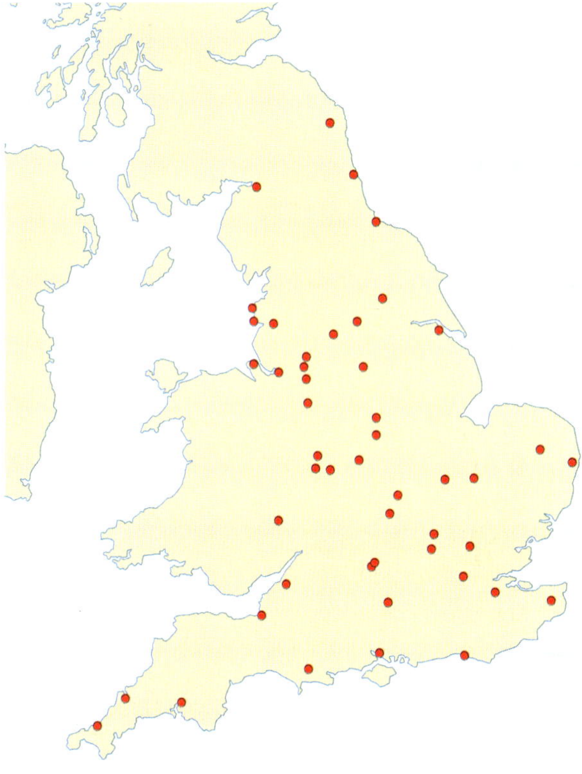

▶ ***Figure 3*** *Enterprise Zones in England in 2020*

Your questions

1 On a map of the UK, mark and name the UK's peripheral areas mentioned in this section.

2 Describe the distribution of the UK's **a** happiest, and **b** least happy regions in Figure 1.

3 Type 'population pyramid for …' into Google for an urban and rural area you know. How similar are their population structures to Figure 2?

4 Complete a table to show advantages and disadvantages of the four incentives for reducing the gap between core and peripheral regions.

Exam-style questions

5 Define the term 'rural periphery'. (1 mark)

6 Compare the two age-sex pyramids in Figure 2. (3 marks)

7 Explain **two** ways in which government policies have attempted to reduce differences between core and peripheral regions of the UK. (4 marks)

5.3 Who we are

In this section, you'll understand why the population geography of the UK is changing.

A fast-changing country

If you are called Olivia or Oliver, your name was the most popular for babies born in England and Wales in 2020; in Scotland, it was Olivia or Jack. Wherever they were, and whatever their names, all those babies were born into a rapidly changing UK. Census estimates show that the population reached 66.7 million that year, and makes the period since 2005 the fastest-ever growth period in UK history (see census data in Figure 1).

There are two causes of this growth: **net immigration**, and a **rising birth rate**.

Net immigration

In 2019, a record 677 000 long-term immigrants arrived to live in the UK. 407 000 emigrated that year too, so net migration (the difference between emigrants and immigrants) was 270 000. Most immigrants were young working adults aged 18–35. This was caused by:

- EU membership until 2020. Anyone was free to move and work in any member state. Even though the UK has left the EU, many European migrants are still attracted to work.
- **Globalisation,** which has revolutionised migration to the UK. London's 'knowledge economy' (see section 5.5) needs highly qualified and skilled people, and the UK cannot provide all it needs.

A rising birth rate

The UK's higher birth rate has been caused by three factors:

- more women in their twenties choosing to have children earlier
- more women born in the 1970s choosing to have children, having previously postponed having them for career reasons
- increasing numbers of overseas-born women who often have higher fertility rates than UK-born women (e.g. for religious reasons).

Date	Population in millions
1965	54.3
1970	55.6
1975	56.2
1980	56.3
1985	56.6
1990	57.2
1995	58.0
2000	58.9
2005	60.4
2010	62.8
2015	65.0
2020	66.7

▲ ***Figure 1*** *The UK's increasing population*

The impacts of immigration on the population

Census data show that the UK's population is now more **multicultural** than at any time in its history (Figure 2). In 2020, 16% of the working population was born outside the UK. Many migrants choose to live in London, so that London has the world's second largest urban immigrant population, after New York, with 35% of its population born overseas.

Like many HICs, the UK has an **ageing population**, which is expensive – as people get older, pension and health costs increase. The increased numbers of people over 50 in 2018 can be seen in Figure 3. The UK needs more people of working age who pay tax to help pay for the ageing population. The government believes that immigration is an economic benefit to the UK, and it helps to balance the ageing effect.

▼ ***Figure 2*** *Census estimates showing the country of birth of the resident UK population who had been born overseas, in 2020*

Rank	Country of birth	Population (thousands)
1	India	863
2	Poland	818
3	Pakistan	547
4	Romania	427
5	Republic of Ireland	360
6	Germany	289
7	Bangladesh	260
8	South Africa	252
9	Italy	233
10	China	217

Many also believe there are social benefits from immigration. For example, multicultural friendships and families, together with the cultural benefits on British sport, music, and food all make the UK a livelier and more creative country.

Changing population distribution

The population of each region of the UK undergoes continuous dynamic change. Three factors are important.

- Every region experiences **inward migration**. Some have greater inward migration than others – London's population increases with greater immigration, and the South West increases as more people retire there. Areas with slower economic growth (the North East, Scotland, Wales and Northern Ireland) have less inward migration.
- To this should be added any **net increase** in population (births minus deaths) – London has the UK's highest net increase.
- Every region also experiences **outward migration**. In London, outward migration (following Brexit and Covid-19) reduced its growth.

As a result, significant differences in population growth emerge between different UK regions (Figure 4).

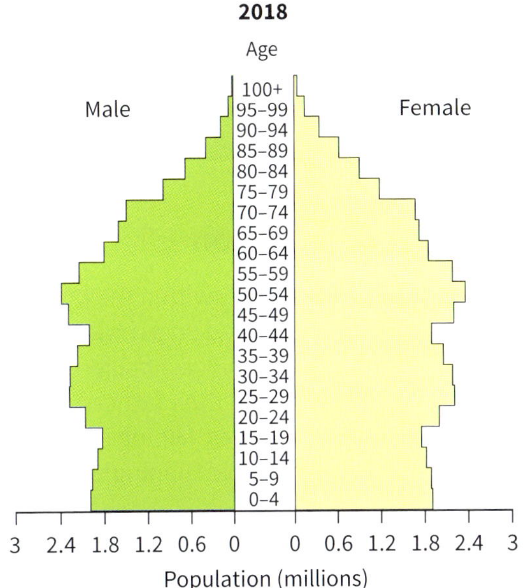

▲ ***Figure 3*** *Census data showing the structure of the UK's population in 2018*

Your questions

1 a Using Figure 1, draw a line graph to show UK population; extend the x-axis to 2050 and the y-axis to 80 million.

b Using a dotted line to project future UK populations, estimate **i)** when the UK population will reach 70 million, **ii)** the UK population in 2050.

2 Draw a spider diagram to show the benefits of immigration to the UK.

3 In pairs, research one country from Figure 2, and explain what past and present links it has to the UK that cause people to move here.

4 Using Figure 4, explain how the UK's ageing population could affect the future distribution of population.

Exam-style questions

5 Define the term 'net immigration'. (1 mark)

6 Using Figure 3, describe the UK population structure in 2018. (3 marks)

7 Explain **two** reasons for the rapid increase in the UK's population since 2000. (4 marks)

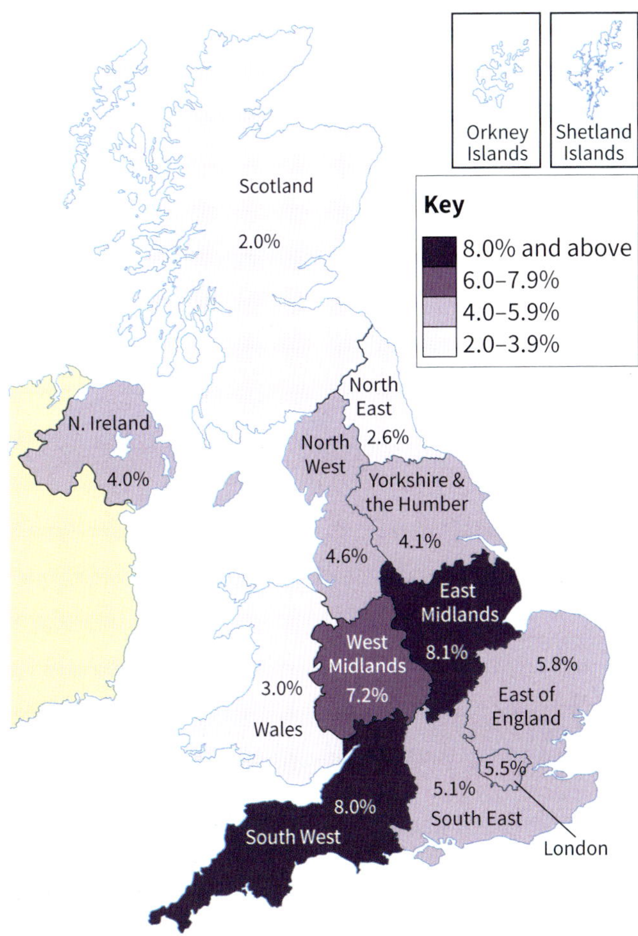

▲ ***Figure 4*** *Percentage increases in UK population using census estimates, 2018–30*

5.4 The decline of the 'old economy'

In this section, you'll understand the relative decline of the primary and secondary sectors of the UK economy.

It's all change in Dinnington

Dinnington is a town of 10 000 people, within the borough of Rotherham in South Yorkshire. Dinnington was based on **primary sector** employment, with one of the Yorkshire coalfield's biggest collieries until 1992, when it closed (Figure 1). Many boys followed their fathers, becoming apprentice engineers and electricians in the mines. Mining jobs were full-time and well paid, and the town thrived – as did Dinnington's Colliery Brass Band.

The colliery band still exists – it has a £1 million recording contract! – but there are no miners. The colliery site is now a large business park, which created 1700 new jobs between 1998 and 2008. Most companies there are tertiary (services), such as home delivery, or sales (e.g. a motorhome dealer). Like many places in the UK, Dinnington has changed from primary and secondary employment to tertiary.

▲ ***Figure 1*** *Dinnington colliery (1a) is now a business park (1b)*

Working in Dinnington now

Losing the mine is not the only change in Dinnington. Most people now have to commute, instead of working locally. In July 2021, of the jobs advertised locally in Dinnington:

- few were in Dinnington itself (most were in Sheffield, Rotherham, or Doncaster)
- some were permanent, but many were temporary contract posts for a few months
- most were full-time, some were part-time, and a few were zero hours
- many were low wage e.g. care workers. Even salaries for skilled workers (e.g. engineers or managers/supervisors) offered just £30 000, little above the average UK salary.

From the old economy to the new

Figure 2 shows changes to UK employment since 1980, which amount to an employment revolution.

- Jobs in the **primary sector** (farming and raw materials) have more than halved. Over 250 000 people were employed in coal mines in the 1970s – but only 1000 by 2020. Farming lost over 100 000 jobs between 1980 and 2015.
- **Secondary sector** employment (manufacturing) has also fallen sharply. These two sectors form the **old** – or traditional – **economy**.
- Meanwhile, the **tertiary/quaternary sector** (services) has increased rapidly, and has become the basis of the **new economy** (see section 5.5). In many rural areas, farming has been replaced by tourism as the main economic activity.

▼ ***Figure 2*** *Changes to primary, secondary and tertiary/quaternary employment in the UK between 1980 and 2020 (in millions of workers)*

Date	Primary	Secondary	Tertiary and Quaternary
1980	0.89	8.9	17.6
1985	0.78	7.3	18.3
1990	0.68	7.5	21.0
1995	0.54	6.3	21.0
2000	0.48	6.2	22.9
2005	0.46	5.6	25.2
2010	0.46	5.0	26.0
2015	0.48	5.1	28.1
2020	0.41	4.8	26.4
Trend	**Down by 54%**	**Down by 46%**	**Up by 50%**

The 'domino effect'

In the 1960s, over 100 000 men in north-east England and Yorkshire worked in coal mining. Coal was supplied to steelworks in Redcar or Sheffield. In the north-east, steel was used in the engineering industry and shipbuilding on the Tyne and Wear rivers (Figure 3a). In Yorkshire, Sheffield made cutlery and steel for machinery. These 'heavy' industries were linked – coal was needed to make steel, which was used to build ships or machinery.

However, each industry suffered problems:

- Coal was expensive to mine, because it was deep. Little coal is produced in the region now.
- Steel suffered from cheap overseas competition. In 2015 the steel works at Redcar was closed.
- Shipbuilding and engineering collapsed when Asian countries began to build cheaper larger ships. Shipbuilding on the Tyne finally ended in 2007.

What happened is a process called the 'domino effect' (also known as the negative multiplier effect) (Figure 3b) – as one industry collapses, it leads to the collapse of others. When this happens, it also damages other local businesses and services. This led to **deindustrialisation** (see section 3.3) across northern England.

▲ ***Figure 3a*** *Shipyards lining the River Tyne in the 1970s*

▲ ***Figure 3b*** *The 'dominoes' – coal, steel, engineering, shipbuilding*

So what's left in the North?

As in the rest of the UK, manufacturing in the north has declined. There is still some manufacturing – Nissan in Sunderland is Europe's largest car factory (it moved there in the 1980s), but most employment is tertiary. Sheffield produces as much steel as ever, but for specialist markets such as the arms trade and construction work.

Other employment in northern England includes:

- transport companies DB Arriva and Go-Ahead
- call centres (e.g. Tesco) – wage rates are 40% lower than London
- financial services, such as Virgin Money.

Part of the region's problem is a lack of high-salary jobs. Leeds is also an important centre for financial services, but is dwarfed by London.

Your questions

1 Complete the table comparing employment in Dinnington before the mine closed with now.

	Dinnington before 1992	Dinnington now
Level of qualification and skill needed?		
Full- or part-time work?		
Temporary or permanent?		
Level of income?		
Benefits men, women, or both?		

2 Make a second copy of the table, comparing farming in rural areas (the 'old') with tourism (the 'new').

3 a Explain how the four industries in Figure 3b were linked before their collapse.

b Explain what is meant by the 'domino effect', and why it happened.

4 Why might it be hard to attract new service industries to Dinnington and northern England?

Exam-style questions

5 Using Figure 2, describe changes to UK employment between 1980 and 2020. (3 marks)

6 Explain **two** reasons for the trends in primary and secondary employment in the UK since 1980. (4 marks)

5.5 The rise of the 'new economy'

In this section, you'll understand the tertiary and quaternary sectors of the UK economy.

Up early – but it's worth it!

6.30am, Canary Wharf, London's Docklands (Figure 1). The working day has started, with banking traders at their desks. By getting in early, they can trade with Singapore and Hong Kong (seven hours ahead of London). Many work long hours until late evening, so they can also trade with New York (five hours behind London).

100 000 people work in Canary Wharf, most commuting from London and south-east England. Their journey is worthwhile, because average salaries are £100 000 a year (four times higher than the UK average), often with bonuses!

Many companies in Canary Wharf work in banking and investment (e.g. HSBC, JP Morgan). They invest money for individuals, companies and pension funds (money saved towards pensions), hoping that its value will rise. They earn commission on every transaction; that's how they make money.

▲ ***Figure 1*** *Canary Wharf in London's Docklands*

The knowledge economy

London is one of a few major cities – called 'world cities' – that trade and invest globally (see section 3.2). That hasn't happened by chance. Faced with declining primary and secondary employment, the UK government in the 1980s encouraged tertiary and quaternary sectors to develop a **'new economy'** (see Figure 2). Both sectors offer services (tertiary), but many are highly specialised (quaternary). The quaternary sector is called the **knowledge economy** – an economy based on specialised knowledge and skill. It requires university degrees, plus specialised training (e.g. law or accountancy).

As well as banks, the knowledge economy in Canary Wharf includes companies working in:

- UK and overseas property development
- law – ensuring investments are legal
- insurance – to protect global shipping and property
- IT – managing financial systems
- creative industries (film, media, advertising).

	IT	Finance, insurance & property	Professional scientific & technical
1980	767 000	1 027 000	1 013 000
2015	1 344 000	1 679 000	2 958 000
% rise	**70%**	**64%**	**192%**

▲ ***Figure 2*** *The growth of tertiary and quaternary employment in the UK since 1980*

The new rural economy

The knowledge economy is no longer tied to cities! Computers now feature in many homes and during the Covid-19 pandemic proved vital to working away from the office, enabled by WiFi. Even in a small flat, laptops enable people to work easily without commuting to the office. Working at home has also become popular for many rural dwellers. Musicians can write and record whole albums at home. New technology means people can work **flexibly** in rural locations without having to travel long distances to cities.

Flexible working means:

- people can use IT to work anywhere, any time
- people work with colleagues elsewhere that they've never met!
- companies avoid paying for expensive city centre offices (annual rents can be £450 per square metre, or £5000 a year for an office the size of a small bedroom!).

Advantages	Disadvantages
• Better health; people take breaks during the day • No commuting – lower stress, less traffic congestion • Less absenteeism and sickness • Parents can work at home, saving money on child-care • It suits those who may struggle to travel by public transport • Allows those who wish to work variable hours to do so easily • Better productivity; people work longer hours instead of commuting	• Isolation from work colleagues and less contact with your boss – being overlooked for promotion? • It's sometimes difficult to motivate and organise home-workers • Work never disappears – when you're working at home, it's always around you

Understanding the 'new economy'

The 'new economy' is a service-sector economy. It's different from the 'old economy', based on traditional industries (section 5.4). Companies in the new economy are called '**footloose**', because they're not tied to location. But it has two sides, as the table shows.

At the high salary end, companies in the 'knowledge economy' need people with top qualifications and skills to work in London, other large cities or science parks near universities.

However, many 'new economy' jobs pay much less, offering low salaries on small industrial estates, or business or retail parks (as in Dinnington, section 5.4).

Low salary 'new economy', e.g. Dinnington and retail centres	High salary 'knowledge economy', e.g. London's Canary Wharf
Sector: Tertiary	**Sector:** Quaternary
Examples: Jobs with delivery firms, in retail parks, or shopping centres. Jobs advertised in Job Centres or online.	**Examples:** Jobs in global banking or law. Jobs advertised globally to get the right people.
Located: On the outskirts of towns for cheaper land and local labour.	**Located:** Where there are highly skilled, educated staff, good IT and broadband.
Qualifications: Mostly unskilled, needing few qualifications.	**Qualifications:** A degree and training, e.g. law, finance, IT.
Part-time or full-time?: A quarter of jobs are part-time; many are temporary, lasting weeks/months (e.g. Christmas).	**Part-time or full-time?:** Mostly full-time; contract jobs also available (with high daily rates of pay).
Earnings: Wages usually low (minimum wage or just above), with more for supervisor roles (e.g. in supermarkets).	**Earnings:** High salaries plus bonuses. Salaries depend on qualifications; many earn six-figure amounts.
Employee mix: Mix of male and female, with women the majority.	**Employee mix:** Mostly male, especially in banks or property companies.

? Your questions

1. Explain the differences between tertiary jobs, quaternary jobs, the 'knowledge economy' and the 'new economy'.
2. Give two specific examples of tertiary and quaternary jobs.
3. **a** Explain, with examples, why most companies in the 'knowledge economy' prefer to locate in city centres.
   b Explain why many companies and individuals changed their opinions about working in cities as a result of the Covid-19 pandemic.
4. Draw a table to show the advantages and disadvantages of the 'old economy' (section 5.4) and the 'new economy'.

Exam-style questions

5. Define the term 'new economy'. (1 mark)
6. State **two** characteristics of quaternary sector employment. (2 marks)
7. Explain **two** reasons why many people prefer to work from home. (4 marks)
8. Explain the positive and negative effects of the 'new economy' on people. (4 marks)

5.6 The impact of globalisation on the UK

In this section, you'll understand how globalisation has affected the UK economy.

It's holiday time!

Manchester Airport, August. A family from Leeds checks in with Virgin Atlantic (Figure 1) to fly to New York. After a week there, they're off for a trip of a lifetime to Las Vegas – with Virgin, of course. Mum and Dad both work for a media company owned and run by Virgin. Daughter Jess works for Virgin Active at weekends. They used to have a Yorkshire Bank account and a Northern Rock mortgage – but both were taken over by Virgin Money. They're the true Virgin family!

▲ ***Figure 1*** *A Virgin Atlantic aircraft taking off from Manchester Airport*

Virgin is one of the world's biggest brand names, and its founder, Richard Branson, one of the best-known industry leaders. From a small beginning in 1968, Virgin now consists of over 400 companies. Originally focused on travel, entertainment and lifestyle, it now also includes financial services, food and drink, and telecommunications. It operates in over 50 countries with over 50 000 employees.

Virgin shows how global the UK's economy has become. Many British companies operate overseas – for example, BP (in almost every country in the world), Body Shop (3000 shops in over 70 countries), British Airways (170 destinations in 70 countries), and Barclays (48 million customers in 50 countries). And that's just a few companies beginning with B!

TNCs, globalisation, and FDI

The UK's economy has become increasingly globalised since the 1980s (see section 2.7). TNCs like Virgin developed globally because of:

- free flows of goods and services between countries, known as **free trade**, without tariffs (section 2.5). So, although the UK left the EU, UK companies locate there because 500 million customers live there.
- their ability to locate in countries with low labour costs, and to **employ people** from almost anywhere in the world.
- investment rules, allowing free movement of capital between countries, known as **Foreign Direct Investment** (FDI) (see section 2.7).

▲ ***Figure 2*** *The St James Court Hotel in London, a luxury hotel owned by Tata. Tata also owns Tetley Tea, Jaguar and Land Rover, amongst its UK brands.*

With FDI, UK companies can invest abroad, but overseas TNCs also invest in the UK. In 2014, the UK received more FDI than any other country in Europe. Overseas TNCs play a big part in our lives. If your family has ever:

- drunk Tetley Tea
- driven on a frozen road treated with salt
- bought or rented a Jaguar or Land Rover
- stayed in luxury hotels in London (see Figure 2)

then you've been helping Tata, an Indian TNC, to earn money for its various companies. TNCs such as Tata are attracted to the UK as one of the world's biggest economies.

TNCs and privatisation

The role of TNCs in the UK has increased with **privatisation.** Privatisation means the change in ownership of services such as rubbish collection from the public sector (run by the government or local councils) to the private sector (owned by shareholders). It's politically sensitive – some people feel it leads to efficiency, others feel that profit should not be made out of basic needs such as water and health.

In the UK, privatisation has taken place in the following areas:

- **Infrastructure**. Many UK energy industries are overseas-owned (e.g. EDF), as are some water and rail companies.
- **Local council services**. You pay Council Tax for rubbish collection, but your rubbish will be collected by a private company, e.g. Aeolia or Serco.
- **The NHS**. Although government funds the NHS, many services are contracted to private companies to do the work, for example as shown in Figure 3.

▲ ***Figure 3*** *Health and care services formerly owned by Virgin Care, and sold in 2021 to another private sector company*

Your questions

1 Explain the difference between globalisation, free trade and FDI.

2 Using examples on these pages, list **a** advantages of British TNCs expanding overseas, **b** disadvantages of overseas TNCs expanding into the UK.

3 Using Figure 2 to guide you, present a six-slide PowerPoint on Indian TNC Tata – three slides on its Indian companies, and three on its UK activities.

4 **a** In pairs, discuss advantages and disadvantages of privatising UK industries and services.

b Why is the idea of private companies (as shown in Figure 3) running services in the NHS politically sensitive?

Exam-style questions

5 Define the term 'TNC'. (1 mark)

6 Explain **two** impacts of globalisation on the UK economy. (4 marks)

7 Analyse the text 'It's holiday time', 'TNCs, globalisation, and FDI', and Figure 2. Assess the impacts of globalisation on the UK economy. (8 marks)

5.7 Understanding London's location

In this section, you'll understand the importance of London's site and situation.

Past meets present

Central London. Construction of new London offices is halted. The reason? Archaeologists need time to assess Roman remains and ancient streets discovered in the foundations. It is further evidence of Roman settlement. Building continues after everything is mapped and photographed.

The Romans bridged the Thames (near London Bridge in Figure 1) after their arrival in Britain in 43 AD. The **site** they chose (i.e. land on which they built) was the last place where the Thames was shallow enough to cross before reaching its estuary. Market traders had always met there, but the new bridge drew more. With the market came houses; within decades a significant town had grown, called Londinium. It was a Roman multiplier effect!

- By 200 A.D. Londinium had become the capital, replacing Colchester.
- By 1300, further growth brought traders by sea as well as land, and people in search of work.

It wasn't an easy city to settle! Figure 1 shows how much land near the river was marsh. But it was ideal for a port – notice West India Docks (where Canary Wharf is now) had been developed by 1801.

▼ ***Figure 1*** *An extract from the first Ordnance Survey map of east London in 1801*

London's situation

An important factor in London's growth is its **situation,** i.e. its location within the UK.

Being close to Europe, London could trade there by sea quickly. Even when large industrial cities of the Midlands and North were growing during the Industrial Revolution, London had a far bigger population, economy and port.

Internationally, London's time zone helps its economic growth today. Those working in London's finance industries (section 5.5) can trade with Asia (5–7 hours ahead), Australia (8–11 hours ahead), and later in the same day with New York (5 hours behind).

The wider world

London is one of two major 'world cities' – the other is New York. However, the world's fastest economic growth recently has been in Asia, so how has London kept its position? The reason is its **connectivity** (i.e. how easy it is to travel or connect with other places).

- **Internationally,** London has the world's second biggest international airport at Heathrow (section 3.2). But add together international passengers at *all* its airports and it is by far the world's largest international air 'hub'. Eurostar also brings European major cities within a few hours travel of London.

- **Nationally,** the UK's fastest rail services link London and major UK cities. Manchester and Birmingham each have three fast services an hour. Future travel times will be quicker still, with HS2. However, these timings distort the UK as Figure 2 shows. While urban core regions are brought closer to London, peripheral areas seem a lot further away.
- **Regionally,** most major A roads and motorways lead to London, linking it with other major cities. It's a **radial** network – roads converge in London like spokes of a wheel. Note in Figure 3 how the A1 and M1 lead north, then other links are numbered clockwise.

Time-distance transformed maps showing fastest journey time by rail from London

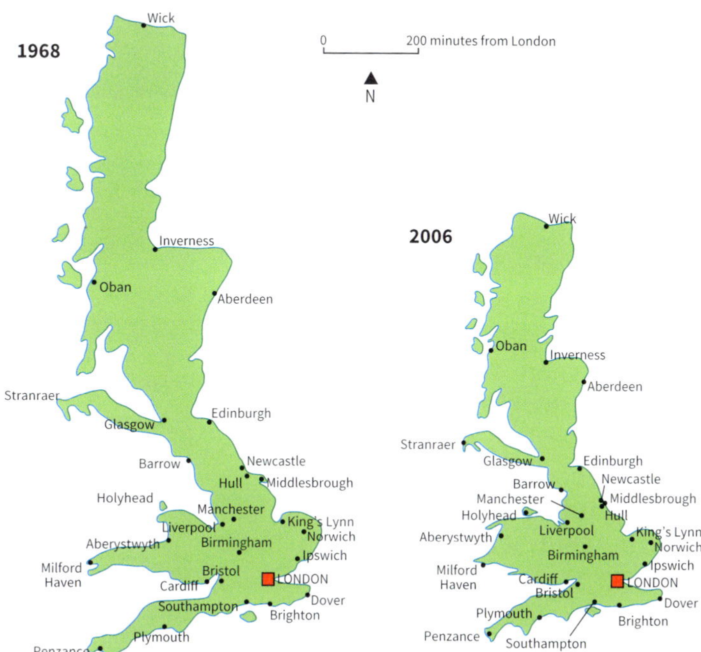

▲ ***Figure 2*** *Maps of the UK distorted to show travel time to London rather than physical distance*

London's cultural diversity

London has wide cultural connections, too. It's always been a diverse city. The late seventeenth century brought Protestants seeking religious freedom and Jews escaping persecution, and many refugee groups have since arrived. Now, London's knowledge economy makes it a global magnet for migrants. London's schools teach students from over 200 countries. It's easy to see why Nelson Mandela felt as he did about London (Figure 4).

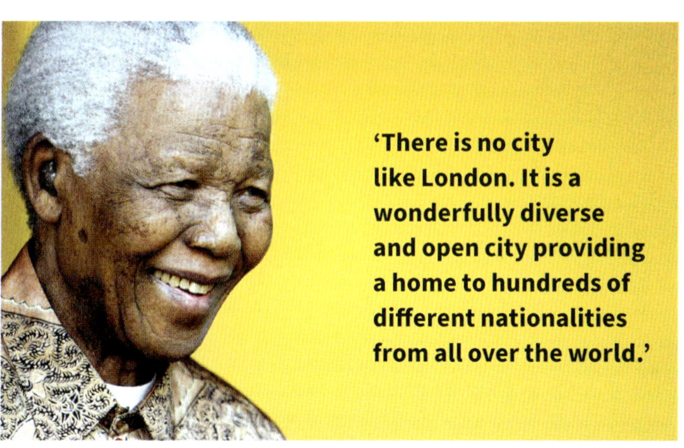

▲ ***Figure 4*** *How Nelson Mandela felt about London when it was bidding to host the 2012 Games*

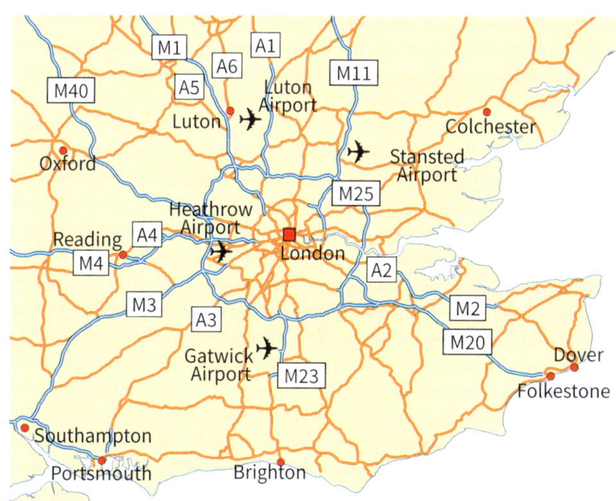

▲ ***Figure 3*** *London's radial road and motorway network*

Your questions

1 Using Figure 1, describe the physical landscape of central and east London before it was settled.

2 **a** Explain how the maps in Figure 2 have been drawn.

b Explain **i)** the benefits of fast rail to the urban core regions of the UK, **ii)** the problems this poses for rural peripheral parts of the UK.

3 Using Figure 3, suggest why A roads go into the centre of London, but motorways do not.

Exam-style question

4 Explain **two** reasons why some UK cities are better connected than others. (4 marks)

5.8 Understanding London's structure

In this section, you'll understand London's structure and functions, and how it varies between one part and another.

Chaos or order?

Seen from the ground, cities seem chaotic – a mass of traffic, buildings and people. You can spend so much effort crossing roads safely, or getting from A to B, that it's easy to ignore what's around you. Seen from the air, things look very different, as Figure 1 shows. There's a **structure** – or arrangement of buildings. The high-rise buildings are clustered together, in an area called the **Central Business District (CBD)** – in which land use is dominated by offices. Here, people are concentrated working for London's banks and financial services.

▲ ***Figure 1*** *Central London and the high-rise offices that make up the City, or CBD – London actually has three CBDs!*

London's CBD

What's special about the CBD? It's the **oldest** part of the city, where it first began (see 5.7). It's also the place where most offices are found – as the term CBD suggests, it's central! London's radial roads mean that it's the most accessible area from all parts of London. This leads to higher land values, which in turn make it the most **densely built** part. Figure 2 shows how land values vary in different areas of cities. The peaks in Figure 2 are high land values – but they're like the photo in Figure 1, because these areas also have the highest buildings! If land is expensive, the only way to maximise its value is to build up. The diagram was designed to explain American cities, but it fits London well.

London's expanding knowledge economy (see section 5.5) means that the CBD has expanded.

- Canary Wharf, 4 km east of the City, has formed a second CBD.
- London's 'West End' – its shopping streets (e.g. Oxford Street) and theatreland – form a third CBD.

In spite of the density, central London's **environmental quality** benefits from its royal parks (e.g. Hyde Park and St. James's Park). But it also has the UK's worst air quality, caused by traffic.

Did you know?

Office rents in central London are some of the most expensive in the world – a square foot in Mayfair costs £110 per year!

▼ ***Figure 2*** *A general pattern of urban land values. Notice how the CBD is the highest value part of the city – others occur where roads intersect (or meet).*

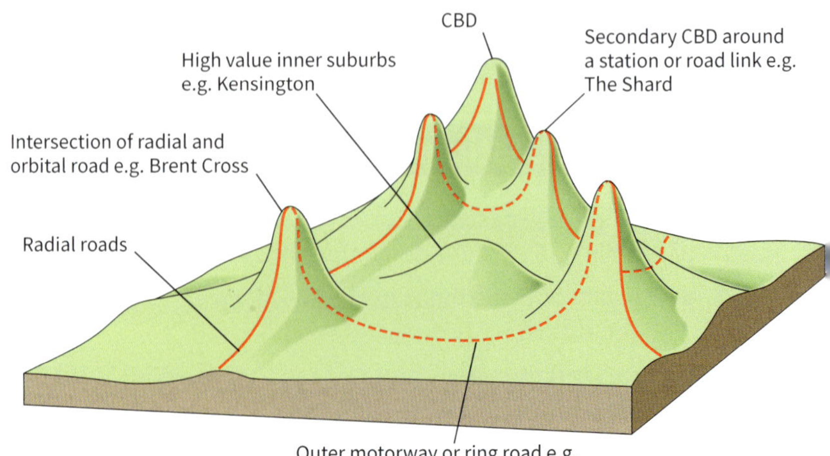

London's inner suburbs

In the Industrial Revolution (18th/19th centuries), factories and densely packed terraced housing (like Manchester housing in TV's 'Coronation Street') were built close to central London. A few high-income suburbs also developed, whose population (e.g. landed gentry) wanted to be close to the city.

The inner suburbs are varied.

- 1 km west from London's West End is Kensington, on the edge of Hyde Park, and one of the world's most expensive suburbs.
- 1 km east from the City is Hackney, an area of older factories and newer flats (Figure 3).

The inner suburbs are changing rapidly. They're close to the City, so large older houses have been divided into flats for rent. Environmental quality varies between areas that are run-down but changing (e.g. parts of Hackney like Figure 3) and smarter areas like Notting Hill near London's parks.

▲ ***Figure 3*** *Hackney, showing old industrial buildings by the River Lea Navigation (a canal) side-by-side with newer flats*

London's urban-rural fringe

Figure 4 shows Loughton, near Epping Forest on London's rural-urban fringe, where city meets the countryside. It's very different from Hackney. Almost every house has a garden, so building density is lower. Most houses were built in the late 20th century. There is some industry towards the edge of town near the underground line – but it's mostly residential. Environmental quality is higher, and Epping Forest is a short walk away.

▲ ***Figure 4*** *Loughton, in north-east London's rural-urban fringe*

Your questions

1 Describe the CBD of London using Figures 1 and 2.

2 Copy and complete the following table to compare the different areas of London in this section.

Area of London	Building age	Building density	Land use(s)	Environmental quality
CBD				
Wealthy inner suburbs				
Poorer inner suburbs				
Outer suburbs				

Exam-style questions

3 Define the term 'Central Business District'. (1 mark)

4 State **two** characteristics of the inner suburbs of a named UK major city. (2 marks)

5 Explain **two** characteristics of the rural-urban fringe around major UK cities. (4 marks)

6 Explain why the inner suburbs of a named UK major city are so varied. (4 marks)

5.9 London and migration

In this section, you'll understand the causes of migration and assess how migration affects the character of different parts of London.

London's changing population

The world's largest cities are booming, caused by concentrated economic growth and jobs. After decades of decline, London's population is growing faster than at any time in its history. The reason is **migration** from within the UK and overseas (see section 5.3). **Overseas immigration** is by far the bigger, with 2 million people settling in London from overseas between 2010 and 2020!

Who are the migrants?

Most migrants are working age adults between 21 and 35 years old. **Internal** migration from within the UK consists mostly of recent graduates from UK universities seeking work and a London lifestyle. International migration consists of two groups of workers – **skilled** and **unskilled**.

- Many **skilled workers** take up well-paid jobs in the knowledge economy in the City (section 5.5). London companies appoint migrants with particular skills from overseas, as there are not enough skilled people in the UK. Most migrants in this category tend to be highly qualified professionals from the EU, USA, South Africa and Australia.
- **Unskilled workers** also find work easy to get. Many do jobs unwanted by UK workers (e.g. refuse collection), or those with unsocial hours (e.g. childcare, pizza delivery). London's construction, hotel, and restaurant companies would find it hard without them. Many of these come from the EU, but also from India, Pakistan, Bangladesh, and increasingly West Africa.

▼ ***Figure 1*** *Clustering of two ethnic groups in London – Asian Indian British (Map 1a) and black Caribbean British (Map 1b)*

Map 1a

Harrow
Wembley
Southall
Hounslow
Richmond upon Thames
London
Newham

Key

Percentage of Asian Indian British residents

- up to 4%
- 4–9%
- 9–16%
- 16–26%
- 26–37%
- over 37%

Migration and ethnic communities

Most recent migrants seek cheap rented accommodation. They are not eligible for social housing, and so take private rented property in inner city areas. Often, clusters of particular ethnic communities develop (Figure 1). These:

- help to defend migrants against discrimination
- support ethnic shops and services (banks, places of worship, etc)
- help to preserve cultural distinctiveness. Some areas have developed cultural festivals, e.g. Notting Hill Carnival.

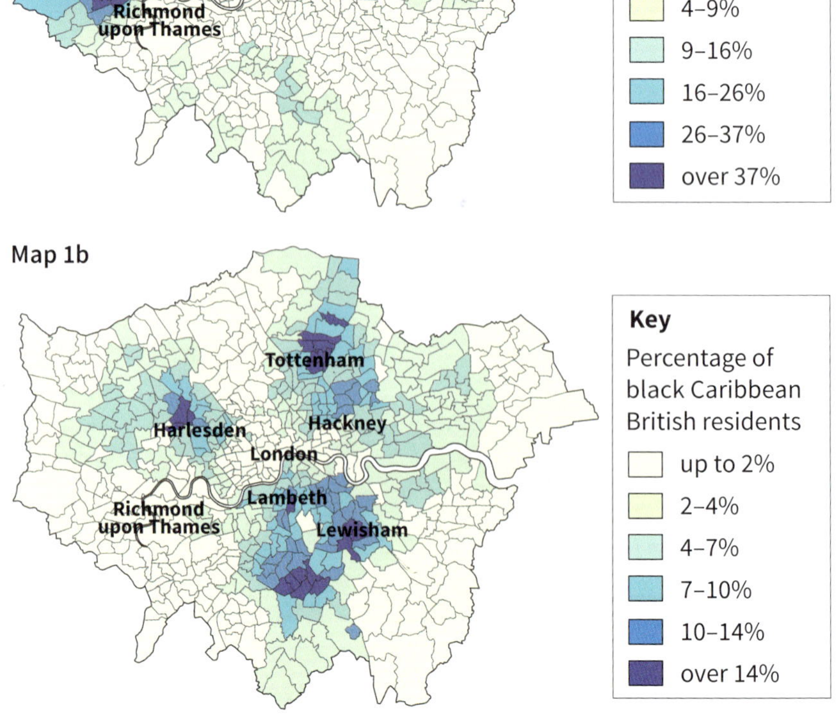

How migration affects three London suburbs

Age-sex structure

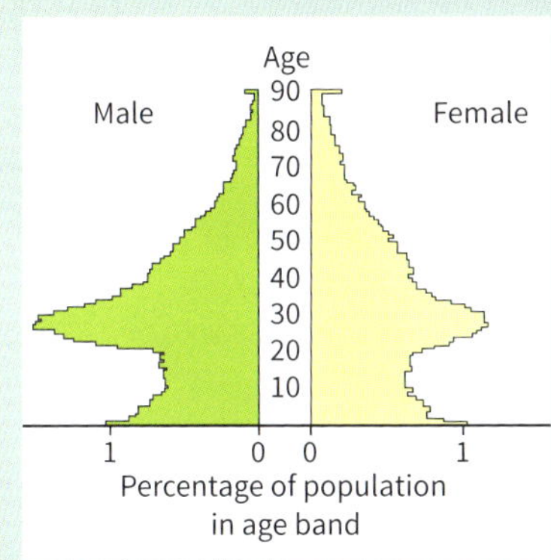

1 Newham

- **Ethnicity.** One of London's most diverse boroughs – 30% White British, 26% Black British of Caribbean and African heritage, 39% Asian heritage (mainly Indian, Bangladeshi and Pakistani).
- **Income.** Low-income area.
- **Housing.** 32% own their property, 35% rent privately, 32% rent from social housing.
- **Services.** Schools under pressure caused by high birth rate. Pressure on social services with 38% of children living in poverty.
- **Culture.** Mainly Asian with food shops, small businesses and several temples and mosques; also several African Anglican churches.

2 Lambeth

- **Ethnicity.** Diverse (38% born outside UK from 152 countries). Black British (25%), White British (55%), small Asian British population (8%).
- **Income.** Average income area with recent inward migration by middle class.
- **Housing.** 44% own their property, 20% rent privately, 34% rent from social housing.
- **Services.** 81% of children in schools from ethnic backgrounds. 140 languages spoken in Lambeth schools; English is a second language for half of school students.
- **Culture.** Varies, from Black Caribbean culture (Caribbean food market stalls and restaurants in Brixton) to White British.

3 Richmond upon Thames

- **Ethnicity.** One of the least diverse boroughs in London – 85% White British, 7% Asian or Asian British. But many residents born overseas, particularly in the USA and EU.
- **Income.** Very high income area – 69% have professional or managerial occupations. Average income £41 000 – much higher than UK average.
- **Housing.** Stable area where affluent people buy expensive property. 69% own their property, 16% rent privately, 15% rent from social housing.
- **Services.** Less pressure on schools with fewer children, but has higher than average percentage in care homes.
- **Culture.** Predominantly middle class.

Your questions

1 Explain why immigration has become an economic necessity for London.

2 Using Figure 1, describe the distribution of **a** Asian Indian British people, **b** Black Caribbean British people.

3 Explain why distinctive areas develop within London of different ethnic groups and nationalities.

4 Compare the three age-sex pyramids in terms of **a** largest age groups, **b** male–female balance, **c** birth rate, d proportions of elderly people (above 70).

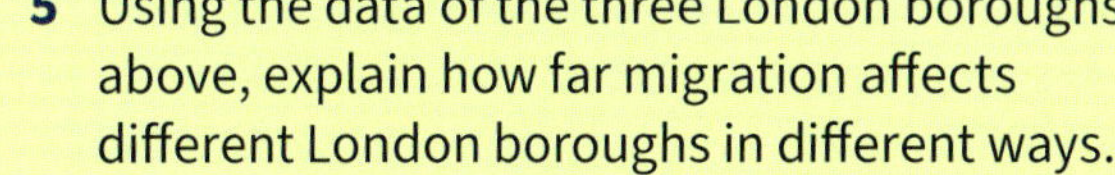

5 Using the data of the three London boroughs above, explain how far migration affects different London boroughs in different ways.

Exam-style questions

6 Using Figure 1, compare the distribution of Asian Indian British and Black Caribbean British people. (3 marks)

7 Analyse Maps 1a and 1b, and the text 'How migration affects three London suburbs'. Assess the impacts of migration on London's suburbs. (8 marks)

5.10 London's inequalities

In this section, you'll understand the reasons for different levels of inequality in different parts of London.

It's only a short commute

In normal times, hundreds of thousands of people use London's underground. They listen to music, read the paper, perhaps even sleep a bit. The journey from London Bridge station on the Jubilee Line towards Canary Wharf (see Figure 1) is only three stations. But in that short distance, life expectancy in local communities falls – from London Bridge (84 years) to Canada Water (78) – then rises again at Canary Wharf (89). Stay on the train for two more stations to Canning Town, and it falls again to 79. The same journey will take you from places with the least child poverty in London to those with the highest levels.

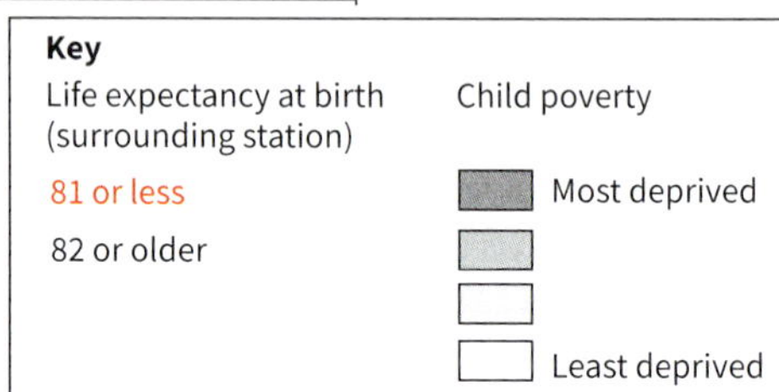

▲ ***Figure 1*** *'Lives on the line' – how even short journeys on London's underground show up differences in life expectancy and child poverty in the city*

While parts of London are booming and wealthy, some are very deprived. **Deprivation** has several causes. The government gathers information on employment, health (how good or poor it is), education, housing and services to produce an **Index of Multiple Deprivation** which shows how deprived places are. There's a close link between deprivation and life expectancy.

London's inequalities

In 2020, over 2.5 million people in London lived in poverty – that's 28% of its population, and 6% higher than the rest of England. That might sound surprising because of London's huge wealth. But incomes in London are more unequal than any other part of the UK. One million of the UK's poorest people and one million of its wealthiest live in London.

Deprivation – a lack of wealth and services. It usually means low standards of living caused by low income, poor health, and low educational qualifications.

Comparing Newham and Richmond upon Thames

Newham in east London is one of London's most deprived boroughs, while Richmond in south-west London is one of its wealthiest. Both are shown on Figure 2. The two show how closely deprivation and health can be linked (see Figure 3). In 2019:

- Incomes are low in Newham. As a result, Newham has far greater numbers of children on free school meals.
- Incomes in Richmond are far higher (see section 5.9). Average household incomes are twice as great as in Newham. The percentage of those with degree qualifications enables people in Richmond to gain employment with higher incomes.

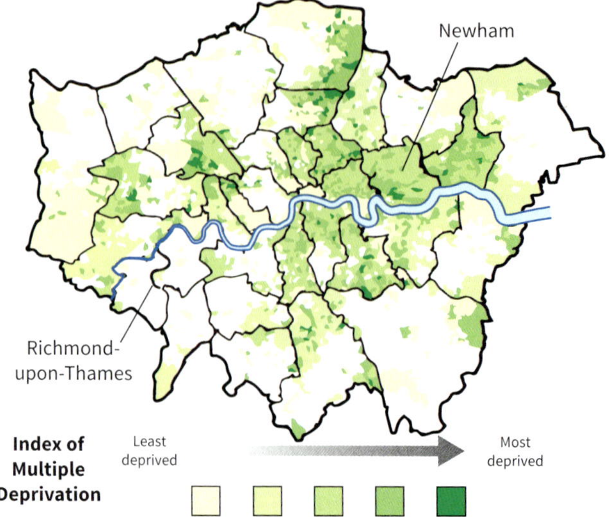

▲ ***Figure 2*** *Map showing Index of Multiple Deprivation for London boroughs (for the borough names, see Figure 2 page 180)*

- The health of those in Newham is worse than in Richmond (Figure 3), especially people with fair, bad, or very bad health. Ill health (e.g. arthritis) is one of the factors that limits people's ability to work and earn.
- In Newham, the percentage of people without qualifications is high. This group is often limited to unskilled jobs with low earnings. However, Newham has had considerable success in school improvement, with the average Attainment 8 GCSE score for students in EBacc subjects in Newham only just below that of Richmond.

	Newham	Richmond
General health		
Infant mortality (per 1000 births)	3.7	2.2
People living in fair, bad, or very bad health (%)	16.9	12.2
Premature deaths (before 75, per 100 000 population)	346	245
Education		
Average Attainment 8 score for students aged 16 at GCSE in EBacc subjects, 2019 (%)	48.6	53.9
% adults aged 16–64 with no qualifications	12.5	2.5
% secondary students eligible for free school meals, 2019	19.3	9.9
% adults educated to degree level	35	58.8

▲ ***Figure 3*** *Deprivation – comparing health and education indicators for Newham and Richmond*

Multiple deprivation in London

In order to understand deprivation, the government compiles census data on incomes, housing, health and services. The result is the Index of Multiple Deprivation (see Figure 4), which helps to assess in which areas a community is deprived ('high' or 'very high') or not ('low' or 'very low'). Crime is linked to deprivation, with highest crime rates occurring in the most deprived urban areas. You can view crime data for any part of the UK – just visit ukcrimestats.com and key in the name of any area or postcode.

Indicator of deprivation	What it measures	Proportion of population living in areas of London with this indicator
Income	People on low incomes	High
Employment	Those unable to work through unemployment, sickness or disability	High
Health and disability	People in poor physical or mental health	High
Education, skills and training	People with low educational attainment	Low
Housing and services	Affordability of housing and within reach of services, e.g. transport, doctor	Housing – Very high Services – Low
Crime	People affected by crime	High
Living environment	Those living in sub-standard housing (e.g. lack of heating, damp)	High

▲ ***Figure 4*** *The Index of Multiple Deprivation (IMD) in Newham. Compare this with Cornwall in section 5.18.*

Your questions

1. Describe the pattern of **a** life expectancy, and **b** child poverty in Figure 1.
2. In a paragraph, suggest why it matters that London is an unequal city.
3. Use Figure 3 to suggest how and why:
   - **a** general health is worse in Newham than in Richmond
   - **b** education achievement is lower in Newham than in Richmond.
4. Suggest how further improvements to exam results in Newham's schools could impact upon future levels of deprivation within the city.
5. Visit the UK Crime Stats website (ukcrimestats.com) to investigate crime in Richmond (postcode TW9) and Newham (postcode E15) for the most recent month. Find **a** which has the greatest total crimes, and **b** which four types of crime occur most.

Exam-style question

6. Analyse Figures 1–4. Assess the reasons for contrasting levels of deprivation within London. (8 marks)

5.11 Facing decline

In this section, you'll understand the reasons why parts of London suffered decline in the past.

A land of dereliction

It's hard to think of London as a decaying city. It's so lively, with restaurants and bars open into the night, and transport busy round the clock. But it wasn't always like this. When London's docks closed in 1981, the riverside between Tower Bridge and the Thames estuary looked like the photo in Figure 1.

The decline of London's docks was caused by the use of containers for transporting goods by sea (see section 2.9). New container ships were larger, so ports moved downstream where water was deeper. The closure of the docks had a huge impact.

- Industries that relied on the port (like the flour mill in Figure 1) moved too. Nearby, industries in London's biggest manufacturing area, the Lea Valley, also closed. It was an example of **deindustrialisation** (section 3.3). By 2001, only 7.5% of people worked in manufacturing in London, down from 30% in 1971.
- The closures had a massive impact on communities. Parts of east London had unemployment rates of over 60%. The area suffered **depopulation** as people left in search of work. Between 1971 and 1981, inner London boroughs lost over 500 000 people – 16% of its population! Nearly 100 000 of those were from areas closest to the docks and the Lea Valley.

▲ ***Figure 1*** *A derelict former flour mill beside the Royal Docks. Imagine 21 square km of derelict land looking like this – all the way upstream to Tower Bridge!*

Suburbanisation

The depopulation of inner London speeded up a process which had been taking place for 60 years. Outer suburbs of London gained people, while inner suburbs lost them. The process is called **suburbanisation**, where people leave the inner city for a house in the outer suburbs, which has space for a garden (see Figure 2). Many left London altogether – the city lost 1.5 million people between 1951 and 1981.

Suburban growth became possible because of transport changes.

- London's **underground** opened in 1863, and by 1930, the network was established. Suburban office workers could be in the City in 30 minutes.
- The **electrification** of surface rail in the 1920s made travel beyond London faster. Commuters in Guildford, 50 km away, could be in London in 30 minutes.

▲ ***Figure 2*** *Late 20th century lower-density suburban housing*

Decentralisation

The shift to the suburbs led people to spend their money there too, instead of in London. Shopping habits changed; people shopped by car, not train. It caused **decentralisation** – shifting the balance of shopping activity and employment away from the CBD. It led to the growth of:

- **out of town shopping centres** (e.g. Croydon's Whitgift Centre) which developed under cover shopping to attract customers. Larger centres developed even further out, close to the M25, such as Bluewater in Kent.
- **retail parks,** built away from suburban shopping centres, but close to major roads like Kew Retail Park on London's South Circular Road in Figure 3.
- **business parks,** which are areas for employment, such as Stockley Park near London's Heathrow Airport, close to the M4 and M25.

Buying online – called **e-commerce** (e.g. Amazon) – has decentralised shopping further, particularly during the Covid-19 pandemic when shops were closed. Online shopping now threatens many high streets and national stores.

▲ ***Figure 3*** *Kew Retail Park – a recently built shopping centre*

The fight back

The problem with decentralisation is that people spend money outside the city. London gains if people spend their money there, instead of at Bluewater in Kent. To attract people back, two shopping centres have been developed in inner London by Australian company Westfield.

- One is at Stratford, east London (Figure 4) – it's Europe's largest shopping centre! Stratford is London's most accessible point by rail and underground outside the centre of the city.
- The second is at Shepherd's Bush in west London. It's close to tube and surface rail, but also to the M40.

▲ ***Figure 4*** *Stratford's Westfield shopping centre – Europe's largest!*

Your questions

1 Explain the difference between deindustrialisation, depopulation, suburbanisation, and decentralisation.

2 Why does it matter to a city if its industries and port close?

3 Explain how the move to the suburbs was caused by changes to transport.

4 Draw a table to compare the advantages and disadvantages of city centre shopping compared to out-of-town or retail park shopping.

Exam-style questions

5 Define the term 'suburbanisation'. (1 mark)

6 For a major UK city, explain **two** impacts of deindustrialisation. (4 marks)

7 For a major UK city, explain how online shopping threatens to change cities. (4 marks)

5.12 Expansion and regeneration!

In this section, you'll understand why London has reversed a period of declining population.

The sprawling city

People who work in London have a difficult choice to make – where to live?

- Choosing London means very high rent or mortgage payments.
- Choosing further out of London means a cheaper house, but a lot of money and time spent commuting.

London isn't an easy city to live in.

Part of the problem is London's size. It's 70 km north–south, and a similar distance east–west. Even though it lost 1.5 million people between 1951 and 1981, it still managed to grow in size! The reasons are:

- **Counter-urbanisation.** Although people moved out of London, it was often to the Home Counties (those which surround London) whose population increased. The boundary between city and countryside has become more blurred.
- **Suburbanisation** (section 5.11)**.** Moving from inner London to the outer suburbs usually means moving from a small house to one with a garden. The same number of people therefore take up more space.
- **Family size** has fallen. Fertility rates (the number of children born per woman) fell from almost 3 in 1961 to 1.6 by 2011. But when those born in the 1960s created their own families, they needed more homes and space!
- **Increasing divorce** and **later marriage** means people are single for longer, increasing the number of homes needed.

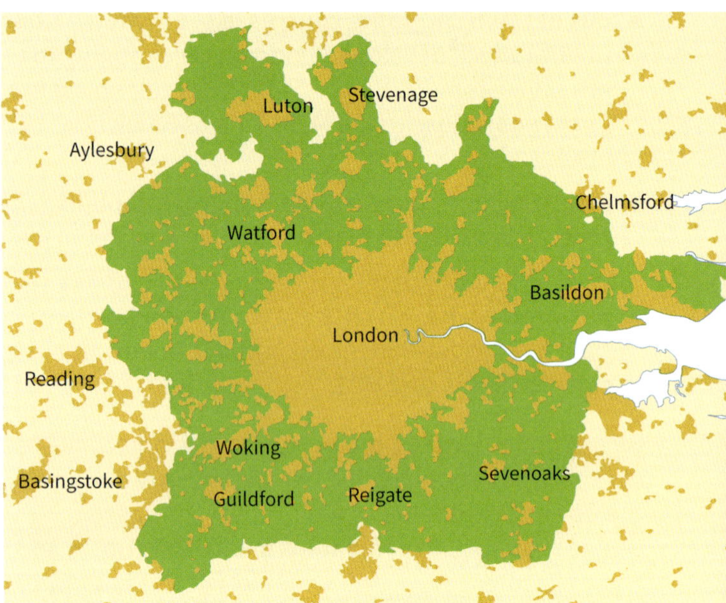

▲ ***Figure 1*** *London's green belt*

Regeneration means redeveloping former industrial areas or housing to improve them.

Brownfield sites are former industrial areas that have been developed before.

London has become a sprawling city, invading surrounding rural areas (known as the rural–urban fringe). To limit its growth, a **green belt** (shown in Figure 1) was introduced by planners to protect the countryside. It created a 'belt' where no major building was allowed. Further expansion could only occur beyond it, in cities such as Chelmsford.

Did you know?
The cost of renting a home in London is more than double the UK average.

Re-urbanisation

Since 1991, the flow of people leaving London has reversed. **Re-urbanisation** (where a city regrows) has taken place. This has been caused by four factors:

- **Space.** The closure of London's docks and industries (section 5.11) created space for redevelopment (known as **regeneration**). New housing (see Figure 2) and offices have been developed on **brownfield sites** around docklands.
- **Investment** by large TNCs created jobs in financial and business services in London's Docklands (especially Canary Wharf – 100 000 people work there!). Bank of America has its European headquarters there, and HSBC its global headquarters.

- **Gentrification.** Many high-income earners now prefer to live closer to work than commute. Many former working-class areas in inner London such as Fulham and Bermondsey have become gentrified – that is, occupied and renewed by middle classes enjoying the lifestyle London offers – theatres, bars and restaurants.
- **Studentification.** University expansion has been caused by demand from overseas students. Universities have a big impact on communities – known as studentification, which refers to parts of cities where students live in large numbers. Universities bring employment (e.g. academic lecturers), and student spending regenerates pubs, shops and buy-to-let properties. In east London, Queen Mary University of London and the University of East London have significantly increased student numbers.

▲ ***Figure 2*** *East Village, one of east London's newest suburbs in Stratford. These flats and houses were the Athletes Village during the 2012 Games.*

Culture and leisure

Regeneration has played its part in culture and sport. London's 2012 Games left a legacy of a huge park, the Queen Elizabeth Olympic Park. Before the Games, it was a largely derelict and industrial area (Figure 3). In 2016, the Olympic Stadium opened as West Ham's new football stadium, and athletics events and open-air music concerts are also held there. Nearby, a new cultural quarter known as East Bank is being developed by the Sadlers Wells Opera Company. This is part of a major development including new music studios for the BBC.

Your questions

1 Explain how the processes of counter-urbanisation, suburbanisation, regeneration, and re-urbanisation have all affected London.

2 Explain the difference between gentrification and studentification.

3 Explain why increasing divorce rates and later marriage have increased the number of homes needed in London.

4 **a** Describe the extent and pattern of London's green belt in Figure 1.

b In pairs, list advantages and disadvantages of having a green belt.

5 Draw a spider diagram to show changes to east London resulting from the 2012 Games.

6 Which would you argue would be better to manage London's housing demand – to build more housing **a** within London, **b** outside the green belt, or **c** within the green belt? Explain your reasons.

Exam-style question

7 Explain **two** impacts of the decision to create a green belt around major UK cities. (4 marks)

Before

After

▲ ***Figure 3*** *The site of the Olympic Stadium before development and after*

5.13 The impacts of rebranding

In this section, you'll assess the impacts of regeneration and rebranding parts of London.

From old to gold

Cities compete to host the Olympic and Paralympic Games. It's not just the attention brought by 27 days of sport; it gives the host city a chance to invest in and regenerate areas that are run down. Where London's 2012 Olympic venues now stand was once London's most derelict land in the Lea Valley in east London. It had been one of Europe's biggest industrial areas. Its decline in the 1970s (section 5.11) led to the abandonment of buildings like the one in Figure 1a. Clearing the land to create the Olympic Park was one of London's biggest projects.

East London's rising population

The Queen Elizabeth Olympic Park (Figure 1b) is just one of a series of regeneration projects in London. The impact on the city's population has been considerable. A declining population between 1951 and 1981 has reversed to become a rapidly growing population of young professionals. Tower Hamlets – one of east London's inner boroughs – lost nearly 40% of its population between 1951 and 1981. Its population has doubled between 1987 and 2019, and more people now live there than ever before.

From a city whose population suffered **counter-urbanisation** for so long, London is now where many people want to live. This is because many inner city suburbs have been **rebranded** – that is, they have had a change of image.

- Land that was once derelict has now changed to housing, offices and hotels (Figure 2).
- New transport links make east London very accessible. Bus routes have been expanded, and run 24 hours. Tube and rail links have been extended, and also run 24 hours a day.

There is a downside, however. London's housing is among the world's most expensive. This is caused by:

- population growth which is faster than the rate at which houses are being built.
- overseas investors who buy London property but leave it vacant. This reduces London's housing stock.

London desperately needs more affordable housing.

▲ ***Figure 1*** *Before and after – derelict land in what became the Queen Elizabeth Olympic Park. Figure 1a was taken in July 2005, and Figure 1b in 2015.*

▲ ***Figure 2*** *Flats and hotels alongside what used to be the Royal Docks in east London. The cranes are no longer in use, but are there to keep a sense of the area's past.*

Changing environmental quality

Rebranding means that areas which were once run-down have become more desirable. Some of London's poorest areas have been gentrified (section 5.12), and environmental quality has improved because of regeneration, e.g. the new Olympic Park.

But little open space has been created. Inner London has many long-established parks, e.g. Victoria Park in Hackney. But the recent Queen Elizabeth Olympic Park was its first major park in 150 years, and many suburbs have little open space nearby. Pressure to build housing means that London's population density is increasing, and more cars on the road means poorer air quality.

London's changing economy

In 2019, before the Covid-19 pandemic, London's economy was almost as large as that of the Netherlands!

- Its economy creates jobs. Estimates claim that London's economy will create 35 000 new jobs annually, ranging from low income (e.g. cleaning) to high (e.g. law).
- Demand for housing and offices means that London's construction industry is booming.

The city creates its own multiplier effect!

But there's a downside.

- It's expensive to live there, which displaces many people away from London. Housing demand drives up prices to buy or rent. Employers pay people more, so the cost of living rises. The lifestyle in Figure 3 requires a salary of over £42 000. The majority don't earn anything like this.
- The Covid-19 pandemic altered views of working in London. 48% of London's employees worked from home. Companies found it worked for them, questioning the costs of office space in central London, while employees no longer faced expensive travel costs.

Item (September 2021 prices)	Monthly cost
Average monthly rent in a shared flat or house	£1000.00
Total household bills (e.g. heating and council tax)	£220.00
Travel: cost of a monthly Oyster travel card (Zones 1–4)	£200.00
Food shopping per month	£160.00
Coffees and lunches at work (£12 a day x 22 working days a month)	£264.00
Going out with friends/colleagues x 3 nights a week costing £30 a night x 4 weeks	£360.00
Mobile phone (£40) and laptop (£20) cost per month	£60.00
Monthly tax, pension, and National Insurance payments	£1300.00
Total per single month	**£3554.00**
Total for a year	**£42 648.00**

▲ ***Figure 3*** *Costs of living in London in 2021*

Your questions

1 Explain how 'rebranding' differs from 'regeneration' in cities.

2 Using Figures 1a and 1b, explain how the 2012 Olympics helped to rebrand east London.

3 Copy the table below, to show the positive and negative impacts of the following changes in London:

Change	Positive impact	Negative impact
Rising population		
Environmental quality		
Economic opportunities		

4 In pairs, design a flow diagram to show how and why London has become so expensive to live in.

5 Explain the potential future impacts of Covid-19 on London as a place to **a** work, and **b** live.

Exam-style questions

6 Define the term 'rebranding'. (1 mark).

7 Explain **two** reasons why inner city areas have seen rapidly increasing populations in recent years. (4 marks)

8 Analyse Figures 1, 2 and 3, and the text on these pages. For a named major UK city, assess the economic and environmental impacts of regeneration on different groups of people. (8 marks)

5.14 Geographical skills: investigating changing environments

In this section, you'll use geographical skills to investigate different land use types in east London.

Identifying land uses in east London

You will find it helpful to use the following:

- page 327 which contains the key for 1:25 000 scale maps
- page 326 which will help you to develop your skills in interpreting maps.

The part of east London shown in Figure 2 contains the Queen Elizabeth Olympic Park. The river that runs from north to south through the centre of the map is the River Lea. The area around the River Lea was London's most industrial area until the 1970s, when it began to decline. Study the map (Figure 2) and identify where a few industries remain. You'll find two:

- one lies between Hackney Wick at grid reference 372847 and the area around the two canal locks to the south between Old Ford and Stratford Marsh.
- a second area lies south of Mill Meads, whose name is at 385831.

You can spot each of these areas by the larger buildings. You can also spot the Aquatics Centre (Figure 1), which replaced many of the former industries there.

Did you know?
Next time you go to east London, take your swimming gear and £2.70 (off peak) so you can swim in the Aquatics Centre.

Figure 1 *London's Aquatics Centre from the 2012 Olympic and Paralympic Games*

Your questions

1 Use 6-figure grid references to identify two examples of each of the following land uses in the map extract (Figure 2):

- **a** industry (shown by the letters 'Wks' – short for 'Works')
- **b** venues from the 2012 Olympic and Paralympic Games
- **c** areas of older terraced housing (shown by parallel straight rows of high density housing)
- **d** education establishments.

2 Use 6-figure grid references to identify the area known as East Village in Stratford (the Athletes Village in the 2012 Games).

3 Identify the most likely location on the map (Figure 2) from which the photo in Figure 1 was taken.

4 Suggest possible reasons why the older industrial areas are all close to the River Lea or canals.

5 Describe four pieces of evidence with grid references to show that this area is very well served by public transport.

6 Suggest three reasons why Westfield (an Australian property company) decided to build the Westfield Stratford City shopping centre (grid square 3885).

7 Hackney has one of London's fastest growing populations and is widely advertised as a pleasant area in which to live. Identify three features that would make Hackney:

- **a** easy to live in (e.g. for day-to-day living)
- **b** pleasant to live in.

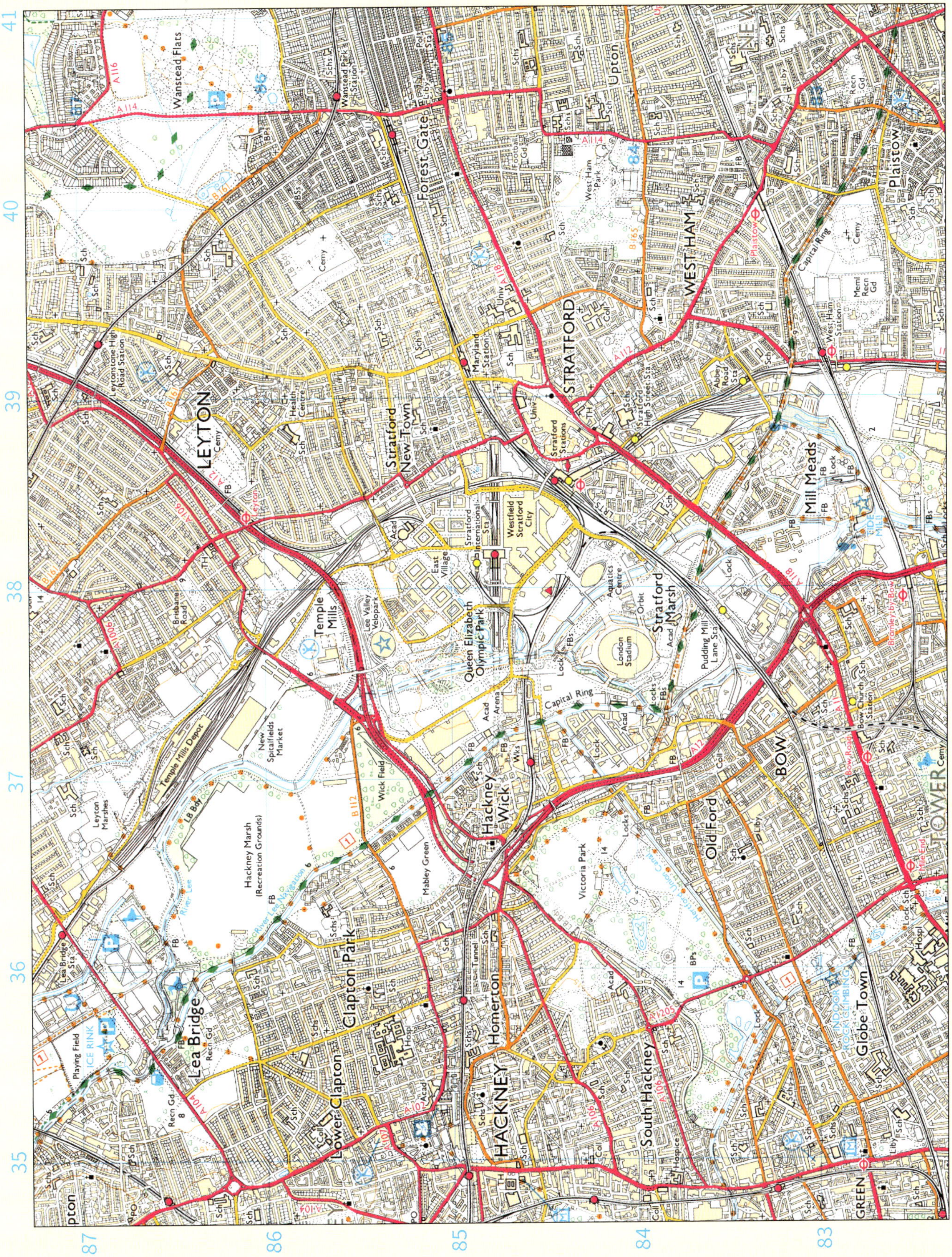

▲ ***Figure 2*** *1:25 000 Ordnance Survey (OS) map extract of the Stratford area of east London. The area contains the Olympic Park, where London's 2012 Games were held.*

5.15 Improving London

In this section, you'll evaluate the attempts to improve quality of life in London and to make the city more sustainable.

The daily grind

Before the Covid-19 pandemic, three million people used to **commute** into London daily. As well as central London, there are large employers elsewhere, such as Heathrow Airport. Commuting takes a lot of time and expense, spent on crowded journeys, where getting a seat may be a luxury. Before the pandemic, London accounted for 90% of commutes made by underground and tram in England, and 75% of those by train. It's not surprising that the pandemic led many to question whether commuting was the best way of spending time, if they could work from home.

▲ ***Figure 1*** *Commuting in London*

Travel to work affects people's **quality of life** – that is, the life that they enjoy apart from the money they earn. How can London become more **sustainable** (see section 3.12) so that quality of life is improved? It has six problems which are related:

- **Transport.** How can all the people in London be transported sustainably? The car is not the answer – it has to be public transport.
- **Affordable housing.** People increasingly face long commutes just so they can live somewhere that's affordable. Cheaper housing lies outside London, which means longer journeys to work.
- **Energy efficiency.** Can affordable houses be cheap to run by being **energy efficient**?
- **Employment.** Can employers be persuaded that people do not necessarily need to travel into the office to do their job?
- **Green space.** London has more open space than most of the world's large cities. Providing more affordable housing might use some of it. Should it be protected?
- **Waste.** Can London control its waste, which in 2015 cost over £600 million to collect and dispose of? London is behind the rest of the UK for recycling – only 34% of its waste is recycled. The national average is 45.5%, and 28 of London's 33 Borough Councils are below this level (see Figure 2).

Some of the challenges in tackling these problems are shown opposite (Figure 3).

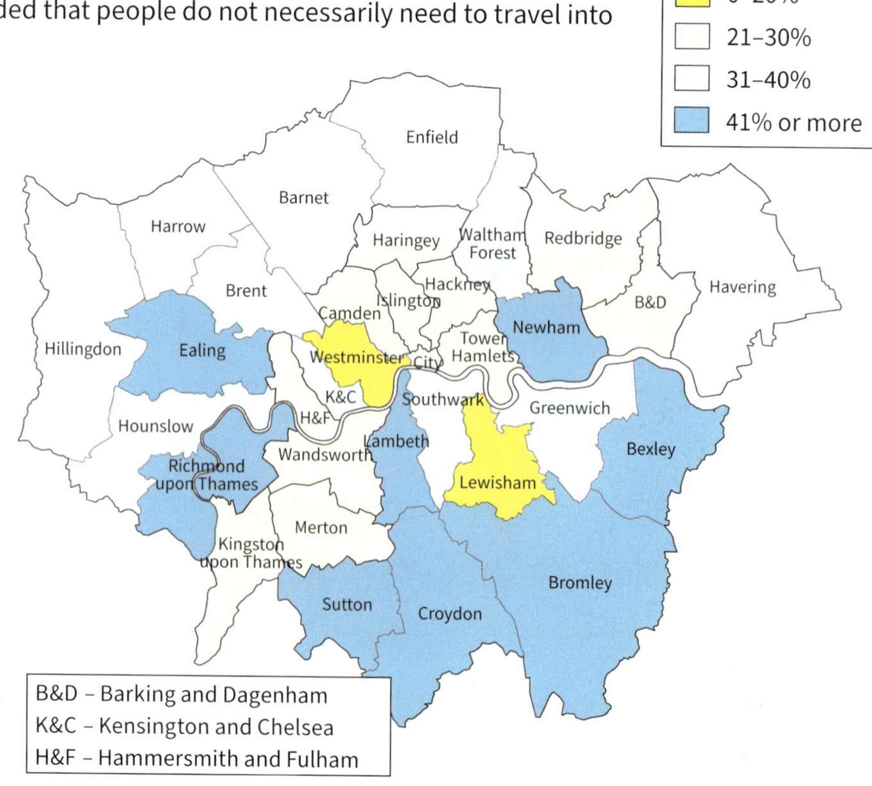

▶ ***Figure 2*** *The percentage of waste which is recycled in each London borough in 2019–20*

A more sustainable London?

Problem and challenges in tackling it

Transport

Tackling transport is one of the main ways in which London aims to reduce its greenhouse gas emissions.

- London's congestion charge charges motorists for daytime travel into central London. £1.2 billion of income was invested in London's buses, initially resulting in a 6% increase in passengers, but increased use of taxis and minicabs had by 2018 actually increased congestion.
- Since 2012, all new London buses have been hybrid (using a conventional engine combined with an electric motor) making buses cleaner and more fuel-efficient.
- 'Source London' provides the UK's first city-wide electric vehicle charging point network, with 1500 charging points in London in 2020 (three times the number of petrol stations!).

Employment

- Especially since the Covid-19 pandemic, many companies and organisations are encouraging people to work at home 1–2 days a week. The number of people who worked mostly from home reached 48% in London in 2020, but many were unable to do so, such as NHS workers, teachers, and supermarket and delivery workers – all known as 'essential workers'.
- Flexible working hours are also more common, which help people to travel more cheaply outside the normal rush hour.
- When at least half the population had been vaccinated against Covid-19, the move to bring workers back to the city centre grew. Many decided to work at home permanently.

Affordable housing

- Many housing projects claim to be affordable! The East Village in Stratford (section 5.12) is 50% affordable housing. But the qualifying salary for affordable housing there is £60 000! Workers on minimum wage find themselves squeezed out by those earning far more but who still qualify!
- Shared Ownership is a programme organised by the Mayor of London's office to help low income Londoners to buy, by offering shared ownership. You buy 25% or 50% of a property, and rent the remainder. As income rises, so you buy more. But some projects, such as Kings Park in Harold Wood in Essex, are still expensive – over £400 000 for a 3-bedroomed property, with mortgage and rent together costing £1300 a month.

Energy efficient housing

BedZED (Beddington Zero Energy Development) in Sutton, south London, is a sustainable community that promotes energy conservation. There are nearly 100 apartments and houses, plus offices and workplaces. BedZED's homes use 81% less energy for heating, 45% less electricity, and 58% less water than an average British home. They also recycle 60% of their waste. But it's London's only project like this!

Green spaces

Green space is essential for quality of life. But some think housing demand in London can only be met by building on greenfield land (section 3.6). This has disadvantages:

- Loss of farmland. The area lost to urban development nationally since 1945 is 750 000 hectares – bigger than London, Berkshire, Hertfordshire and Oxfordshire combined.
- Loss of rural scenery. The Council for the Protection of Rural England (CPRE) believes that this is as serious as the loss of farmland.
- Many people question whether London's 'Green Belt' (section 5.12) can survive – it is close to London and would be ideal for housing.

Recycling

By 2030, London aims to recycle 50% of all household waste across the city, and even more from businesses. It plans to achieve this by:

- reusing waste
- providing accessible recycling and composting services
- providing recycling bins all over the city
- developing waste-burning power stations to generate heat and power. One third of fuel used in the Energy Centre in the Olympic Park is household waste, which heats water to generate energy for the whole park.

▲ ***Figure 3*** *Problems and challenges in making London more sustainable*

Your questions

1 In pairs, design a diagram (e.g. spider diagram, or Venn diagram) to show how London's six problems are related.

2 a In pairs, rank the six problems from 1 (most serious) to 6.

b Justify your rankings and compare with others.

3 a Study the 'sustainability stool' in Section 3.12. Take each of the six problems above ('A more sustainable London?') and rank which of the six is being tackled 1 (most effectively) to 6 (least effectively).

b Justify your rankings.

c Based on this, which two problems do you think London's Mayor should be most focused on? Justify your ideas.

Exam-style questions

4 Define the term 'quality of life'. (1 mark)

5 Study Figure 2. Describe the pattern of recycling in London's boroughs. (3 marks)

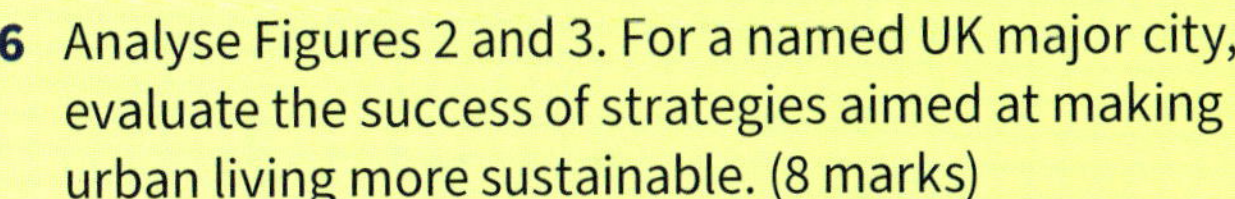

6 Analyse Figures 2 and 3. For a named UK major city, evaluate the success of strategies aimed at making urban living more sustainable. (8 marks)

5.16 Beyond the capital

In this section, you'll understand why accessible rural areas near London are dependent on the city.

Life in Terling

Terling is a traditional-looking English village, near Chelmsford in Essex. The church and village green, shown in Figure 1, are near the cricket meadow, where local teams play in summer. Nearby is Terling Place, the home of Lord Rayleigh, whose family owns most of the land for miles around – and has for centuries.

But, although all looks fine, services in many villages like Terling are struggling. The shop fights for survival, the doctor's surgery has recently closed and been taken over by a cafe, the bus runs twice a week, and the pub has closed. The dairy farms around Terling no longer provide jobs. The cows were sold when milk prices fell, and farm workers were sacked. The price of grain soared, so now land is used for arable farming. But outside contractors are brought in to plough, sow and harvest – then leave. Old farmhouses have been sold off and cow milking sheds and barns lie unused.

▲ ***Figure 1*** *The green and church at Terling in Essex*

Despite the problems, property prices are booming, and the village primary school is full! Traditional cottages can sell for up to £1 million, depending on their size. This is because Terling is very **accessible** (i.e. easy to get to and from) for people who commute to London (see Figure 2).

- The railway station at Hatfield Peverel is five minutes' drive away, with trains taking 45 minutes to London.
- A five-minute drive also takes you to the A12, which links London and Colchester.
- Chelmsford is just seven miles away. Its supermarkets, shops and services have meant reductions in Terling's own services. Terling now depends on Chelmsford and London for goods, services and work.

Flows of people

The flow of people moving from London to towns and villages on the rural–urban fringe has been continuous. Chelmsford's population rose from 58 000 in 1971 to about 180 000 in 2020. Its growth has been caused mainly by people migrating out of London (counter-urbanisation). It's a process that works well for many. High London salaries make the cost of rail season tickets affordable, and housing is cheaper in Essex. London actually depends on the rural–urban fringe.

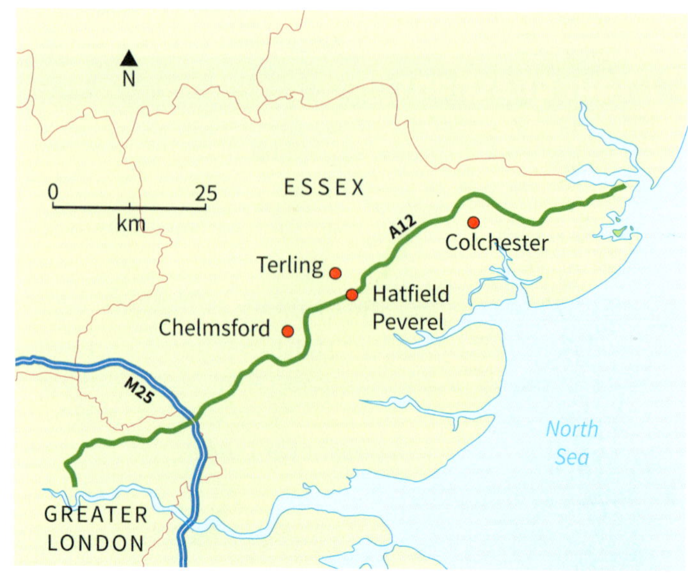

▶ ***Figure 2*** *The location of Terling and Chelmsford on London's rural–urban fringe*

There are nowhere near enough people to work in London within the city itself, so it needs employees from the surrounding region. The benefits work the other way too, with places like Chelmsford and Terling benefiting from higher salaries paid in London. The downside is crowded commuter trains and congested roads.

Because of this flow, the environment changes constantly across the rural–urban fringe. As housing estates are built on the edges of London, the fringe shifts outwards. So Greater London gradually changes from:

- outer suburbs, to ...
- ... green belt, where little development is allowed to ...
- ... dormitory towns like Chelmsford to ...
- ... villages like Terling to ...
- ... the rural landscape, where farming takes place.

The rural–urban fringe

Terling and Chelmsford are part of a bigger area surrounding London, called the **rural–urban fringe.** It lies beyond the suburbs and is mainly rural – but depends on London for work and services. Growing towns and roads expand into the countryside. Before the Covid-19 pandemic, 650 000 commuters left places like Terling to work in London. Settlements in the rural–urban fringe become **dormitory towns and villages** – places where people sleep but are away during the day.

The rural–urban fringe is the change between rural and urban land uses. As you move out from urban London:

- the urban landscape of Greater London gradually changes to a rural one but there are towns like Chelmsford along the way.
- bigger places like Chelmsford, shown in Figure 3, gradually change to rural villages such as Terling.

Figure 3 *Aerial view of the rural–urban fringe around Chelmsford*

Thinking beyond

What's the future for rural life and traditional country skills if all villages fill up with city commuters?

Your questions

1 a In pairs, use a property search website (such as rightmove) to research and compare prices for a 3-bedroomed house in Terling with those in central London.

b Use a website (such as trainline) to find the cost of an annual season ticket from Hatfield Peverel to London.

c Using your research and information here, list the economic, social and environmental costs and benefits for Terling of being so close to London.

2 Using Figure 3, identify evidence that the area shown is part of the rural–urban fringe.

3 How might population increase in villages like Terling affect services there?

4 Explain why the rural–urban fringe is often a difficult place to live for low wage earners or the elderly.

Exam-style questions

5 Explain how cities and accessible rural areas depend on each other. (4 marks)

6 Explain how living in accessible rural areas can have positive and negative impacts for people. (4 marks)

5.17 Off to Devon!

In this section, you'll understand why some accessible rural areas experience economic and social change.

The biggest IT move in history!

If you've ever lived through the mess of moving house, think what it's like moving an office of 1200 people. Imagine too that it's the Met Office, responsible for weather forecasting and analysis of weather patterns – with powerful supercomputers. That's what happened when it moved from its offices outside London to the edge of Exeter in East Devon. It was the biggest IT move in history!

It was a result for East Devon! Not everyone moves when a company relocates, so it meant new jobs for this mainly rural region. The local council estimates that the move has brought an extra £74 million annually to rural East Devon. Their calculations are based on:

- money spent moving house. Every employee had grants towards moving costs – e.g. to pay for things like carpets
- regular weekly spending – e.g. supermarket shopping, leisure.

The Met Office's move had a huge multiplier effect on the region.

People are still moving out!

London's rapid population increase (section 5.9) masks one big contradiction – that never before have people **left** the city in such numbers! Between 2017 and 2020, almost a million people left London. But the population kept rising because nearly a million babies replaced them, and over a million migrants arrived from overseas.

Figure 1 *The Met Office in Exeter*

Why East Devon?

Exeter is an attractive location for companies. Because the area is mainly rural, land costs are cheaper – office rental costs £9 per square foot annually compared to £90 in central London! And although East Devon is 170 miles from central London, it's very accessible.

- Met Office buildings are 3 km from Exeter Airport, with daily flights to London, northern UK cities, and Europe.
- There are over 40 train services daily to London (fastest journey two hours).
- The Met Office is 0.5 km from the M5 Junction 29.

However, this kind of relocation puts three pressures on East Devon – population change, pressure on housing, and pressure on leisure and recreation.

Did you know?

With computing technology, does it really matter where the Met Office is located?

◀ ***Figure 2*** *Map showing the location and landscape of East Devon*

In or out migration	2014–15	2015–16	2016–17
Inward	8243	8318	9285
Outward	5877	5848	6665
Net	2366	2470	2620

▲ ***Figure 3*** *Table showing migration in and out of East Devon, 2014–17*

▲ ***Figure 4*** *Rural Devon*

▲ ***Figure 5*** *Sidmouth, on Devon's coast*

1 Population change

Devon is in demand. In 2017 alone, it gained over 9000 migrants from other parts of the UK. And it's sunny, accessible East Devon that most people want – 40% of Devon's arrivals move here (Figure 3). Part of the increase is due to an inward flow of retired people, but there is also a significant increase in family migrants as well.

2 Pressure on housing

Increasing populations mean one thing – East Devon needs more housing. There are two housing problems:

- Two-thirds of East Devon is classified as 'AONB' (Area of Outstanding Natural Beauty). Looking at Figures 4 and 5, it is easy to see why. Demand for housing here pushes up prices, but it is hard to increase supply; planning permission for new housing is hard to get because of its impact on the scenery.
- Average incomes in east Devon (£520 per household per week in 2019) are 10% below the UK average (£585). Yet housing is only 3% cheaper than average. More affordable housing is needed.

3 Pressure on leisure and recreation

East Devon has no shortage of leisure attractions. It has its own stunning coastline around Exmouth and Sidmouth (Figure 5). To the east, the World Heritage Jurassic Coast stretches into Dorset. 30 minutes to the west is Dartmoor National Park. The accessibility of the area brings huge numbers of visitors. Most visits to all three areas are day trips – and there are an estimated 15 million of those each year! This puts pressure on roads, air quality and environmentally sensitive areas.

Your questions

1. Using material in this section, name four things that attract businesses to Exeter and East Devon.
2. Explain why the Met Office move had such a big multiplier effect on East Devon.
3. Draw a spider diagram to show how population change leads to pressures on **a** housing, and **b** leisure and recreation in rural areas like East Devon.
4. Explain in 300 words the arguments **a** in favour, and **b** against building more housing in East Devon.

Exam-style questions

5. State **two** reasons why companies are attracted to accessible rural areas. (2 marks)
6. Explain **two** pressures that accessible rural areas experience from inward migration. (4 marks)

5.18 Challenges facing rural areas

In this section, you'll understand the challenges facing changing rural areas.

Changing Cornwall

To the west of Devon lies the county of Cornwall (Figure 1). It has a strong image – Poldark, pasties, clotted cream, seafood. About 570 000 people live there, but four million tourists visit in August! Together with Devon, it's the UK's most popular holiday destination. On a sunny day in summer, that's no surprise – it has a 700 km coastline with sandy beaches, small fishing harbours (e.g. Padstow, see Figure 2), and isolated coves. People love it – which explains why it has one of the UK's fastest growing populations.

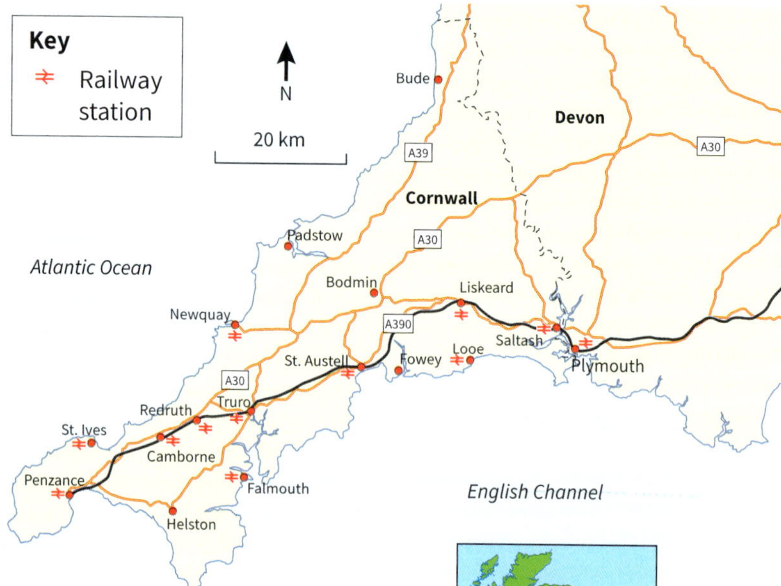

▲ **Figure 1** *Cornwall's location in south-west England*

But it's not perfect

In spite of its holiday image, Cornwall has some problems.

- It's a long county – nearly 140 km from one end to the other. Road and rail transport is slow. There's no motorway, and trains take two hours between Plymouth and Penzance – about 120 km!
- There are no large population centres – St Austell is the largest with 35 000 people. There are several small towns but none large enough to attract large employers.
- There's no knowledge economy (except teleworkers) that could raise incomes. As a result, Cornwall has the UK's lowest average salaries – £24 500, compared to a UK mean of £31 000, and London's £39 500.
- Like most tourist areas, most jobs are seasonal, part-time, and low-wage. In 2020, £9 per hour was typical for hotel or restaurant work.

The decline in primary employment

Cornwall's biggest problem is employment. The decline of its traditional **primary** economy – farming, fishing, china clay quarrying and tin mining – has left the county with few permanent, full-time jobs.

- **Farming.** The number of dairy cattle has fallen by 60% since 2000. Falling milk prices (caused by supermarkets driving down prices) have forced many farmers to give up dairying.
- **Fishing.** Fishing has suffered serious decline, caused by previous overfishing of the UK's fish stocks by UK and EU fishing boats.

▲ **Figure 2** *Padstow, a traditional Cornish fishing village*

- **China clay quarrying.** The St Austell area has some of the world's best china clay. The quarries are owned by Imerys, a French TNC. In the 1960s, over 10 000 people were employed there. Now, cheaper clay overseas has reduced the number of jobs to under 1000. The quarries have left Cornwall with an ugly legacy of wasteland (Figure 3).
- **Tin mining.** Tin is hard to mine, and needs a high global price to make it worth mining. The collapse of tin prices led to the closure of Cornwall's last mine in 1998.

▲ ***Figure 3*** *The legacy of china clay quarrying*

Health and services

West Cornwall is one of the UK's most **deprived** areas. Not only does it have the UK's lowest average incomes and a high percentage of elderly people, it also has few services.

- Only 38% of villages have a doctor's surgery, and most of those only open one morning a week. Buses serve 70% of villages, but there may be only 3–4 a day.
- The main hospital in Truro provides a wide range of treatments, but for many parts of west Cornwall it's 50 km away, which can make a difference between life and death.
- Young people sometimes have to travel over 50 km for sixth form education or training, and travel costs are high.

Indicator of deprivation	What it measures	Proportion of population living in parts of Cornwall with this indicator
Income	People on low incomes	High
Employment	Those unable to work through unemployment, sickness or disability	Medium
Health and disability	People in poor physical or mental health	High
Education, skills and training	People with low educational attainment	Low
Housing and services	Affordability of housing and within reach of services e.g. transport, doctor	Very high
Crime	People affected by crime	Very low
Living environment	Those living in sub-standard housing (e.g. lack of heating, damp)	Very high

▲ ***Figure 4*** *The Index of Multiple Deprivation (IMD) in Cornwall*

So deprivation is not just about income. The government uses an **Index of Multiple Deprivation** (IMD) in order to get an idea of deprivation in different parts of the UK (explained in section 5.10). Cornwall's indicators are shown in Figure 4.

Your questions

1 Explain two reasons why salaries in Cornwall are lower than **a** the rest of the UK, and **b** London.

2 Explain why the decline of Cornwall's primary industries has made its economic problems worse.

3 Explain why rural areas often have poorer services (school, doctor, buses) than urban.

4 Using Figure 4, compare the IMD scores for London (section 5.10) with those for Cornwall.

5 Using Figure 4, explain why the elderly and the young are most affected by the indicators shown.

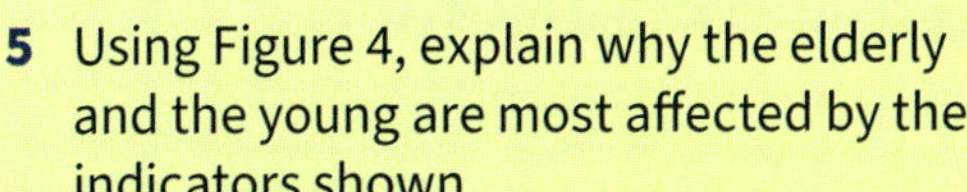

Exam-style questions

6 For a named rural area, explain how economic change has affected it. (4 marks)

7 Explain why deprivation can be high in rural areas. (4 marks)

5.19 New opportunities?

In this section, you'll assess how far rural diversification has brought benefits and costs.

A day out in a quarry?

▲ ***Figure 1*** *The Eden Project*

By 10.00 a.m. on most summer mornings, a queue of cars builds up outside the village of Trethurgy, three miles from St Austell, Cornwall. They're queuing outside Cornwall's biggest all-weather attraction, the Eden Project (Figure 1). At first, a day out in what used to be a china clay quarry may not sound appealing. But the 21st century landscape of plastic domes – called 'biomes' because they contain plant displays from rainforests and Mediterranean regions – offers much more. There's a zip wire, winter ice-skating, summer outdoor concerts, art projects – all in addition to the biomes that are the star attraction.

The Eden Project is Cornwall's only all-year-round indoor visitor attraction. Tourists can visit whatever the weather, and local people gain full-time work. It was designed on sustainable principles, and an education centre runs courses about sustainable living. In the Eden Project's first ten years:

- 13 million people visited.

- Visitor spending on accommodation and meals generated an extra £1 billion to the Cornish economy.
- It employed 700 people, and generated 3000 jobs elsewhere.

However, it has two problems:

- Its sustainable aims have fallen short. While it promotes local cycle paths, and offers reduced admission for anyone arriving by public transport, 97% of visitors arrive by car.
- Visitor numbers are falling; there is evidence that, having been once, few return.

Did you know?

The Eden Project now employs as many people as the whole of the china clay industry in Cornwall.

Farming and diversification

With farm incomes falling, farmers look for other ways of making an income. It's called **rural diversification** – finding a wider range of activities in rural areas to enable farms and other rural businesses to survive. Farm shops and tourist accommodation are two ways.

1 Farm shops

Until 2003, three brothers from the Lobb family in south Cornwall were earning just £30 000 from their 800 acres. That had to be shared between three families.

Their farm is near the Lost Gardens of Heligan, south Cornwall's second biggest tourist attraction (after the Eden Project). They developed a farm shop to sell their beef and lamb to the 218 000 people who visit the Gardens each year (Figure 2). The project – Lobb's Farm Shop – was financed using £200 000 grant funding from the EU and UK government.

The shop sells meat and vegetables produced on the farm, and local products such as Cornish wine and cheese. It has created 12 full-time jobs, and 8 part-time, and most are year-round. Its turnover is now over £700 000 per year. The shop also includes a visitor centre with information about welfare and environmental farming, as well as holding craft fairs.

What helps locally is that £10 spent in farm shops becomes worth £23 in the local economy through the multiplier effect (e.g. by paying employees and suppliers). That same £10 spent in a supermarket is worth only £13 locally.

2 Tourist accommodation

Many farms now supplement their income with that from tourism. These include:

- barn conversions, turning farm buildings into holiday cottages
- camp sites.

A few have invested in leisure complexes (see Figure 3), including log cabins, health spas, swimming pools and play areas in order to encourage families. These are economically beneficial, but the increasing number of barn conversions has led to a reduction in nesting places for birds, such as swallows and owls.

▲ ***Figure 2*** *Lobb's Farm Shop near Gorran in Cornwall*

▲ ***Figure 3*** *Flear Farm Cottages near Kingsbridge in South Devon, a luxury complex of 13 holiday cottages and leisure complex*

Your questions

1 Copy and complete the following table to show the benefits and costs of the Eden Project, Lobb's Farm Shop, and tourist accommodation (like that in Figure 3).

Project	Benefits	Costs
Eden Project		
Lobb's Farm Shop		
Tourist accommodation		

2 Which of the three would you argue has the greatest benefits for the local economy? Explain.

Exam-style questions

3 Define the term 'rural diversification'. (1 mark)

4 Explain **two** reasons why projects are needed to diversify the economy of rural areas. (4 marks)

5 Analyse the information on these pages. Assess the extent to which rural diversification has brought benefits to rural areas. (8 marks)

5.20 Geographical skills: investigating tourism

In this section, you'll use geographical skills to investigate tourism in south Cornwall.

1 Identifying features of south Cornwall

You will find it helpful to use the following:

- page 327 which contains the key for 1:25 000 scale maps
- page 326 which will help you to develop your skills in interpreting maps.

The part of south Cornwall shown in Figure 2 contains the Eden Project, a former china clay quarry. It also contains evidence of other primary employment (see section 5.18). The map in Figure 2 contains evidence of where this employment took place.

Your questions

1 Use 6-figure grid references to identify evidence of the following primary employment in this area: **a** mining, **b** farming, **c** quarrying for china clay, **d** a fishing port. 

2 Use 4-figure grid references to identify that this area is important as **a** place for tourists to a stay in, **b** eat in, and **c** go for a day visit.

3 **a** Using the map, identify one safe cycle route from the hotel at 053521 to the Eden Project. Justify your choice of route.

b In which direction would you be cycling from the hotel towards the Eden Project? 

c Using a piece of string or similar, measure the distance of this route to the Eden Project. Give your answer to the nearest 0.5 km. 

d Identify three features that might make this cycle route challenging for cyclists.

e Explain why the cycle ride back to the hotel is likely to be easier than the one going.

2 Tourism in the local economy

There is a proposal to build about 500 homes on the site of Cornwall Coliseum at 058522. It will extend between the shore and the cliff. Find this location. Figure 1 shows the site of the proposed development.

Your questions

1 Using Figure 2, identify in which direction the camera was pointing to take Figure 1.

2 Describe the site of the proposed housing development.

3 In pairs, use Figure 2 and Figure 1 to identify three advantages and three disadvantages of this location as a place to live.

4 Research the progress of the development at Carlyon beach online.

5 Discuss whether you would give planning permission for this development. Justify your decision.

▼ ***Figure 1*** *The site of the development proposal for Carlyon Bay. The large stack is Crinnis Island (056520 on the map in Figure 2)*

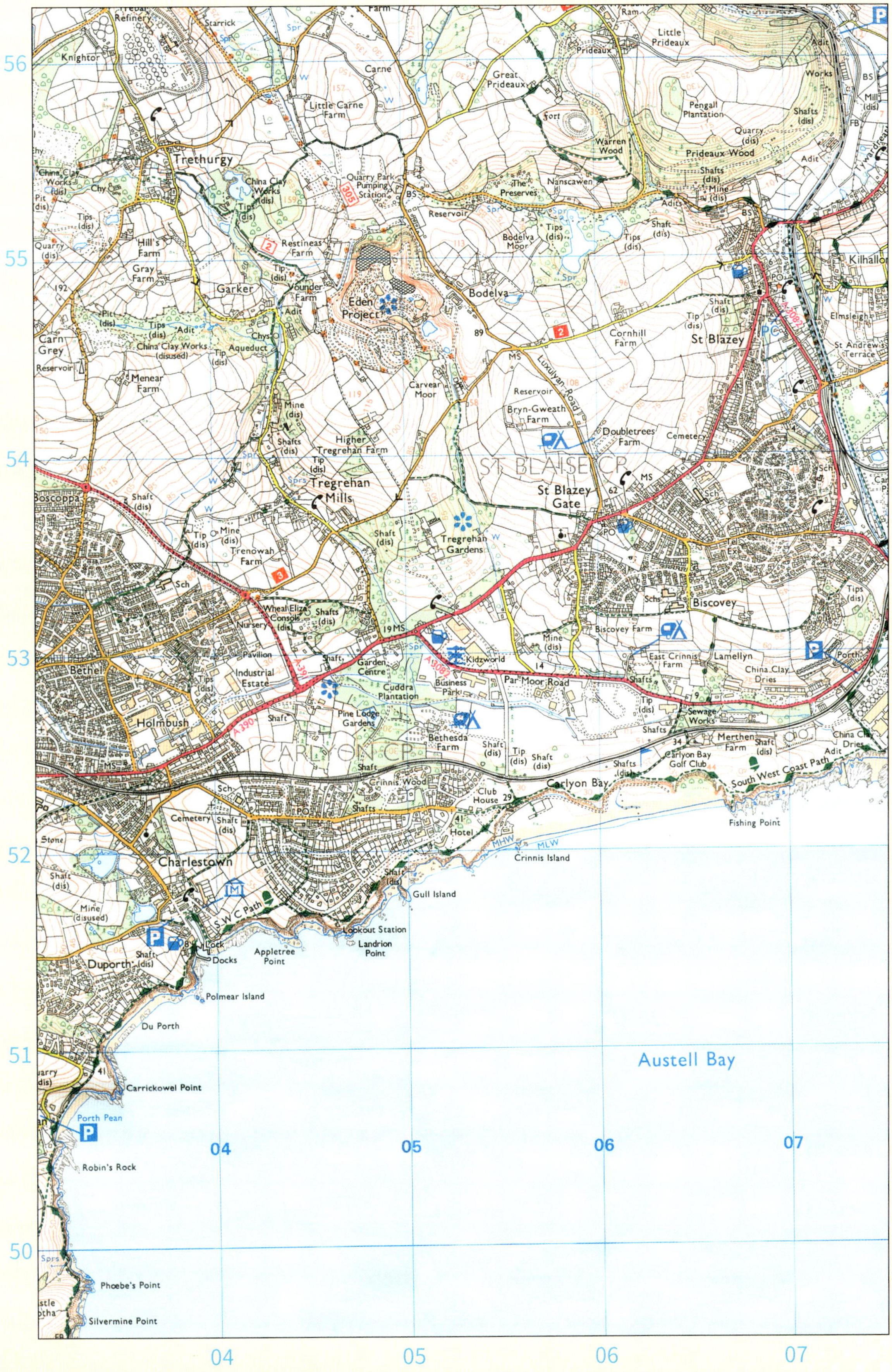

▲ ***Figure 2*** *1:25 000 Ordnance Survey (OS) map extract of part of south Cornwall*

6.1 Investigating coastal processes and management

In this section, you'll understand how to investigate coastal processes and management using fieldwork and research.

Designing an enquiry

Coasts are popular places for people as a whole and for geographers! Figure 1 shows two locations on Britain's coast. Have a look at these. Think about the kind of questions geographers might ask when they see places such as this.

For example, they might ask:

- What has happened here?
- How did it happen?
- What might happen to this place in the future, and why?

Many coastal locations are good places for fieldwork since there are lots of questions like these to investigate. This is the starting point for any **enquiry**. An enquiry is a series of stages that start with a question (see Figure 2, Stage 1) and end up with an answer or conclusion (Stage 5). You will probably have completed an enquiry in geography (or science) before and will have used **fieldwork** and practical work in the same way.

Each stage is equally important, right from the initial question, to the research and context, through to the overall evaluation. Only at the end can you have an opportunity to reflect on what you have found and what it means.

▲ ***Figure 1*** *Two locations on Britain's coast – 1a) the Isle of Wight (top), and 1b) Porthcurno in Cornwall (bottom)*

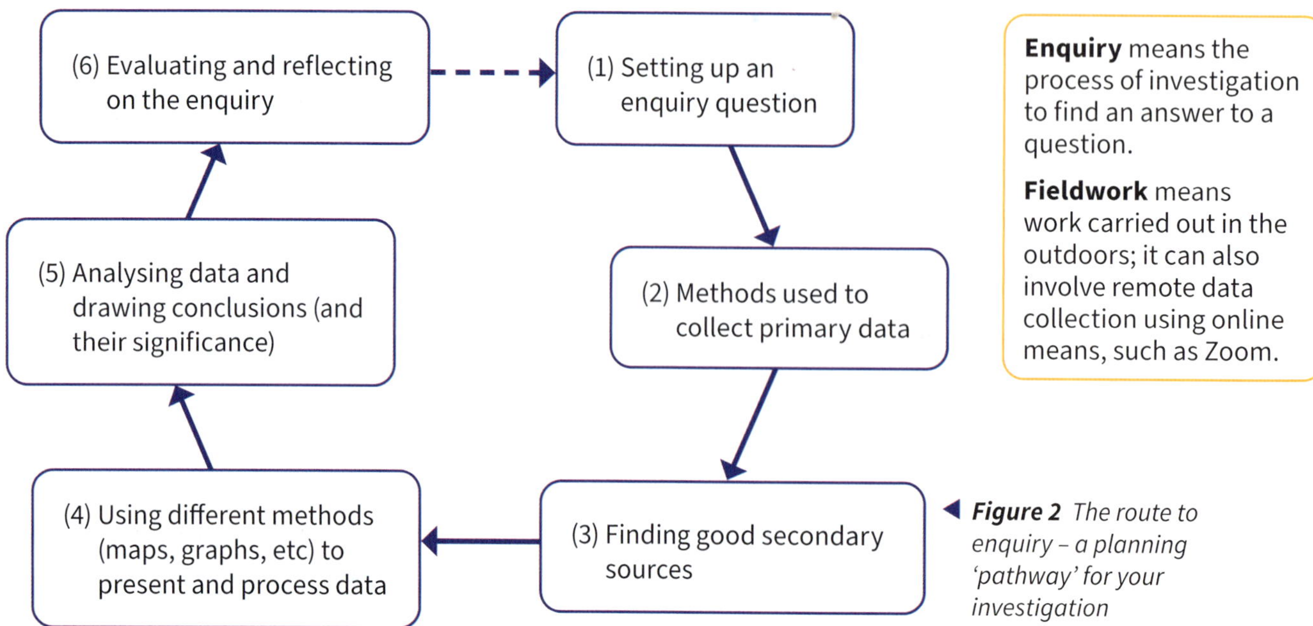

◄ ***Figure 2*** *The route to enquiry – a planning 'pathway' for your investigation*

Enquiry means the process of investigation to find an answer to a question.

Fieldwork means work carried out in the outdoors; it can also involve remote data collection using online means, such as Zoom.

Developing an enquiry question

A good enquiry depends upon having a good question. A good question must be directly linked to the overall theme – i.e. coasts! An example is:

What impacts do different methods of coastal management have on coastal processes and communities at Barton-on-Sea?

But that's very broad – so it needs breaking down into questions that are simpler and more workable. To carry out this investigation you could subdivide the main question, for example:

1. What are the problems facing the coast at Barton-on-Sea?
2. What types of coastal management techniques are used?
3. What impacts have these techniques made on coastal processes?
4. What impacts have these techniques made on communities?

To complete the enquiry you will need to use both primary and secondary data as sources.

Primary data

Fieldwork data which you collect yourself (or as part of a group) are called 'primary data' which are first-hand information. There are many different types of primary data covered in section 6.2.

Secondary data

Secondary data have been collected by someone else. They are important in giving background information and a context for your enquiry. Two examples that you will find useful are:

- A geology map (use the British Geological Survey bgs.ac.uk website)
- A Shoreline Management Plan (SMP), shown in Figure 3. These are published for stretches of coastline with plans a) about how to manage and protect the coast, and b) to help decide on which methods of protection are best used for which places.

▼ ***Figure 3*** *An extract from a local SMP. This information helps you to develop background information for the enquiry.*

Your questions

1. Copy and complete the following table, which refers to Figures 1 and 2.

Stage	
(1) An enquiry question we could investigate for either Photo 1a or Photo 1b.	
(2) Three methods we could use to collect primary data.	
(3) Two sources of secondary data we could use.	
(4) How we could present our data.	

2. Explain how the Shoreline Management Plan shown in Figure 3 could help you in your enquiry.

Exam-style question

3. For the coastal location in which you carried out fieldwork, explain **two** reasons why particular enquiry aims or questions were chosen. (4 marks)

6.2 Primary data collection in coastal fieldwork

In this section, you'll understand different techniques for collecting primary data in a coastal enquiry.

Enquiry design

The point of fieldwork is to collect your own data.
The student in Figure 1 is using measuring instruments called callipers – like a ruler – to find out the size of stones on a beach. There's a point to this – the student is investigating longshore drift and wants to find out whether stone size changes along a beach.

Data are essential! They help you to understand what is happening along a stretch of coast. You can also compare your fieldwork along a coast to what textbooks tell you. It makes good teamwork.

It's important to consider what data you need when you design your investigation, so that any data collected are as reliable and accurate as possible. In particular, you should think about:

▲ *Figure 1 Measuring pebble size along a beach, using callipers*

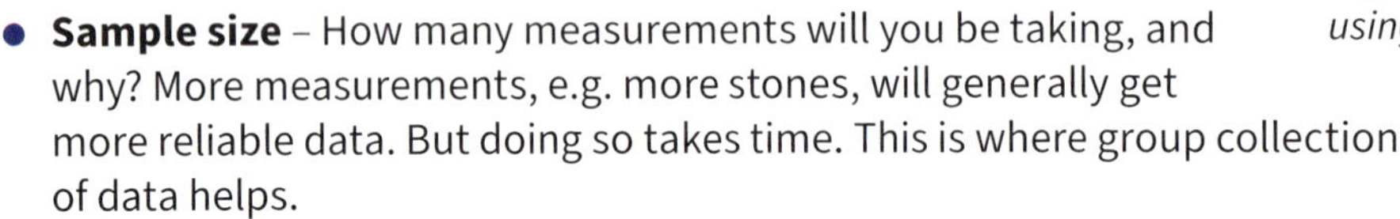

- **Sample size** – How many measurements will you be taking, and why? More measurements, e.g. more stones, will generally get more reliable data. But doing so takes time. This is where group collection of data helps.

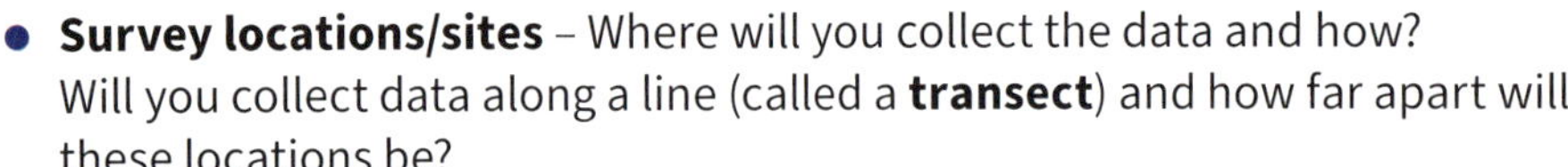

- **Survey locations/sites** – Where will you collect the data and how? Will you collect data along a line (called a **transect**) and how far apart will these locations be?
- **Accuracy** – How can you ensure that your data are accurate? Will you need to measure several samples of stones and calculate an average?

Different types of primary data: quantitative and qualitative

Any coastal investigation is likely to involve a range of data. There are two types – **quantitative** and **qualitative data**. Whatever data you collect must link directly to the enquiry question that you have set yourself (see section 6.1).

Quantitative data

Beach studies include a number of quantitative fieldwork techniques. The table in Figure 3 shows data that are commonly collected in beach investigations. All quantitative techniques need equipment, like the clinometer shown in Figure 4.

Sampling method is important to quantitative techniques. Three types are used in collecting quantitative data:

- **Random** – where samples are chosen using a random number generator, and every pebble has an equal chance of being selected.
- **Systematic** – means working to a system to collect data, for example every ten metres along a beach.
- **Stratified** – means collecting a sample made up of different parts; for example, deliberately selecting samples of different pebble sizes from a point on a beach so that you include the whole range of pebble sizes found there.

▲ *Figure 2 Collecting quantitative data for a coastal investigation*

Data required	Equipment needed	Brief description and reasons for doing this
Beach gradient (in degrees)	Clinometer, tapes and ranging poles	The gradient is measured at distances up the beach from the water mark, often at 10 or 20 m intervals. The steepness of a beach can help us understand more about the processes operating and types of waves that often reach the coast.
Pebble shape (using a scale of roundness)	Identification chart	Measures how round or angular a sample of about 10–20 stones is. This tells us how eroded pebbles are – the more rounded they are, the more they have been smoothed off by abrasion.
Pebble size in cm or mm	Ruler or calliper (Figure 1)	Measures the length of the long axis of a sample of stones. This helps us know whether longshore drift is taking place, with the furthest material along a beach being the smallest.

▲ ***Figure 3*** *Examples of quantitative data used in beach investigations*

Qualitative data

Qualitative data include a number of techniques that don't involve numbers or counting. They are subjective and involve the judgment of the person collecting. Techniques for collecting qualitative data with coastal areas include:

- written site descriptions
- taking photographs
- recording videos
- field sketches (shown in Figure 5).

They can be used to record use of fieldwork equipment as well as capture coastal landscapes and management. Always think carefully about the frame of your photo, and if necessary, use an object such as a coin for scale if you are taking close-up pictures of things such as beach pebbles.

▲ ***Figure 4*** *Using a clinometer to measure beach gradient*

◀ ***Figure 5*** *A good field sketch – but do annotate it! The annotations will help explain the geography of the landscape around you.*

Your questions

1. Explain the differences between quantitative and qualitative data.
2. In pairs, draw up a table of the advantages and disadvantages of quantitative and qualitative data.
3. Make a blank copy of Figure 3. Complete it with three **qualitative** types of data for your coastal fieldwork.
4. For each **qualitative** technique, decide whether you would use random, stratified or systematic sampling to collect your data. Explain your reasons.

Exam-style questions

5. For your chosen coastal location, explain **two** methods that you used to collect primary fieldwork data. (4 marks)
6. Explain **one** way in which you attempted to make your data collection as reliable as possible. (2 marks)

6.3 Processing and presenting coastal fieldwork data

In this section, you'll understand how to present your data using graphs, photos or maps.

Getting it all together

Students often overlook the importance of managing and organising data. It's vital to make sure that you organise both your own and any group data that you have collected. Usually, a spreadsheet which you can complete and share is the best way of doing this.

It's sometimes a good idea to keep your individual data separately coloured (e.g. within the rows of the spreadsheet) and separated from group data. You might want to note down how and why you selected your individual and group data, and how they link to the enquiry aim or focus.

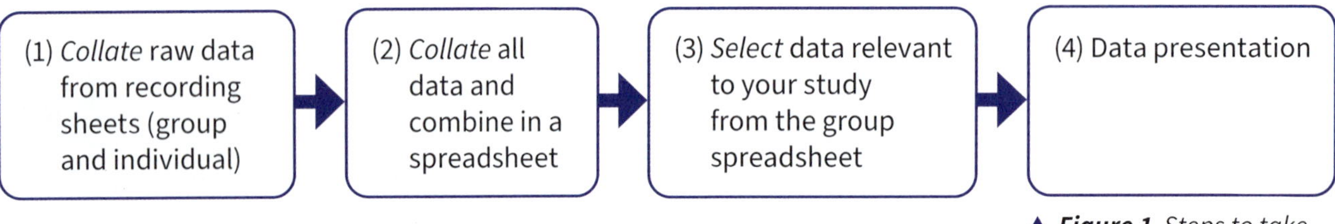

▲ ***Figure 1*** *Steps to take before data presentation*

Presenting your data

When it comes to data presentation, think more widely than bar charts, histograms and pie charts. Figure 2 shows a range of approaches to data presentation.

Maps / Cartography	GIS and photographs	Table(s) of data	Graphs and charts
• Used to show locations and patterns. • Mini-graphs and charts can be located on maps. • This makes it easier to compare patterns at locations.	• Used to show historic maps or sites which have been lost to erosion. • Useful for aerial photos of the coast to show land use. • Helps to show how places have changed after being affected by storms.	• Can be used to present raw data that you and your group collected. • Useful to highlight patterns and trends. • Can be highlighted and annotated, and can help to identify anomalies (any data which look unusual).	• There is a wide range of graphs and charts available. (Hint: make sure you choose the right chart, e.g. do you know when to use a pie chart or bar chart?) • Can show data and patterns clearly – easier to read than a table of data.

▲ ***Figure 2*** *A range of data techniques*

Think about not just **how** you might present a particular set of data, but also **why** a particular technique might be the most suitable. For example, are you dealing with **continuous data** or **categories**? Are you dealing with numbers or percentages? How can you present your data geographically?

- **Continuous data** show change along a line of study – e.g. in sediment along a beach. Beach gradients and sediment data are continuous, so are best presented using a line graph. An example is shown in Figure 3a.
- **Categories** show classifications – e.g. measuring pebble size or long axes and grouping them into sizes. A bar chart would be the best chart here. An example is shown in Figure 3b.
- Where your **sample sizes** are different (e.g. 15 pebbles at one location, 17 at another), turn raw numbers into percentages of different sizes. Then you should use a pie chart.

- Instead of just presenting graphs, locate them on a **map or aerial photo** (e.g. using Google maps or GIS, see Figure 4). This makes change easy to spot, and turns simple data into a geographical display.

You will find other techniques to use in presenting your fieldwork data. For example:

- **annotated photographs** show evidence of coastal processes, e.g. wave erosion at the cliff base
- **field sketches** highlight the way in which people and property are vulnerable to coastal erosion.

GIS is another good way of presenting information since it allows you to overlay some data (as in Figure 4) as well as begin to do some analysis of more complex data. GIS has a number of geo-processing tools that allow you to create specialised maps, as well as look for patterns and relationships.

Thinking beyond

Secondary data can really add to your own fieldwork because the data have been compiled by experts in their field.

▼ ***Figure 3a*** *A line graph showing a beach profile of slope angle up the beach, using a clinometer*

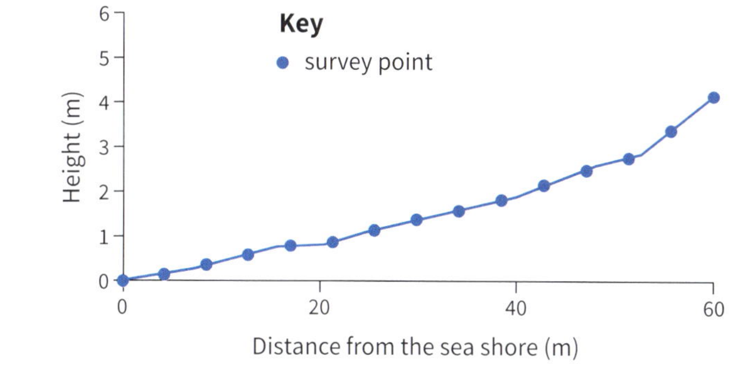

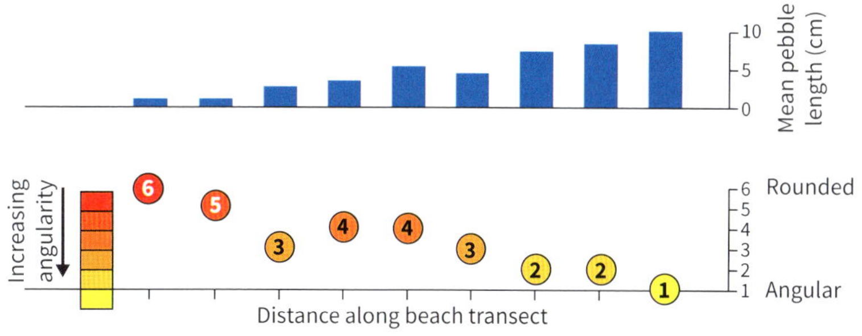

▲ ***Figure 3b*** *These beach pebble data show pebble length (top graph) and shape (bottom graph). The data are plotted on two graphs, one above the other, to aid comparison.*

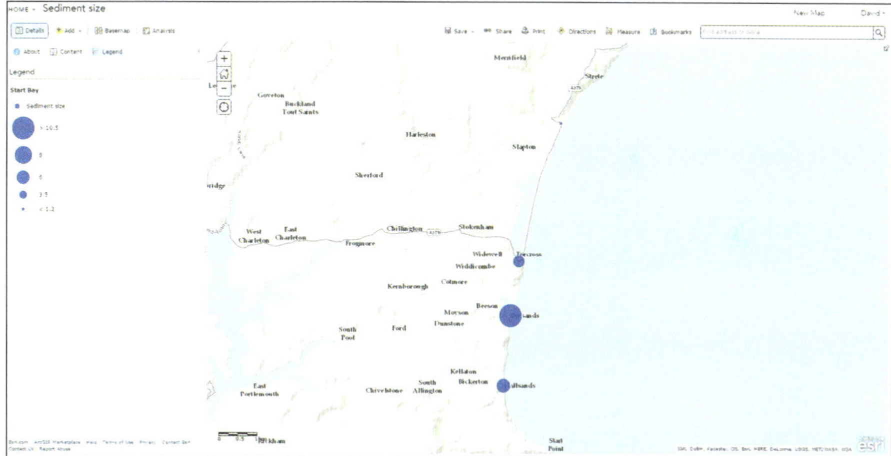

▶ ***Figure 4*** *The data are plotted on a simple base map to aid comparison between the different sites*

Your questions

1. List when you should use the following: bar charts, pie charts, line graphs and histograms.
2. In pairs, research and identify two ways in which you could use GIS to research coastal processes.
3. Study the data in Figures 3a and 3b. Explain reasons why the methods used to present the data in each graph are **a** effective, and **b** mathematically accurate.
4. Study Figures 3a and 3b.
   - **a** Identify any anomalies.
   - **b** Explain possible reasons for these.

Exam-style questions

5. Explain **one** advantage of using a line graph to show a beach gradient cross section. (2 marks)
6. Explain **one** technique that you used to present your beach sediment data. (2 marks)

6.4 Analysis and conclusions – coastal enquiry

In this section, you'll understand how to analyse and draw conclusions.

What is analysis?

Analysis should be done for both primary and secondary data. To analyse, you need to:

- identify patterns and trends in your results, and describe them
- make links between different sets of data – for example, how sediment size and roundness seem to change at the same time
- identify **anomalies** – unusual data which do not fit the general pattern of results
- explain reasons for patterns you are sure about – for example, data that might show a process operating along the coast, such as longshore drift
- suggest possible reasons for patterns you are unsure about – for example, why results suddenly change in a way that you can't explain.

Cause and effect	Emphasis	Explaining	Suggesting
as a result of...	above all...	this shows...	could be caused by...
this results in...	mainly...	because...	this looks like...
triggering this...	mostly...	similarly...	points towards...
consequently...	most significantly...	therefore...	tentatively...
the effect of this is...	usually...	as a result of...	the evidence shows...

▲ ***Figure 1*** *The language of analysis – these words and short phrases are useful to use in analysis*

Writing your analysis

When you write your analysis, you should have a clear and logical format. Start with an introductory statement, and then write about each point in more detail. Good analysis also:

- uses the correct geographical terminology
- uses the past tense
- is written in the third person
- avoids the use of 'I' or 'we'.

Figure 1 has some handy phrases you can use, depending on your results.

Analysing data

You need to be able to use both quantitative and qualitative techniques.

1 Using quantitative techniques

Quantitative techniques are about handling numerical data from different sites, like that shown in Figure 2a. These can be analysed using statistical techniques – for example, you should be able to calculate the **mean** (the average of the values in the data).

Site A – western part of the beach	Site B – centre of the beach	Site C – eastern part of the beach
Long axis (mm)		
95	24	10
68	19	12
48	16	64
49	15	32
90	29	34
82	18	55
86	6	37
56	10	18
80	19	19
49	20	19
69	13	12
42	9	8
68	15	63
57	18	62
70	19	15
59	21	9

▲ ***Figure 2a*** *Beach sediment data from three sites A, B and C*

You can also use a dispersion diagram (see Figure 2b) which will help to find the following values:

- **Median** – to find the median you need to order the data (like the dispersion diagram) and then find the middle value. This divides the data set into two.
- **Mode** – the number that appears most frequently in a data set.
- **Range** – the difference between the highest and lowest values.
- **Quartiles** – dividing a list of numbers into four equal groups – two above and two below the median. You could use quintiles (five groups) as well.

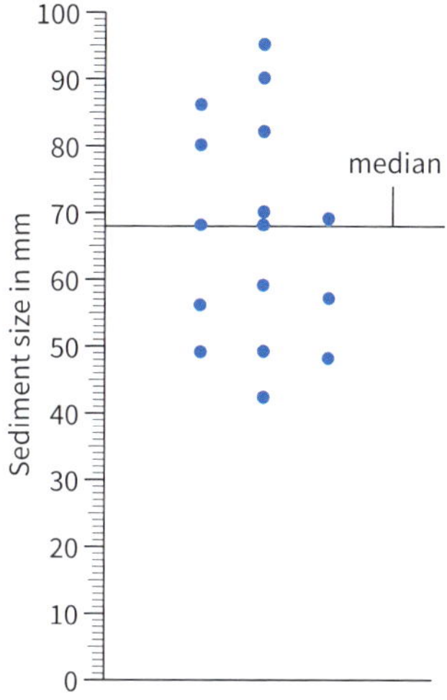

▲ ***Figure 2b*** *A dispersion diagram for Site A*

2 Using qualitative techniques

Just like sketches, photographs are far more than mere space-fillers. They are examples of qualitative techniques that provide vital clues and evidence about the fieldwork experience, as well as help with the analysis of information. Figure 3 is an example of a photograph that could be analysed by using annotations. Remember, annotations are good explanations and can also be numbered to show a sequence. Similarly, field sketches can be used to analyse processes as well as change over time.

Writing a good conclusion

A conclusion is almost the end point in your enquiry, with several important ingredients. It is shorter than the analysis, because it is focused. Make sure you:

- Refer to the main aim of your investigation – see section 6.1. What did you find out? Give an answer to the question!
- State the most important data that support your conclusion – both primary and secondary.
- Comment on any anomalies and/or unexpected results.
- Comment on the wider geographical significance of your study, e.g. why it might be important, whether your results could be useful to others, or whether you think all coastlines are like this.

▲ ***Figure 3*** *A stretch of coast with many physical features*

Your questions

1 Using Figure 2a, calculate the mean sediment size for each of the sites, A, B and C.

2 **a** Draw a dispersion diagram like the one in Figure 2b for sites B and C.

b Use the diagrams to calculate the median, mode and range for each site.

c Divide each diagram into quartiles. Calculate the mean of the upper quartile and of the lower quartile.

3 Which site has **a** the largest sediment overall, **b** the smallest, and **c** the most variation in size?

4 Is it likely that longshore drift operates on this coast? If so, which way does it go? Explain your answer.

5 Is there a part of the beach which is an anomaly? What possible suggestions could there be for this?

6 Draw an annotated field sketch of Figure 3.

Exam-style question

7 Explain **two** ways in which you analysed your beach sediment data. (4 marks)

6.5 Evaluating your coastal enquiry

In this section, you'll understand how to evaluate, reflect and think critically about the different parts of the enquiry process.

The importance of an evaluation

The evaluation is the last part of the enquiry process. Many find it very difficult. This is the part of the enquiry which aims to both evaluate and reflect on a number of things, including:

- the process of collecting data
- the overall quality of the results and conclusion.

It is much more than a list of things that simply might have gone wrong (see Figure 1).

Many approaches to fieldwork and research have limitations and errors which can affect their findings. It is very important that you accept that no study is perfect, even those carried out by university academics and professionals! It is sensible to highlight where you think the shortcomings of your work might be.

▲ **Figure 1** *Some key questions to ask in an evaluation*

What might have affected your results?

Reliability is the extent to which your investigation produced consistent results. In other words, if you were to repeat the enquiry, would you get the same results? Are your results valid – i.e. did your enquiry produce a trustworthy outcome?

Several factors can influence the reliability, validity and therefore the overall quality of your enquiry, as shown in Figure 3.

Sample size	Reliability
Sampling method	Time of day and year
Accuracy of data collection methods	Weather conditions
Accuracy of data	Equipment used

▲ **Figure 2** *Eight memory joggers to help you evaluate*

Possible sources of error	Impact on quality
Sample size	Smaller sample sizes usually means lower quality data (less reliability).
Frequency of sample (e.g. every 10 metres instead of every 100 metres)	Fewer sites reduces frequency, which then reduces quality.
Type of sampling	Sampling approaches may create 'gaps' and introduce bias in the results.
Equipment used	The wrong / inaccurate equipment can affect overall quality by producing incorrect results.
Time of survey	Different tides might influence beach accessibility and its measurable width.
Location of survey	Big variations in beach profiles and sediment characteristics can occur in locations close to each other.
Quality of secondary data	Age and reliability of secondary data affect their overall quality.

▲ **Figure 3** *Sources of error in a geographical enquiry*

Being critical

Being critical means, simply, whether any shortcomings affected the quality of the results. In a geographical enquiry, researchers generally try to have the most reliable outcomes possible. They accept the limitations of their study. They often consider these types of questions as part of their evaluation:

1. How much do I trust the overall patterns and trends in my results?
2. What is the chance that these outcomes could have been generated randomly (or by chance)?
3. Which of my conclusions are most reliable, compared to other conclusions?
4. Which part of my enquiry caused the most unreliable results?

Such questions, although complex, can often be useful in considering how to write an evaluation.

How do my results affect my understanding of coastal processes?

Another part of the process of evaluation is to link any knowledge you gained from the enquiry back to a theoretical model or idea. It may be a good idea to think about the key factors, as in Figure 4, and to then try to develop your own model.

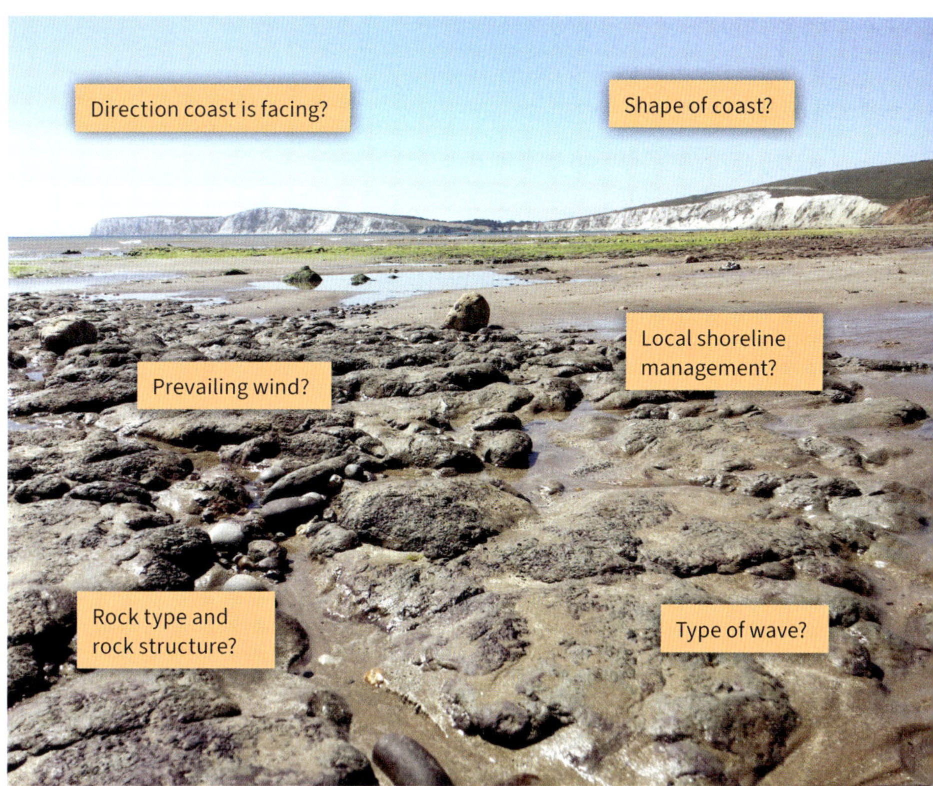

▶ ***Figure 4*** *Factors that may influence coastal processes*

Your questions

1. Explain why an evaluation is the last part of the enquiry process.
2. Explain why an evaluation of one enquiry can be helpful in setting up a second enquiry.
3. **a** In pairs, copy the factors listed in Figure 3. Based on your own fieldwork, rank them in terms of importance.
   b Explain your final rank order.
   c Now add an extra column to the table in Figure 3. Write a title 'The most significant sources of error in our enquiry', and add points about your own coastal fieldwork.
4. For either Christchurch Bay (see sections 4.6 to 4.14) or your own fieldwork, make a copy of Figure 4 and add notes to it to explain how these factors affect the coast.

Exam-style questions

5. Explain **one** factor about your own primary data which could have affected your results. (2 marks)
6. Evaluate the reliability of your coastal fieldwork conclusions. (8 marks)

6.6 Investigating river processes and management

In this section, you'll understand how to investigate river processes and management using fieldwork and research.

Designing an enquiry

Rivers are popular places for people to live near and to enjoy. Figure 1 shows two contrasting river locations. Have a look at these. Think about the kind of questions geographers might ask when they see places such as this.

For example, they might ask:

- What has happened here?
- How did it happen (short- and longer-term reasons)?
- What might happen to this place in future, and why?

Many river (or fluvial) locations are good places for fieldwork since there are lots of questions like these to investigate. This is the starting point for any **enquiry**. An enquiry is a series of stages that start with a question (see Figure 2, Stage 1) and end up with an answer or conclusion (Stage 5). You will probably have completed an enquiry in geography (or science) before and have used **fieldwork** and practical work in the same way.

Each stage is equally important, right from the initial question, to the research and context, through to the overall evaluation. Only at the end can you have an opportunity to reflect on what you have found and what it means.

▲ ***Figure 1*** *Two contrasting river landscapes – 1a (top) is the River Tone at Taunton, and 1b (bottom) is an area of the Peak District*

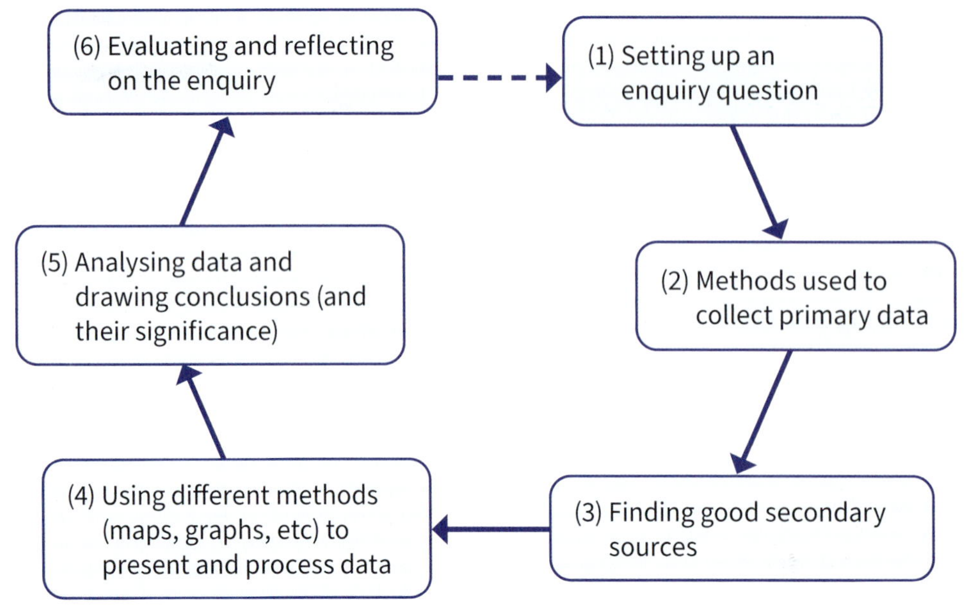

◀ ***Figure 2*** *The route to enquiry – a planning 'pathway' for your investigation*

Enquiry means the process of investigation to find an answer to a question.

Fieldwork means work carried out in the outdoors; it can also involve remote data collection using online means, such as Zoom.

Developing an enquiry question

A good enquiry depends upon having a good question. A good question must be directly linked to the overall theme – i.e. rivers! An example is:

How do different drainage basin and river channel characteristics influence flood risk for people and property along the River Exe?

But that's very broad – so it needs breaking down into simpler and more workable questions. In this investigation, you could subdivide the main question, for example:

1. What are the main characteristics of the river's channel in its upper course?
2. What places are most at risk along the River Exe?
3. What impacts does channel shape have on flood risk?
4. What impacts does valley shape and land use have on flood risk?

To complete the enquiry you will need to use both primary and secondary data as sources.

Primary data

Fieldwork data which you collect yourself (or as part of a group) are called 'primary data' which are first-hand information. Different types of primary data are covered in section 6.7.

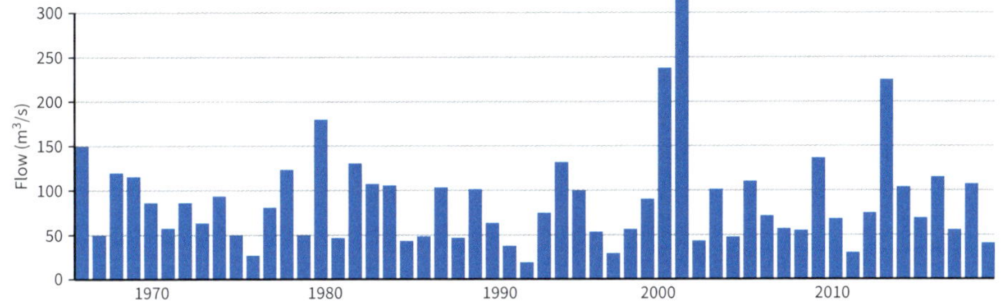

◀ ***Figure 3*** *Maximum yearly peak flows for the River Creedy (near Exeter) 1966–2019. This information helps you ask questions to get a context for the enquiry.*

Secondary data

Secondary data have been collected by someone else. They are important in giving background information and a context for your enquiry. Two examples are:

- Past river flow data, such as those in Figure 3.
- A flood risk map such as those from the Environment Agency.

Your questions

1 Copy and complete this table, which refers to Figures 1 and 2.

Stage	
(1) An enquiry question we could investigate for either Photo 1a or Photo 1b.	
(2) Three methods we could use to collect primary data.	
(3) Two sources of secondary data we could use.	
(4) How we could present our data.	

2 Explain how the data in the hydrograph shown in Figure 3 could help you in your enquiry.

Exam-style questions

3 Study the OS map in section 4.19. Explain **one** question or aim that could be used to investigate physical processes in this river landscape. (2 marks)

4 For the river along which you carried out fieldwork, explain **two** reasons why particular enquiry aims or questions were developed. (4 marks)

6.7 Primary data collection in river fieldwork

In this section, you'll understand different techniques for collecting primary data in a rivers enquiry.

Enquiry design

The point of fieldwork is to collect your own data. The student in Figure 1 is using a measuring instrument – a 'pebbleometer'– to find out the size of stones collected in a river. There's a point to this – the student is investigating river processes and wants to find out about stone size changes along the river.

Data are essential! They help you to understand what is happening in a drainage basin. You can also compare your fieldwork along a river to what textbooks tell you. It makes good teamwork.

It's very important to consider what data you need when you design your investigation, so that any data collected are as reliable and accurate as possible. In particular, you should think about:

- **Sample size** – How many measurements will you be taking, and why? More measurements, e.g. more stones, will generally get more reliable data. But doing so takes time. This is where group collection of data helps.
- **Survey locations/sites** – Where will you collect the data and how? Will you collect data along a line (called a **transect**) and how far apart will these locations be?
- **Accuracy** – How can you ensure that your data are accurate? Will you need to measure several samples of stones and calculate an average?

▲ ***Figure 1*** *Measuring pebble size from a river, using a homemade pebbleometer*

Different types of primary data: quantitative and qualitative

Any river investigation is likely to involve a range of data. There are two types – **quantitative** and **qualitative data**. Whatever data you collect must link directly to the enquiry question that you have set yourself (see section 6.6).

Quantitative data

River studies include a number of quantitative fieldwork techniques. The table in Figure 3 shows data that are commonly collected in river investigations. All quantitative techniques need equipment, like the metre stick shown in Figure 2.

Sampling method is important to quantitative techniques. Three types are used in collecting quantitative data:

- **Random** – where samples are chosen using a random number generator, and every pebble has an equal chance of being selected.
- **Systematic** – means working to a system to collect data, for example measuring depth every 10 cm across the river.

▲ ***Figure 2*** *Using a tape measure to measure out a five-metre stretch over which to test the river's speed (or velocity)*

Data required	Equipment needed	Brief description and reasons for doing this
River gradient (in degrees)	Clinometer, tapes and ranging poles	The gradient is measured at sites along the river and is measured over 10 or 20 m. The gradient of a river can, for example, help us understand more about the processes operating and the influence of geology.
River speed (velocity)	Flow meter or float	Measures how fast the river is flowing. This tells us about the amount of energy in a river and we can investigate whether it conforms to a model.
Pebble size in cm or mm	Ruler or pebbleometer/ calliper (Figure 1)	Measures the length of the long axis of a sample of stones. This helps to link processes of river erosion with position within a river catchment.
River width and depth	Tape measure, ruler	Measure width of the river in cm. Measure depth in cm by taking five readings – one at each bank, plus at a quarter, half, and three-quarters of the width.

▲ ***Figure 3*** *Examples of quantitative data used in river investigations*

- **Stratified** – means collecting a sample made up of different parts; for example, deliberately selecting samples of different pebble sizes from a point in the river so that you include the whole range of pebble sizes found there.

Qualitative data

Qualitative data include a number of techniques that don't involve numbers or counting. They are subjective and involve the judgment of the person collecting. Techniques for collecting qualitative data with rivers areas include:

1 written site descriptions

2 taking photographs

3 recording videos

4 field sketches (shown in Figure 4).

▲ ***Figure 4*** *A good field sketch – but do annotate it! The annotations will help explain the geography of the landscape around you.*

They can be used to record use of fieldwork equipment as well as capture river landscapes and management. Always think carefully about the frame of your photo, and if necessary, use an object such as a coin for scale if you are taking close-up pictures of things such as river sediment.

Your questions

1 Explain the differences between quantitative and qualitative data.

2 In pairs, draw up a table of the advantages and disadvantages of quantitative and qualitative data.

3 Make a blank copy of Figure 3. Complete it with three **qualitative** types of data for your river fieldwork

4 For each **qualitative** technique, decide whether you would use random, stratified or systematic sampling to collect your data. Explain your reasons.

Exam-style questions

5 For your chosen river location, explain **two** ways that you collected quantitative fieldwork data. (4 marks)

6 Explain **one** way in which you attempted to make your data collection as reliable as possible. (2 marks)

6.8 Processing and presenting river fieldwork data

In this section, you'll understand how to present your data using graphs, photos or maps.

Getting it all together

Students often overlook the importance of managing and organising data. It's vital to make sure that you organise both your own and any group data that you have collected. Usually, a spreadsheet which you can complete and share is the best way of doing this.

It's sometimes a good idea to keep your individual data separately coloured (e.g. within the rows of the spreadsheet) and separated from group data. You might want to note down how and why you selected your individual and group data, and how they link to the enquiry aim or focus.

▼ ***Figure 1*** *Steps to take before data presentation*

(1) *Collect* raw data from recording sheets (group and individual) → (2) *Collate* all data and combine in a spreadsheet → (3) *Select* data relevant to your study from the group spreadsheet → (4) Data presentation

Presenting your data

When it comes to data presentation, think more widely than bar charts, histograms and pie charts. Figure 2 shows a range of approaches to data presentation that might be considered.

▼ ***Figure 2*** *A range of data techniques*

Maps / Cartography	GIS and photographs	Table(s) of data	Graphs and charts
• Used to show locations and patterns. • Mini-graphs and charts can be located on maps. • This makes it easier to compare patterns at locations.	• Used to show historic maps or sites which have been lost to erosion. • Useful for aerial photos of rivers to show land use. • Helps to show how places have changed after being affected by storms.	• Can be used to present raw data that you and your group collected. • Useful to highlight patterns and trends. • Can be highlighted and annotated, and can help to identify anomalies (any data which look unusual).	• There is a wide range of graphs and charts available. (Hint: make sure you choose the right chart, e.g. do you know when to use a pie chart or bar chart?) • Can show data and patterns clearly – easier to read than a table of data.

Think about not just **how** you might present a particular set of data, but also **why** a particular technique might be the most suitable. For example, are you dealing with **continuous data** or **categories**? Are you dealing with numbers or percentages? How can you present your data geographically?

- **Continuous data** show change along a line of study – e.g. in sediment along a stream course. River gradients, for example, are continuous, so are best presented using a line graph. An example is shown in Figure 3a, which has been created with GIS (ArcGIS Online).
- **Categories** show classifications – e.g. measuring pebble size or long axes and grouping them into sizes. A bar chart would be the best chart here. An example is shown in Figure 3b.
- Where your **sample sizes** are different (e.g. 15 pebbles at one location, 17 at another), turn raw numbers into percentages of different sizes. Then you should use a pie chart.
- Instead of just presenting graphs, locate them on a **map or aerial photo** (e.g. using Google Maps or GIS, See Figure 4). This makes change easy to spot, and turns simple data into a geographical display.

You will find other techniques to use in presenting your fieldwork data. For example:

- **annotated photographs** show evidence of river processes, e.g. erosion on the outer edge of a meander
- **field sketches** highlight the way in which people and property are vulnerable to river flooding.

GIS is another good way of presenting information since it allows you to overlay some data (as in Figure 4) as well as begin to do some analysis of more complex data. GIS has a number of geo-processing tools that allow you to create specialised maps, as well as look for patterns and relationships.

▼ ***Figure 3a*** *A line graph showing a river valley long profile (created with GIS) which helps us to understand the drainage basin*

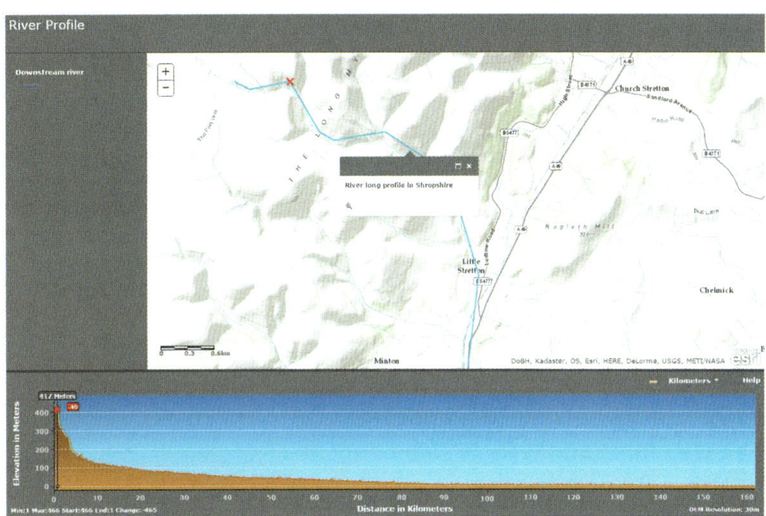

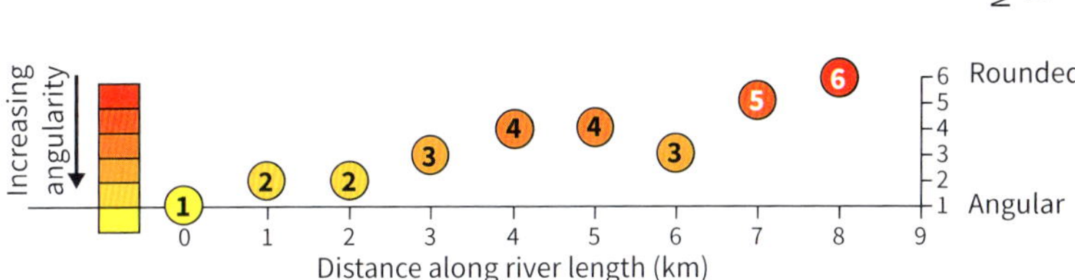

▶ ***Figure 3b*** *These river pebble data show pebble length (top graph) and shape (bottom graph). They are category data (where there are gaps between plot points). The data are plotted on two graphs, one above the other, to aid comparison.*

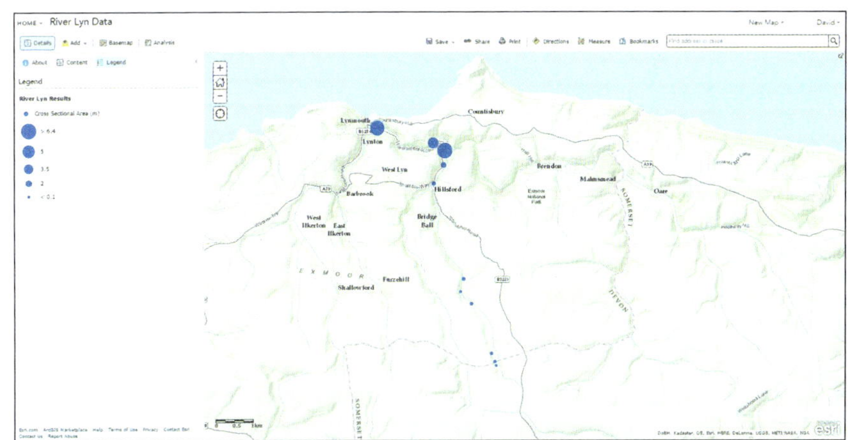

▶ ***Figure 4*** *The data are plotted on a simple base map to aid comparison between the different sites*

Thinking beyond

Secondary data can really add to your own fieldwork because the data have been compiled by experts in their field.

Your questions

1. List when you should use the following: bar charts, pie charts, line graphs and histograms.
2. In pairs, research and identify two ways in which you could use GIS to research river processes.
3. Study the data in Figures 3a and 3b. Explain reasons why the methods used to present the data in each graph are **a** effective, and **b** mathematically accurate.
4. Study Figures 3a and 3b.
   - **a** Identify any anomalies.
   - **b** Explain possible reasons for these.

Exam-style questions

5. Explain **one** advantage of using a line graph to show the long profile of a river. (2 marks)
6. Explain **one** technique that you used to present your river sediment data. (2 marks)

6.9 Analysis and conclusions – river enquiry

In this section, you'll understand how to analyse and draw conclusions.

What is analysis?

Analysis should be done for both primary and secondary data. To analyse, you need to:

- identify patterns and trends in your results, and describe them
- make links between different sets of data – for example, how sediment size and roundness seem to change at the same time
- identify **anomalies** – unusual data which do not fit the general pattern of results
- explain reasons for patterns you are sure about – for example, data that might show a process operating along a river, such as deposition
- suggest possible reasons for patterns you are unsure about – for example, why results suddenly change in a way that you can't explain.

Cause and effect	Emphasis	Explaining	Suggesting
as a result of...	above all...	this shows...	could be caused by...
this results in...	mainly...	because...	this looks like...
triggering this...	mostly...	similarly...	points towards...
consequently...	most significantly...	therefore...	tentatively...
the effect of this is...	usually...	as a result of...	the evidence shows...

▲ ***Figure 1*** *The language of analysis – these words and short phrases are useful to use in analysis*

Writing your analysis

When you write your analysis, you should have a clear and logical format. Start with an introductory statement, and then write about each point in more detail. Good analysis also:

- uses the correct geographical terminology
- uses the past tense
- is written in the third person
- avoids the use of 'I' or 'we'.

Figure 1 has some handy phrases you can use, depending on your results.

Analysing data

You need to be able to use both quantitative and qualitative techniques.

1 Using quantitative techniques

Quantitative techniques are about handling numerical data from different sites, like that shown in Figure 2a. These can be analysed using statistical techniques – for example, you should be able to calculate the **mean** (the average of the values in the data).

Site A – upstream section	Site B – middle section	Site C – downstream section
Long axis (mm)		
95	24	10
68	19	12
48	16	64
49	15	32
90	29	34
82	18	55
86	6	37
56	10	18
80	19	19
49	20	19
69	13	12
42	9	8
68	15	63
57	18	62
70	19	15
59	21	9

▲ ***Figure 2a*** *River sediment data from three sites A, B and C*

You can also use a dispersion diagram (see Figure 2b) which will help to find the following values:

- **Median** – to find the median you need to order the data (like the dispersion diagram) and then find the middle value. This divides the data set into two halves.
- **Mode** – the number that appears most frequently in a data set.
- **Range** – the difference between the highest and lowest values.
- **Quartiles** – dividing a list of numbers into four equal groups – two above and two below the median. You could use quintiles (five groups) as well.

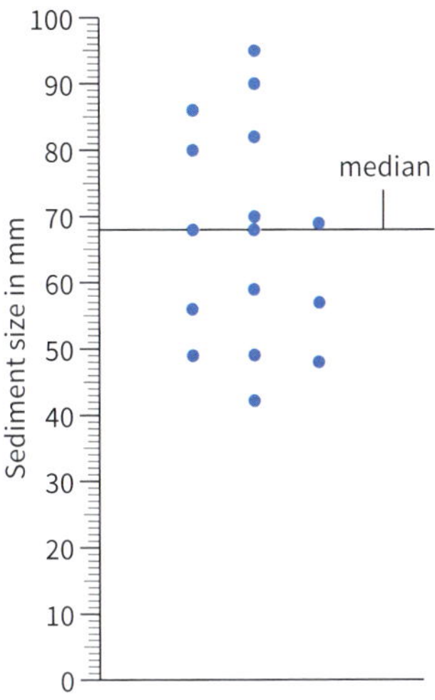

▲ ***Figure 2b*** *A dispersion diagram for Site A*

2 Using qualitative techniques

Just like sketches, photographs are far more than mere space-fillers. They are examples of qualitative techniques that provide vital clues and evidence about the fieldwork experience, as well as help with the analysis of information. Figure 3 is an example of a photograph that could be analysed by using annotations. Remember, annotations are good explanations and can also be numbered to show a sequence. Similarly, field sketches can be used to analyse processes as well as change over time.

Writing a good conclusion

A conclusion is almost the end point in your enquiry, with several important ingredients. It is shorter than the analysis, because it is focused. Make sure you:

- Refer to the main aim of your investigation – see section 6.6. What did you find out? Give an answer to the question!
- State the most important data that supports your conclusion – both primary and secondary.
- Comment on any anomalies and/or unexpected results.
- Comment on the wider geographical significance of your study, e.g. why it might be important, whether your results could be useful to others, or whether you think all rivers are like this.

▲ ***Figure 3*** *A section of river with many physical features*

Your questions

1 Using Figure 2a, calculate the mean sediment size for each of the sites, A, B and C.

2 **a** Draw a dispersion diagram like the one in Figure 2b for sites B and C.

b Use the diagrams to calculate the median, mode and range for each site.

c Divide each diagram into quartiles. Calculate the mean of the upper quartile and of the lower quartile.

3 Which site has **a** the largest sediment overall, **b** the smallest, and **c** the most variation in size?

4 Using the data in Figure 2b, try and explain the overall changes in sediment sizes from upstream to downstream.

5 Are there any anomalies in the river data across the three sites? What possible suggestions could there be for this?

6 Draw an annotated field sketch of Figure 3.

Exam-style question

7 Explain **two** ways in which you analysed your river sediment data. (4 marks)

6.10 Evaluating your river enquiry

In this section, you'll understand how to evaluate, reflect and think critically about the different parts of the enquiry process.

The importance of an evaluation

The evaluation is the last part of the enquiry process. Many find it very difficult. This is the part of the enquiry which aims to both evaluate and reflect on a number of things, including:

- the process of collecting data
- the overall quality of the results and conclusion.

It is much more than a list of things that simply might have gone wrong (see Figure 1).

Many approaches to fieldwork and research have limitations and errors which can affect their findings. It is very important that you accept that no study is perfect, even those carried out by university academics and professionals! It is sensible to highlight where you think the shortcomings of your work might be.

▲ ***Figure 1*** *Some key questions to ask in an evaluation*

Sample size	Reliability
Sampling method	Time of day and year
Accuracy of data collection methods	Weather conditions
Accuracy of data	Equipment used

▲ ***Figure 2*** *Eight memory joggers to help you evaluate*

What might have affected your results?

Reliability is the extent to which your investigation produced consistent results. In other words, if you were to repeat the enquiry, would you get the same results? Are your results valid – i.e. did your enquiry produce a trustworthy outcome?

Several factors can influence the reliability, validity and therefore the overall quality of your enquiry, as shown in Figure 3.

Possible sour ces of error	Impacts on quality
Sample size	Smaller sample sizes usually means lower quality data (less reliability).
Frequency of sample (e.g. every 10 metres instead of every 100 metres)	Fewer sites reduces frequency, which then reduces quality.
Type of sampling	Sampling approaches may create 'gaps' and introduce bias in the results.
Equipment used	The wrong / inaccurate equipment can affect overall quality by producing incorrect results.
Time of survey	Different times of the year will significantly influence the amount of water in the river and may not be representative.
Location of survey	Big variations in river channel depth and width, as well as sediment characteristics, can occur in locations close to each other.
Quality of secondary data	Age and reliability of secondary data affect their overall quality.

▲ ***Figure 3*** *Sources of error in a geographical enquiry*

Being critical

Being critical means, simply, whether any shortcomings affected the quality of the results. In a geographical enquiry, researchers generally try to have the most reliable outcomes possible. They accept the limitations of their study. They often consider these types of questions as part of their evaluation:

1. How much do I trust the overall patterns and trends in my results?
2. What is the chance that these outcomes could have been generated randomly (or by chance)?
3. Which of my conclusions are most reliable, compared to other conclusions?
4. Which part of my enquiry caused the most unreliable results?

Such questions, although complex, can often be useful in considering how to write an evaluation.

How do my results affect my understanding of river processes?

Another part of the process of evaluation is to link any knowledge you gained from the enquiry back to a theoretical model or idea. It may be a good idea to think about the key factors, as in Figure 4, and to then try to develop your own model.

Figure 4 Factors that may influence river processes

Your questions

1. Explain why an evaluation is the last part of the enquiry process.
2. Explain why an evaluation of one enquiry can be helpful in setting up a second enquiry.
3. a In pairs, copy the factors listed in Figure 3. Based on your own fieldwork, rank them in terms of importance.
   b Explain your final rank order.
   c Now add an extra column to the table in Figure 3. Write a title 'The most significant sources of error in our enquiry', and add points about your own fieldwork.
4. For either the River Wharfe (see sections 4.15 to 4.19) or your own rivers fieldwork, make a copy of Figure 4 and add notes to it to explain how these factors affect the river.

Exam-style questions

5. Explain **one** factor about your own primary data which could have affected your results. (2 marks)
6. Evaluate the reliability of your river fieldwork conclusions. (8 marks)

6.11 Investigating variations in urban quality of life

In this section, you'll understand how to investigate differences in urban quality of life using fieldwork and research.

Designing an enquiry

Urban areas are popular for fieldwork as they are familiar and often close to us. Figure 1 shows two very different urban locations. Have a look at these. Think about the kind of questions geographers might ask when they see places such as this.

For example, they might ask:

- Which place offers a higher quality of life?
- Why are they so different?
- What might happen to each place in future, and why?

Many urban locations are good places for fieldwork since there are lots of questions like these to investigate. This is the starting point for any **enquiry**. An enquiry is a series of stages that start with a question (see Figure 2, Stage 1) and end up with an answer or conclusion (Stage 5). You will probably have completed an enquiry in geography (or science) before and have used **fieldwork** and practical work in the same way.

Each stage is equally important, right from the initial question, to the research and context, through to the overall evaluation. Only at the end can you have an opportunity to reflect on what you have found and what it means.

▲ ***Figure 1*** *Two contrasting urban locations – 1a (top) is from a street in central Sheffield, and 1b (bottom) is Portobello Road in west London*

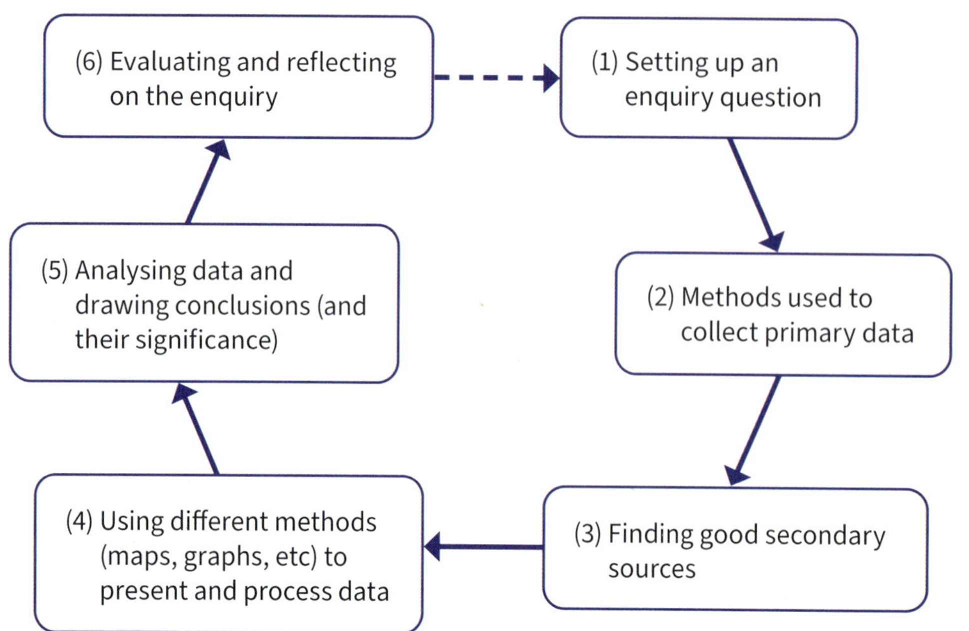

Enquiry means the process of investigation to find an answer to a question.

Fieldwork means work carried out in the outdoors; it can also involve remote data collection using online means, such as Zoom.

◀ ***Figure 2*** *The route to enquiry – a planning 'pathway' for your investigation*

Developing an enquiry question

A good enquiry depends upon having a good question. A good question must be directly linked to the overall theme – i.e. urban environments! An example is:

How and why are there variations in quality of life for different census output areas within Leeds?

But that's very broad – so it needs breaking down into questions that are simpler and more workable. To carry out this investigation, you could subdivide the main question, for example:

1. What are the challenges facing some areas of Leeds?
2. What is the primary evidence of differences in quality of life between areas?
3. How far does the secondary evidence support the primary fieldwork findings?
4. What impacts have these differences had on communities?

To complete the enquiry you will need to use both primary and secondary data as sources.

Primary data

Fieldwork data which you collect yourself (or as part of a group) are called 'primary data' which are first-hand information. There are many different types of primary data covered in section 6.12.

Secondary data

Secondary data have been collected by someone else. They are important in giving background information and a context for your enquiry. Two examples that you will find useful are:

- Census data from the Office for National Statistics (ONS)
- Data from the Index of Multiple Deprivation (IMD) like the graph in Figure 3.

▼ ***Figure 3*** *Different weightings given to the elements of the IMD. This information helps you get a context for the enquiry.*

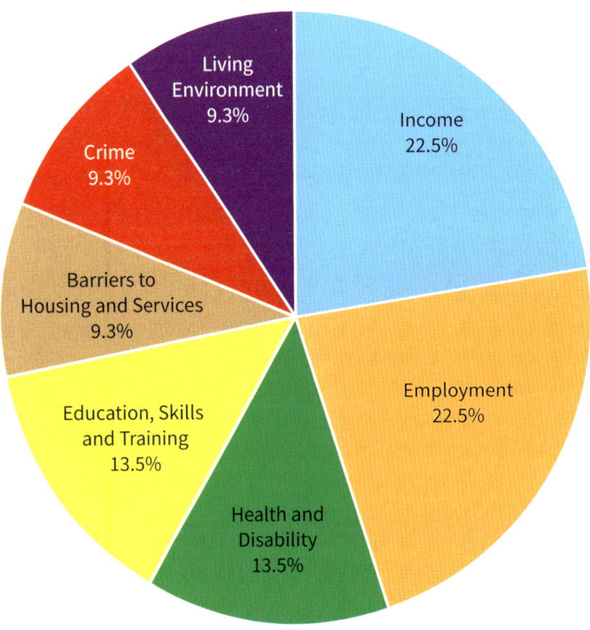

Thinking beyond

Secondary data can really add to your own fieldwork because the data have been compiled by experts in their field.

Your questions

1. Copy and complete this table, which refers to Figures 1 and 2.

Stage	
(1) An enquiry question we could investigate for either Photo 1a or Photo 1b.	
(2) Three methods we could use to collect primary data.	
(3) Two sources of secondary data we could use.	
(4) How we could present our data.	

2. Explain how IMD data (Figure 3) could help you in your enquiry.

Exam-style questions

3. Study the OS map in section 5.14. Explain **one** question or aim that could be used to investigate variations in quality of life in this area. (2 marks)
4. For your chosen urban area, explain **two** reasons why particular enquiry aims or questions were developed. (4 marks)

6.12 Primary data collection for urban fieldwork

In this section, you'll understand different techniques for collecting primary data in an urban enquiry.

Enquiry design

The point of fieldwork is to collect your own data. The student in Figure 1 is using a camera to record the buildings and architecture in a town. There's a point to this – she is investigating urban deprivation and wants to use photographic evidence to document the change between different parts of the town.

Data are essential! They help you to understand what is happening in an urban area. You can also compare your fieldwork in a town or city to what textbooks tell you. It makes good teamwork.

It's very important to consider what data you need when you design your investigation, so that any data collected are as reliable and accurate as possible. In particular you should think about:

- **Sample size** – How many measurements will you be taking, and why? More measurements, e.g. more questionnaires, will generally get more reliable data. But doing so takes time. This is where group collection of data helps.
- **Survey locations/sites** – Where will you collect the data and how? Will you collect data along a line (called a **transect**) and how far apart will these locations be?
- **Accuracy** – How can you ensure that your data are accurate? Will you need to conduct a number of questionnaires and calculate an average, median or mode?

▲ ***Figure 1*** *Using a camera to record the built environment*

Different types of primary data: quantitative and qualitative

There are two types of primary data - **quantitative** and **qualitative**. Both will be used in an urban enquiry. Whatever data you collect must link directly to the enquiry question that you have set yourself (see section 6.11).

Quantitative data

Urban studies include a number of quantitative fieldwork techniques. The table in Figure 2 shows data that are commonly collected in urban investigations. Most quantitative techniques need equipment or recording sheets, like the one shown in Figure 3.

Data required	Equipment needed	Brief description and reasons for doing this
Land use map	Large-scale base map and land use map key	Categories of land use are recorded either along a line (e.g. sides of a road) or in areas to produce a spatial picture of urban land use.
Shopping / environmental quality survey	Pre-prepared environmental quality survey (see Figure 3)	Measures different characteristics of a place based on numerical judgements, with a simple scoring system that can be tallied at the end.
'Local or visitor' coded questionnaire	Questionnaire with 'closed' questions and specific answers	Specific questions are used to gauge the perceptions of people (respondents), either visitors or locals, about how they 'feel' about area.

▲ ***Figure 2*** *Examples of quantitative data used in urban investigations*

Sampling method is important to quantitative techniques. Three types are used in collecting quantitative data:

- **Random** – where samples are chosen using a random number generator and every person in a questionnaire survey has an equal chance of being selected.
- **Systematic** – means working to a system to collect data, for example, every 20 metres or paces along a road to record land use.
- **Stratified** – means collecting a sample made up of different parts; for example, deliberately selecting samples of different people within the town/city so you include the whole range of people found there.

	Qualities being assessed	High +2	Good +1	Average 0	Fairly poor −1	Very poor −2	
Building design and quality	1. Well designed / pleasing to the eye						Poorly designed / ugly
	2. In good condition – e.g. paintwork, woodwork						In poor condition
	3. Houses well maintained or improved						Poorly maintained / no improvement
	4. Outside, gardens are kept tidy / in good condition						Outside gardens, or land / open space in poor condition
	5. No vandalism, or any graffiti has been cleaned up						Extensive vandalism or graffiti in large amounts
Traffic noise and parking	6. Roads have no traffic congestion						Streets badly congested with traffic
	7. Parking is easy; garages or spaces provided						Parking is difficult; no parking provided / on the street
	8. No road traffic, rail or aircraft noise						High noise volume from road, rail, and air traffic
	9. No smell from traffic or other pollution						Obvious smell from traffic or other pollution

▲ ***Figure 3*** *An example of an environmental quality survey that could be used in an urban area*

Qualitative techniques

Qualitative data include a number of techniques that don't involve numbers or counting. They are subjective and involve the judgment of the person collecting. Techniques for collecting qualitative data in urban areas include:

1. written site descriptions
2. taking photographs and videos
3. field sketches
4. interviews (shown in Figure 4).

They can be used to record use of fieldwork equipment as well as capture different aspects of the urban environment. Always think carefully about the frame of your photo. In urban areas, 360° panoramas work well to illustrate contrasts between areas.

▶ ***Figure 4*** *Extended interviews in an urban area to find out people's attitudes and opinions*

Your questions

1. Explain the differences between quantitative and qualitative data.
2. In pairs, draw up a table of the advantages and disadvantages of quantitative and qualitative data.
3. Make a blank copy of Figure 2. Complete it with three **qualitative** types of data for your urban fieldwork.
4. For each **qualitative** technique, decide whether you would use random, stratified or systematic sampling to collect your data. Explain your reasons.

Exam-style questions

5. For your chosen urban location, explain **two** ways that you collected quantitative fieldwork data. (4 marks)
6. Explain **one** way in which you attempted to make your data collection as reliable as possible. (2 marks)

6.13 Processing and presenting urban fieldwork data

In this section, you'll understand how to present your data using graphs, photos or maps.

Getting it all together

Students often overlook the importance of managing and organising data. It's vital to make sure that you organise both your own and any group data that you have collected. Usually, a spreadsheet which you can complete and share is the best way of doing this.

It's sometimes a good idea to keep your individual data separately coloured (e.g. within the rows of the spreadsheet) and separated from group data. You need to be able to explain:

- how and why you selected your individual and group data
- how the data will help you answer your enquiry question.

▼ **Figure 1** *Steps to take before data presentation*

(1) *Collect* raw data from recording sheets (group and individual) → (2) *Collate* all data and combine in a spreadsheet → (3) *Select* data relevant to your study from the group spreadsheet → (4) Data presentation

Presenting your data

When it comes to data presentation, think more than bar charts, histograms and pie charts. Figure 2 shows a range of approaches that you might consider.

▼ **Figure 2** *A range of data techniques*

Maps / Cartography	GIS and photographs	Table(s) of data	Graphs and charts
• Used to show locations and patterns. • Mini-graphs and charts can be located on maps. • This makes it easier to compare patterns at locations.	• Used to show historic maps to show change in an urban area. • Useful for aerial photos of the town / city to show land use. • Helps to show deprivation and / or 'health' of a place.	• Can be used to present raw data that you and your group collected. • Useful to highlight patterns and trends. • Can be highlighted and annotated, and can help to identify anomalies (any data which look unusual).	• There is a wide range of graphs and charts available. (Hint: make sure you choose the right chart, e.g. do you know when to use a pie chart or bar chart?) • Can show data and patterns clearly – easier to read than a table of data.

Think about not just **how** you might present a particular set of data, but **why** a particular technique might be the most suitable. E.g., are you dealing with **continuous data** or **categories**, with numbers or percentages? How can you present your data geographically?

- **Continuous data** show change along a line of study – e.g. in land use along a road. Pedestrian flows might be continuous, for example, so are best presented using a line graph.
- **Categories** show classifications – e.g. putting environmental quality scores into classified groups. Proportional circles have been used in Figure 3.

▼ **Figure 3** *Locations in Ipswich on a GIS map where EQA (Environmental Quality Assessment) data have been displayed as proportional circles, colour coded with a 'colour-ramp' (values are in the centre).*

- Where your **sample sizes** are different (e.g. 15 quality scores at one location, 17 at another), turn raw numbers into percentages of different sizes. Then you should use a pie chart.
- Instead of just presenting graphs, locate them on a **map or aerial photo** (e.g. using Google Maps or GIS, see Figure 3 and Figure 5). This makes differences easy to spot, and turns simple data into a geographical display.

You will find other techniques to use in presenting your fieldwork data. For example:

- **annotated photographs** show evidence of dereliction and decay, possibly indicating a lower quality of life
- **field sketches** highlight the way in which people and property are influenced by areas of changing environmental quality.

We have already seen that GIS is another good way of presenting information since it allows comparisons (as in Figures 3 and 5). It also allows more sophisticated presentation tools to be used, such as digital choropleth maps. GIS also has a number of geo-processing tools that allow you to create specialised maps, as well as look for patterns and relationships.

Respondent: 65-year-old retired person, discussing changes in their local town.

"I have lived in Tiverton, Devon all my life and there has been a lot of change. For a start, many of the smaller local shops have gone; the big supermarket near the river was to blame. I'm less mobile than I was and there are also fewer bus services so I need to rely on my car (but parking is free for a short period in the town). But the town has been improved I think. It's better for people as they have stopped cars driving through the middle like they used to, plus I like the coffee shops where I can relax and they have outside seating. I don't like the fact that there are fewer banks and book shops but that's probably just an age thing. I don't really do internet shopping!"

▲ ***Figure 4*** *A coding technique is useful for a variety of text-based resources, whether primary or secondary information. Simple highlighting is used to show positive (yellow) and negative (blue) comments. The example here helps to analyse results from a questionnaire about attitudes to quality of life.*

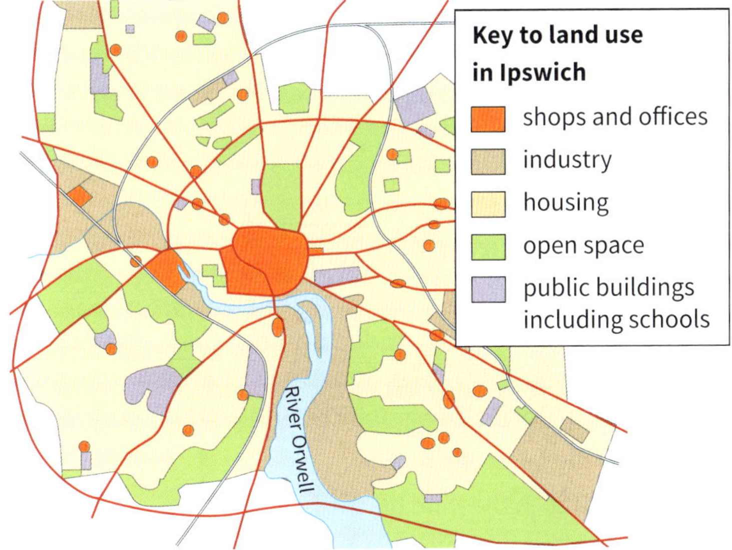

▶ ***Figure 5*** *Land use maps are a useful technique for fieldwork in urban areas – they can help explain quality of life or differences in environmental quality*

Your questions

1. List when you should use the following: bar charts, pie charts, line graphs and histograms.
2. In pairs, research and identify two ways in which you could use GIS to research urban quality of life.
3. Choose two ways of presenting your urban fieldwork data. In pairs, draw a table of the advantages and disadvantages of each of these methods.
4. Study collated data from your class fieldwork. How far have you been able to spot anomalies?

Exam-style questions

5. Explain **one** advantage of using a line graph to show changes in environmental quality along a transect in an urban area. (2 marks)
6. Explain **one** technique that you used to present your secondary IMD data. (2 marks)

6.14 Analysis and conclusions – urban enquiry

In this section, you'll understand how to analyse and draw conclusions.

What is analysis?

Analysis should be done for both primary and secondary data. To analyse, you need to:

- identify patterns and trends in your results, and describe them
- make links between different sets of data – for example, how sediment size and roundness seem to change at the same time
- identify **anomalies** – unusual data which do not fit the general pattern of results
- explain reasons for patterns you are sure about – for example, data that might show a process operating in a town or city, such as the spatial change in land use
- suggest possible reasons for patterns you are unsure about – for example, why results suddenly change in a way that you can't explain.

Cause and effect	Emphasis	Explaining	Suggesting
as a result of…	above all…	this shows…	could be caused by…
this results in…	mainly…	because…	this looks like…
triggering this…	mostly…	similarly…	points towards…
consequently…	most significantly…	therefore…	tentatively…
the effect of this is…	usually…	as a result of…	the evidence shows…

▲ ***Figure 1*** *The language of analysis – these words and short phrases are useful to use in analysis*

Writing your analysis

When you write your analysis, you should have a clear and logical format. Start with an introductory statement, and then write about each point in more detail. Good analysis also:

- uses the correct geographical terminology
- uses the past tense
- is written in the third person
- avoids the use of 'I' or 'we'.

Figure 1 has some handy phrases you can use, depending on your results.

Analysing data

You need to be able to use both quantitative and qualitative techniques.

1 Using quantitative techniques

Quantitative techniques are about handling numerical data from different sites, like that shown in Figure 2a. These can be analysed using statistical techniques – for example, you should be able to calculate the **mean** (the average of the values in the data).

Site A	Site B	Site C
95	24	10
68	19	12
48	35	14
49	45	32
90	29	34
82	18	21
86	48	67
56	55	12
80	45	19
49	35	19
69	13	12
42	28	8
68	15	18
57	8	22
70	19	15
59	21	9

▲ ***Figure 2a*** *Totalled environmental quality scores for three different urban locations: A, B and C. A variety of different indicators have been measured in each location; higher scores indicate a better environmental quality.*

You can also use a dispersion diagram (Figure 2b) to help to find the following:

- **Median** – to find the median you need to order the data (like the dispersion diagram) and then find the middle value. This divides the data set into two halves.
- **Mode** – the number that appears most frequently in a data set.
- **Range** – the difference between the highest and lowest values.
- **Quartiles** – dividing a list of numbers into four equal groups – two above and two below the median. You could use quintiles (five groups) as well.

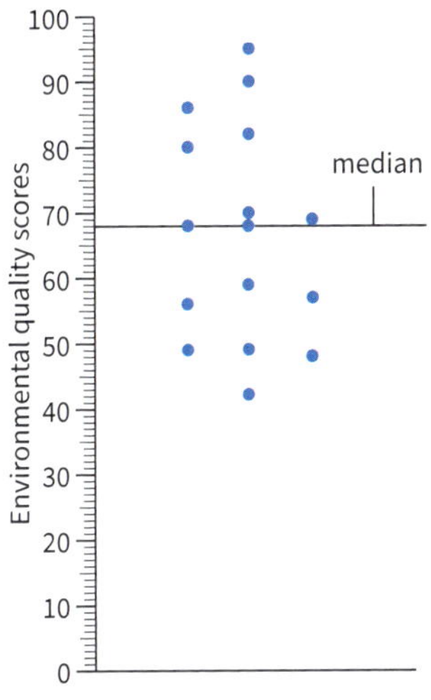

▲ ***Figure 2b*** *A dispersion diagram for Site A*

2 Using qualitative techniques

Just like sketches, photographs are more than mere space-fillers. They are qualitative techniques that provide vital evidence about the fieldwork experience, as well as helping with the analysis of information. Figure 3 is a photograph that could be analysed using annotations. Remember, annotations are good explanations and can also be numbered to show a sequence. Similarly, field sketches can be used to analyse processes as well as change over time.

Writing a good conclusion

The conclusion is almost the end point in your enquiry, with several important ingredients. It is shorter than the analysis. Make sure you:

- Refer to the main aim of your investigation (section 6.11). What did you find out? Answer the question!
- State the primary and secondary data that support your conclusion.
- Comment on any anomalies.
- Comment on the wider geographical significance of your study, e.g. whether your results could be useful to others, or whether you think all urban places are like this.

▲ ***Figure 3*** *An urban environment showing how vacant shops and betting shops are common to many high streets*

Your questions

1 Draw an annotated field sketch of Figure 3.

2 Using Figure 2a, calculate the mean environmental quality survey (EQS) scores for each of the locations, A, B and C.

3 **a** Draw a dispersion diagram like the one in Figure 2b for sites B and C.
b Use the diagrams to calculate the mean, median, mode and range for each site.
c Divide each diagram into quartiles. Calculate the mean of the upper quartile and of the lower quartile.

4 Which site has **a** the highest overall EQS score, **b** the smallest, and **c** the most variation?

5 Using the data in Figure 2a, suggest possible reasons for the variations in EQS between locations.

6 Are there any anomalies in the EQS data between the three locations? What possible suggestions could there be for this?

Exam-style question

7 Explain **two** ways in which you analysed your urban fieldwork data. (4 marks)

6.15 Evaluating your urban enquiry

In this section, you'll understand how to evaluate, reflect and think critically about the different parts of the enquiry process.

The importance of an evaluation

The evaluation is the last part of the enquiry process. Many find it very difficult. This is the part of the enquiry which aims to both evaluate and reflect on a number of things, including:

- the process of collecting data
- the overall quality of the results and conclusion.

It is much more than a list of things that simply might have gone wrong (see Figure 1).

Many approaches to fieldwork and research have limitations and errors which can affect their findings. It is very important that you accept that no study is perfect, even those carried out by university academics and professionals! It is sensible to highlight where you think the shortcomings of your work might be.

▲ ***Figure 1*** *Some key questions to ask in an evaluation*

What might have affected your results?

Reliability is the extent to which your investigation produced consistent results. In other words, if you were to repeat the enquiry, would you get the same results? Are your results valid – i.e. did your enquiry produce a trustworthy outcome?

Several factors can influence the reliability, validity and therefore the overall quality of your enquiry, as shown in Figure 3.

Sample size	Reliability
Sampling method	Time of day and year
Accuracy of data collection methods	Weather conditions
Accuracy of data	Equipment used

▲ ***Figure 2*** *Eight memory joggers to help you evaluate*

Possible sources of error	Impact on quality
Sample size	Smaller sample sizes usually means lower quality data (less reliability).
Frequency of sample (e.g. every 10 metres instead of every 100 metres)	Fewer sites reduces frequency, which then reduces quality.
Type of sampling	Sampling approaches may create 'gaps' and introduce bias in the results.
Equipment used	The wrong / inaccurate equipment can affect overall quality by producing incorrect results.
Time of survey	Different days or times of day might influence perceptions and pedestrian flow, for example.
Location of survey	Big variations in environmental quality can occur between places very close to each other.
Quality of secondary data	Age and reliability of secondary data affect their overall quality.

▲ ***Figure 3*** *Sources of error in a geographical enquiry*

Being critical

Being critical means, simply, whether any shortcomings affected the quality of the results. In a geographical enquiry, researchers generally try to have the most reliable outcomes possible. They accept the limitations of their study. They often consider these types of questions as part of their evaluation:

1. How much do I trust the overall patterns and trends in my results?
2. What is the chance that these outcomes could have been generated randomly (or by chance)?
3. Which of my conclusions are most reliable, compared to other conclusions?
4. Which part of my enquiry caused the most unreliable results?

Such questions, although complex, can often be useful in considering how to write an evaluation.

How do my results affect my understanding of urban deprivation?

Another part of the process of evaluation is to link any knowledge you gained from the enquiry back to a theoretical model or idea. It may be a good idea to think about the key factors, as in Figure 4, and to then try to develop your own model.

▶ ***Figure 4*** *Factors that may influence urban deprivation and quality of life for people living in this area of a major UK city*

Your questions

1. Explain why an evaluation is the last part of the enquiry process.
2. Explain why an evaluation of one enquiry can be helpful in setting up a second enquiry.
3. **a** In pairs, copy the factors listed in Figure 3. Based on your own fieldwork, rank them in terms of importance.
   b Explain your final rank order.
   c Now add an extra column to the table in Figure 3. Write a title 'The most significant sources of error in our enquiry', and add points about your own urban fieldwork.
4. For either London (see sections 5.7 to 5.15) or your own fieldwork, make a copy of Figure 4 and add notes to it to explain how these factors influence quality of life.

Exam-style questions

5. Explain **one** factor about your own primary data which could have affected your results. (2 marks)
6. Evaluate the reliability of your urban fieldwork conclusions. (8 marks)

6.16 Investigating variations in rural deprivation

In this section, you'll understand how to investigate differences in rural deprivation using fieldwork and research.

Designing an enquiry

Rural areas have a number of interesting problems and challenges that affect people who live there. Figure 1 shows two different rural locations. Have a look at these. Think about the kind of questions geographers might ask when they see places such as this.

For example, they might ask:

- What might the challenges be for people living in these areas?
- Why are these challenges different?
- What might happen to these places in future, and why?

Many rural locations are good places for fieldwork since there are lots of questions like these to investigate. This is the starting point for any **enquiry**. An enquiry is a series of stages that start with a question (see Figure 2, Stage 1) and end up with an answer or conclusion (Stage 5). You will probably have completed an enquiry in geography (or science) before and have used **fieldwork** and practical work in the same way.

Each stage is equally important, right from the initial question, to the research and context, through to the overall evaluation. Only at the end can you have an opportunity to reflect on what you have found and what it means.

▲ ***Figure 1*** *Two contrasting rural locations – 1a (top) shows one aspect of rural areas in the 21st century - migrant labour working the land, and 1b (bottom) is a view of Keswick, in the Lake District*

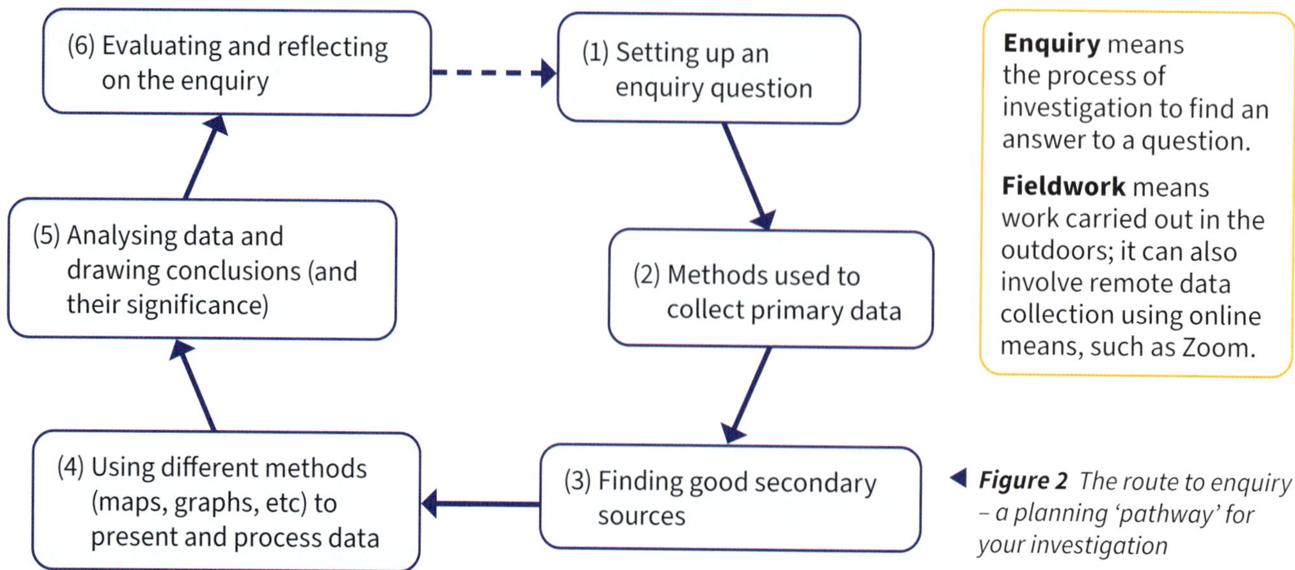

◀ ***Figure 2*** *The route to enquiry – a planning 'pathway' for your investigation*

Enquiry means the process of investigation to find an answer to a question.

Fieldwork means work carried out in the outdoors; it can also involve remote data collection using online means, such as Zoom.

Developing an enquiry question

A good enquiry depends upon having a good question. A good question must be directly linked to the overall theme – i.e. rural environments! An example is:

How and why are there variations in quality of life for different villages in North Norfolk?

But that's quite broad – so it needs breaking down into questions that are simpler and more workable. To carry out this investigation, you could subdivide the main question, for example:

1 What problems does rural deprivation bring to villages in North Norfolk?
2 What is the primary evidence of differences in quality of life between villages?
3 What impacts have these differences had on communities?
4 How far does the secondary evidence support the primary fieldwork findings?

To complete the enquiry you will need to use both primary and secondary data as sources.

Primary data

Fieldwork data which you collect yourself (or as part of a group) are called 'primary data' which are first-hand information. There are many different types of primary data covered in section 6.17.

Secondary data

Secondary data have been collected by someone else. They are important in giving background information and a context for your enquiry. Two examples that you will find useful are:

- Census data from the Office for National Statistics (ONS)
- Data from the Index of Multiple Deprivation (IMD) like the graph in Figure 3.

Figure 3 *Different weightings given to the elements of the IMD. This information helps you get a context for the enquiry.*

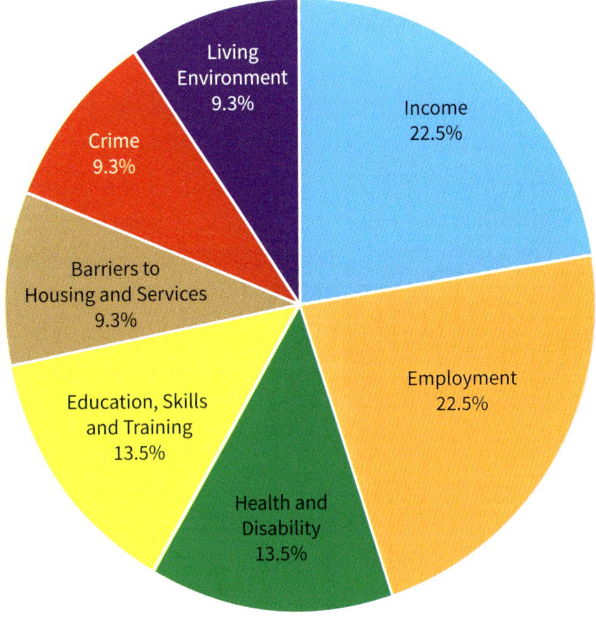

Thinking beyond

Secondary data can really add to your own fieldwork because the data have been compiled by experts in their field.

Your questions

1 Copy and complete this table, which refers to Figures 1 and 2.

Stage	
(1) An enquiry question we could investigate for either Photo 1a or Photo 1b.	
(2) Three methods we could use to collect primary data.	
(3) Two sources of secondary data we could use.	
(4) How we could present our data.	

2 Explain how IMD data (Figure 3) could help you in your enquiry.

Exam-style questions

3 Study the OS map in section 5.20. Explain one question or aim that could be used to investigate variations in deprivation in this area. (2 marks)

4 For your chosen rural area, explain **two** reasons why particular enquiry aims or questions were developed. (4 marks)

6.17 Primary data collection for rural fieldwork

In this section, you'll understand different techniques for collecting primary data in a rural enquiry.

Enquiry design

The point of fieldwork is to collect your own data. Figure 1 shows how speeding traffic can be an issue for rural areas. There's a point to this – students could document rural problems using photographic evidence. Challenges might be between different villages.

Data are essential! They help you to understand what is happening in a rural area. You can also compare your fieldwork in a village to what textbooks tell you. It makes good teamwork.

It's important to consider what data you need when you design your investigation, so that any data collected are as reliable and accurate as possible. In particular, you should think about:

- **Sample size** – How many measurements will you be taking, and why? More measurements, e.g. more questionnaires or environmental quality scores, will generally get more reliable data. But doing so takes time. This is where group collection of data helps.
- **Survey locations/sites** – Where will you collect the data and how? Will you collect data along a line (called a **transect**) and how far apart will these locations be?
- **Accuracy** – How can you ensure that your data are accurate? Will you need to conduct a number of questionnaires and calculate an average, median or mode?

▲ ***Figure 1*** *Fast traffic can be a hazard and lower the quality of life for people living in a village*

Different types of primary data: quantitative and qualitative

There are two types of primary data - **quantitative** and **qualitative**. Both will be used in a rural enquiry. Whatever data you collect must link directly to the enquiry question that you have set yourself (see section 6.16).

Quantitative data

Rural and village studies include a number of quantitative fieldwork techniques. The table in Figure 2 shows data that are commonly collected in rural investigations. Most quantitative techniques need equipment or recording sheets, like the one shown in Figure 3.

Sampling method is important to quantitative techniques. Three types are used in collecting quantitative data:

- **Random** – where samples are chosen using a random number generator and every person in a questionnaire survey has an equal chance of being selected.
- **Systematic** – means working to a system to collect data, for example, every three houses along a road to record building age.
- **Stratified** – means collecting a sample made up of different parts; for example, deliberately selecting samples of different people within the village/town so you include the whole range of people found there.

Data required	Equipment needed	Brief description and reasons for doing this
Land use map	Large-scale base map and land use map key	Categories of land use are recorded either along a line (e.g. sides of a road) or in areas to produce a spatial picture of rural land use.
Environmental quality survey	Pre-prepared environmental quality survey (see Figure 3)	Measures different characteristics of a place based on numerical judgements, with a simple scoring system that can be tallied at the end.
'Local or visitor' coded questionnaire	Questionnaire with 'closed' questions and specific answers	Specific questions are used to gauge the perceptions of people (respondents), either visitors or locals, about how they 'feel' about area.

▲ ***Figure 2*** *Examples of quantitative data used in rural investigations*

	Qualities being assessed	High +2	Good +1	Average 0	Fairly poor -1	Very poor -2	
Building design and quality	1. Well designed / pleasing to the eye						Poorly designed / ugly
	2. In good condition – e.g. paintwork, woodwork						In poor condition
	3. Houses well maintained or improved						Poorly maintained / no improvement
	4. Outside, gardens are kept tidy / in good condition						Outside gardens, or land / open space in poor condition
	5. No vandalism, or any graffiti has been cleaned up						Extensive vandalism or graffiti in large amounts
Traffic noise and parking	6. Roads have no traffic congestion						Roads or lanes badly congested with traffic
	7. Parking is easy; garages or spaces provided						Parking is difficult; no parking provided / on the street
	8. No road traffic, rail or aircraft noise						High noise volume from road, rail, and air traffic
	9. No smell from traffic or other pollution						Obvious smell from traffic or other pollution

▲ ***Figure 3*** *An example of an environmental quality survey that could be used in a rural area*

Qualitative techniques

Qualitative data include a number of techniques that don't involve numbers or counting. They are subjective and involve the judgment of the person collecting. Techniques for collecting qualitative data in rural areas include:

1 written site descriptions

2 taking photographs and videos

3 field sketches

4 interviews (shown in Figure 4).

They can be used to record use of fieldwork equipment as well as capture different aspects of the environment. Always think carefully about the frame of your photo. In villages and towns, 360° panoramas work well to illustrate contrasts between areas.

▶ ***Figure 4*** *Extended interviews in a rural area to find out people's attitudes and opinions*

Your questions

1 Explain the differences between quantitative and qualitative data.

2 In pairs, draw up a table of the advantages and disadvantages of quantitative and qualitative data.

3 Make a blank copy of Figure 2. Complete it with three **qualitative** types of data for your rural fieldwork

4 For each **qualitative** technique, decide whether you would use random, stratified or systematic sampling to collect your data. Explain your reasons.

Exam-style questions

5 For your chosen rural location, explain **two** ways that you collected quantitative fieldwork data. (4 marks)

6 Explain **one** way in which you attempted to make your data collection as reliable as possible. (2 marks)

6.18 Processing and presenting rural fieldwork data

In this section, you'll understand how to present your data using graphs, photos or maps.

Getting it all together

Students often overlook the importance of managing and organising data. It's vital to make sure that you organise both your own and any group data that you have collected. Usually, a spreadsheet which you can complete and share is the best way of doing this.

It's sometimes a good idea to keep your individual data separately coloured (e.g. within the rows of the spreadsheet) and separated from group data. You need to be able to explain:

- how and why you selected your individual and group data
- how the data will help you answer your enquiry question.

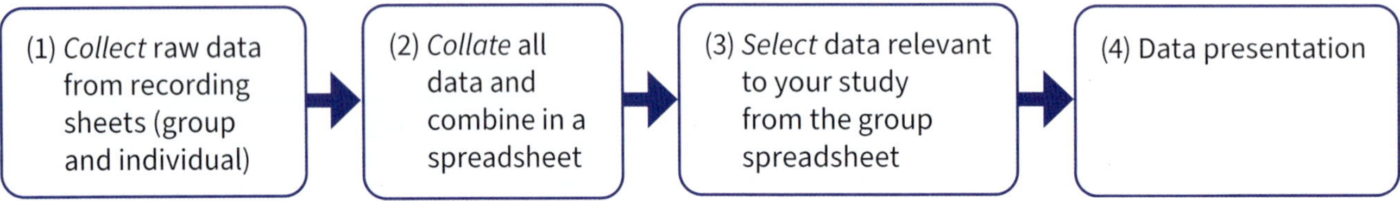

▲ ***Figure 1*** *Steps to take before data presentation*

Presenting your data

When it comes to data presentation, think more widely than bar charts, histograms and pie charts. Figure 2 shows a range of approaches to data presentation that might be considered.

▼ ***Figure 2*** *A range of data techniques*

Maps / Cartography	GIS and photographs	Table(s) of data	Graphs and charts
• Used to show locations and patterns. • Mini-graphs and charts can be located on maps. • This makes it easier to compare patterns at locations.	• Used to show historic maps to show change in a rural village. • Useful for aerial photos of the village(s) to show land use. • Helps to show deprivation and/or 'health' of a place.	• Can be used to present raw data that you and your group collected. • Useful to highlight patterns and trends. • Can be highlighted and annotated, and can help to identify anomalies (any data which look unusual).	• There is a wide range of graphs and charts available. (Hint: make sure you choose the right chart, e.g. do you know when to use a pie chart or bar chart?) • Can show data and patterns clearly – easier to read than a table of data.

Think about not just **how** you might present a particular set of data, but also **why** a particular technique might be the most suitable. For example, are you dealing with **continuous data** or **categories**? Are you dealing with numbers or percentages? How can you present your data geographically?

- **Continuous data** show change along a line of study – e.g. in land use along a road. Pedestrian tourist flows might be continuous, for example, so are best presented using a line graph.

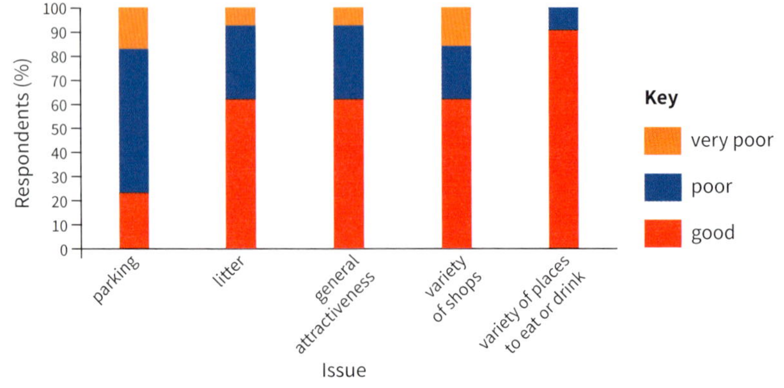

▲ ***Figure 3*** *A compound bar chart showing the results of a closed-question survey about different issues that can affect village communities*

- **Categories** show classifications – e.g. responses from a questionnaire survey.
- Where your **sample sizes** are different (e.g. 15 quality scores at one location, 17 at another), turn raw numbers into percentages of different sizes. Then you could use a compound bar chart, as shown in Figure 3.
- Instead of just presenting graphs, locate them on a **map or aerial photo** (e.g. using Google maps or GIS – see Figure 5). This makes change easy to spot, and turns simple data into a geographical display.

You will find other techniques useful in presenting data. For example:

- **annotated photographs** show evidence of poor quality buildings, possibly indicating a lower quality of life
- **field sketches** highlight the way in which people and property are influenced by areas of changing environmental quality.

Using GIS is a useful way of presenting information, since it allows comparisons (as discussed in Figure 2). It also allows more sophisticated presentation tools to be used, e.g. digital choropleth maps. GIS also has a number of tools that allow you to create specialised maps, as well as look for patterns and relationships (Figure 5).

Interview response from a 45-year-old who has moved to a rural area

"We recently moved to the parish of Morpeth from a large city in the north-east. Wow, what a change! You need a car almost every day so we had to purchase another small vehicle to get around. There is not much work locally so I still commute 45 minutes each way. The shops are great around here with lots of healthy local produce, but it comes at a price. The supermarket is nearly always cheaper for some types of food. And there is the price of local houses. Very expensive, especially if you are on a local wage around here. But overall it's very beautiful, the community is great, and we wouldn't move back to the city.

▲ ***Figure 4*** *This technique helps to analyse results from a questionnaire about attitudes to quality of life. Simple highlighting is used to show positive (yellow) and negative (blue) comments.*

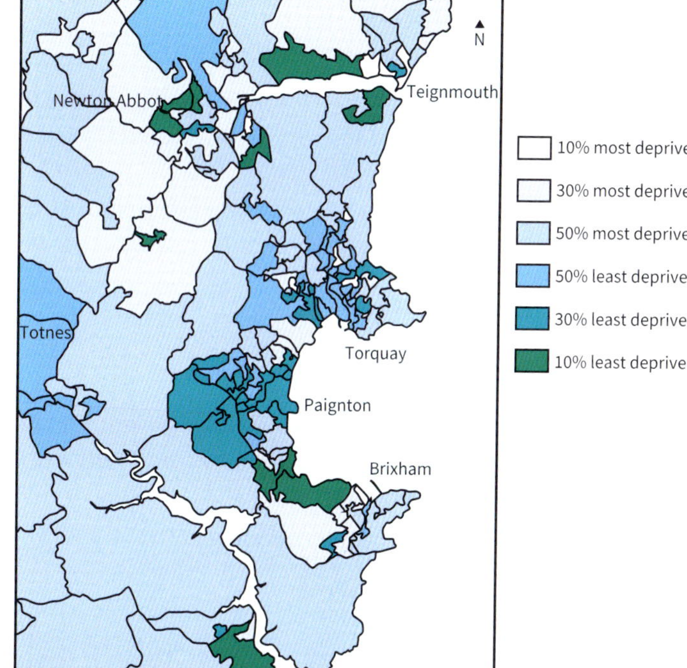

▶ ***Figure 5*** *This GIS map shows secondary IMD data for the area around Torbay in south Devon, colour coded according to IMD score.*

Your questions

1. List when you should use the following: bar charts, pie charts, line graphs and histograms.
2. In pairs, research and identify two ways in which you could use GIS to research rural quality of life.
3. Choose two ways of presenting your rural fieldwork data. In pairs, draw a table of the advantages and disadvantages of each of these methods.
4. Study the collated data from your class fieldwork. How far have you been able to spot anomalies?

Exam-style questions

5. Explain **one** advantage of using a line graph to show changes in environmental quality along a transect in a rural area. (2 marks)
6. Explain **one** technique that you used to present your secondary IMD data. (2 marks)

6.19 Analysis and conclusions – rural enquiry

In this section, you'll understand how to analyse and draw conclusions.

What is analysis?

Analysis should be done for both primary and secondary data. To analyse, you need to:

- identify patterns and trends in your results, and describe them
- make links between different sets of data – for example, how quality of life and environmental quality seem to change at the same time
- identify **anomalies** – unusual data which do not fit the general pattern of results
- explain reasons for patterns you are sure about – for example, data that might show a process operating in a village such as its growth and expansion
- suggest possible reasons for patterns you are unsure about – for example, why results suddenly change in a way that you can't explain.

Cause and effect	Emphasis	Explaining	Suggesting
as a result of...	above all...	this shows...	could be caused by...
this results in...	mainly...	because...	this looks like...
triggering this...	mostly...	similarly...	points towards...
consequently...	most significantly...	therefore...	tentatively...
the effect of this is...	usually...	as a result of...	the evidence shows...

▲ **Figure 1** *The language of analysis – these words and short phrases are useful to use in analysis*

Writing your analysis

When you write your analysis, you should have a clear and logical format. Start with an introductory statement, and then write about each point in more detail. Good analysis also:

- uses the correct geographical terminology
- uses the past tense
- is written in the third person
- avoids the use of 'I' or 'we'.

Figure 1 has some handy phrases you can use, depending on your results.

Analysing data

You need to be able to use both quantitative and qualitative techniques.

1 Using quantitative techniques

Quantitative techniques are about handling numerical data from different sites, like that shown in Figure 2a. These can be analysed using statistical techniques – for example, you should be able to calculate the **mean** (the average of the values in the data).

Village A	Village B	Village C
95	24	10
68	19	12
48	35	14
49	45	32
90	29	34
82	18	21
86	48	67
56	55	12
80	45	19
49	35	19
69	13	12
42	28	8
68	15	18
57	8	22
70	19	15
59	21	9

▲ **Figure 2a** *Totalled environmental quality scores for three different villages: A, B and C. A variety of different indicators have been measured in each place; higher scores indicate a better environmental quality.*

You can also use a dispersion diagram (Figure 2b) to help to find the following:

- **Median** – to find the median you need to order the data (like the dispersion diagram) and then find the middle value. This divides the data set into two halves.
- **Mode** – the number that appears most frequently in a data set.
- **Range** – the difference between the highest and lowest values.
- **Quartiles** – dividing a list of numbers into four equal groups – two above and two below the median. You could use quintiles (five groups) as well.

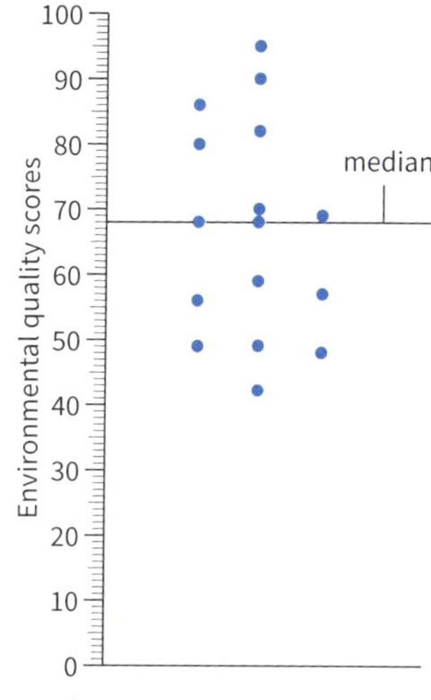

▲ ***Figure 2b*** *A dispersion diagram for Village A*

2 Using qualitative techniques

Just like sketches, photographs are far more than mere space-fillers. They are qualitative techniques that provide vital clues and evidence about the fieldwork experience, as well as helping with the analysis. Figure 3 is an example of a photograph that could be analysed by using annotations. Remember, annotations are good explanations and can also be numbered to show a sequence. Similarly, field sketches can be used to analyse processes as well as change over time.

Writing a good conclusion

A conclusion is almost the end point in your enquiry, with several important ingredients. It is shorter than the analysis. Make sure you:

- Refer to the main aim of your investigation (see section 6.16). What did you find out? Answer the question!
- State the primary and secondary data that support your conclusion.
- Comment on any anomalies.
- Comment on the wider geographical significance of your study, e.g. whether your results could be useful to others, or whether you think all rural places are like this.

▲ ***Figure 3*** *Rural areas often have high environmental quality but also several problems linked to inaccessibility*

Your questions

1 Draw an annotated field sketch of Figure 3.

2 Using Figure 2a, calculate the mean environmental quality survey (EQS) scores for each of the villages, A, B and C.

3 **a** Draw a dispersion diagram like the one in Figure 2b for villages B and C.
 b Use the diagrams to calculate the mean, median, mode and range for each site.
 c Divide each diagram into quartiles. Calculate the mean of the upper quartile and of the lower quartile.

4 Which place has **a** the highest overall EQS score, **b** the smallest, and **c** the most variation?

5 Using the data in Figure 2a, suggest possible reasons for the variations in EQS scores between different places.

6 Are there any anomalies in EQS scores between the three villages? What possible reasons could there be for this?

Exam-style question

7 Explain **two** ways in which you analysed your rural fieldwork data. (4 marks)

6.20 Evaluating your rural enquiry

In this section, you'll understand how to evaluate, reflect and think critically about the different parts of the enquiry process.

The importance of an evaluation

The evaluation is the last part of the enquiry process. Many find it very difficult. This is the part of the enquiry which aims to both evaluate and reflect on a number of things, including:

- the process of collecting data
- the overall quality of the results and conclusion.

It is much more than a list of things that simply might have gone wrong (see Figure 1).

Many approaches to fieldwork and research have limitations and errors which can affect their findings. It is important that you accept that no study is perfect, even those carried out by university academics and professionals! It is sensible to highlight where you think the shortcomings of your work might be.

▲ ***Figure 1*** *Some key questions to ask in an evaluation*

What might have affected your results?

Reliability is the extent to which your investigation produced consistent results. In other words, if you were to repeat the enquiry, would you get the same results? Are your results valid – i.e. did your enquiry produce a trustworthy outcome?

Several factors can influence the reliability, validity and therefore the overall quality of your enquiry, as shown in Figure 3.

Sample size	Reliability
Sampling method	Time of day and year
Accuracy of data collection methods	Weather conditions
Accuracy of data	Equipment used

▲ ***Figure 2*** *Eight memory joggers to help you evaluate*

Possible sources of error	Impact on quality
Sample size	Smaller sample sizes usually means lower quality data (less reliability).
Frequency of sample (e.g. every 10 metres instead of every 100 metres)	Fewer sites reduces frequency, which then reduces quality.
Type of sampling	Sampling approaches may create 'gaps' and introduce bias in the results.
Equipment used	The wrong / inaccurate equipment can affect overall quality by producing incorrect results.
Time of survey	Different days or time of day might influence perceptions and tourist numbers, for example.
Location of survey	Big variations in environmental quality can occur between places very close to each other.
Quality of secondary data	Age and reliability of secondary data affect their overall quality.

▲ ***Figure 3*** *Sources of error in a geographical enquiry*

Being critical

Being critical means, simply, whether any shortcomings affected the quality of the results. In a geographical enquiry, researchers generally try to have the most reliable outcomes possible. They accept the limitations of their study. They often consider these types of questions as part of their evaluation:

1 How much do I trust the overall patterns and trends in my results?

2 What is the chance that these outcomes could have been generated randomly (or by chance)?

3 Which of my conclusions are most reliable, compared to other conclusions?

4 Which part of my enquiry caused the most unreliable results?

Such questions, although complex, can often be useful in considering how to write an evaluation.

How do my results affect my understanding of rural deprivation?

Another part of the process of evaluation is to link any knowledge you gained from the enquiry back to a theoretical model or idea. It may be a good idea to think about the key factors, as in Figure 4, and to then try to develop your own model.

▲ ***Figure 4*** *Factors that may influence rural deprivation and quality of life*

Your questions

1 Explain why an evaluation is the last part of the enquiry process.

2 Explain why an evaluation of one enquiry can be helpful in setting up a second enquiry.

3 **a** In pairs, copy the factors listed in Figure 3. Based on your own fieldwork, rank them in terms of importance.

b Explain your final rank order.

c Now add an extra column to the table in Figure 3. Write a title 'The most significant sources of error in our enquiry', and add points about your own rural fieldwork.

4 For either Devon and Cornwall (see sections 5.17 to 5.20) or your own fieldwork, make a copy of Figure 4 and add notes to it to explain how these factors influence quality of life.

Exam-style questions

5 Explain **one** factor about your own primary data which could have affected your results. (2 marks)

6 Evaluate the reliability of your rural fieldwork conclusions. (8 marks)

COMPONENT THREE
People and Environment Issues

Brown bear in the taiga forests of northern Europe

What is Component Three?

- Pearson Edexcel's GCSE Geography specification B consists of three Components.
- Each Component consists of three Topics, making nine in all.
- Each Component is assessed by its own exam paper – so Component Three is assessed by Paper 3.

What Topics will I study in Component Three?

- **Topic 7 People and the biosphere** is about the global distribution and characteristics of biomes, the importance of the biosphere to human wellbeing, and how people use and modify it to obtain resources.
- **Topic 8 Forests under threat** is about two biomes – tropical rainforests and the taiga – including processes and issues linked to their biodiversity and to their sustainable use and management.
- **Topic 9 Consuming energy resources** is about renewable and non-renewable energy resources, their supply and demand, access and energy security issues, and their sustainable use and management.

You'll also learn several **geographical skills** such as interpreting maps, satellite images, diagrams, statistics, and photos. In particular, you'll learn decision-making skills which form the basis of Paper 3.

What is Paper 3 like?

- **Time:** 1 hour 30 minutes
- **It has four Sections: A, B, C and D**
- **It's worth 64 marks:** 60 across the four sections, with another 4 for Spelling, Punctuation, Grammar and use of specialist geographical terminology (SPaG) which is assessed on the final 12-mark question in Section D.
- **It counts for:** 25% of your final grade.

Where can I get help in preparing for Paper 3?

Chapter 11 gives you all the guidance that you need on all three exam papers. It includes advice relevant to Paper 3 about:

- the exam format, handling different sections, and how exam papers will be marked (sections 11.1 and 11.2)
- how to answer shorter questions worth 1–4 marks (sections 11.3 and 11.4)
- how to answer longer 8-mark questions in Paper 3 (section 11.9)
- how to answer the 12-mark question in Paper 3 (section 11.10).

7.1 What, and where, are biomes?

In this section, you'll understand how the global distribution of biomes is affected by climate.

Getting to know the biosphere

In the 1960s, people marvelled at the first pictures of Earth seen from space, as shown in Figure 1. It was the planet's beauty – its ocean blues, its brilliance in sunlight, and its green forests and yellow deserts. These land surfaces form part of the **biosphere**. The biosphere is the living layer of Earth – a thin smear of life on its surface, between rocks (lithosphere) and air (atmosphere). All plants and animals are found here.

The biosphere is divided into large regions called **biomes**. There are nine major terrestrial (land-based) biomes, shown in Figure 2, each with its own climate, plant species and animal species. Tropical rainforests in Brazil, Cameroon and Indonesia have similar vegetation, but different individual species and ecosystems.

▲ ***Figure 1*** *Earth seen from space*

Figure 2 shows the distribution of biomes. Notice that they are, roughly, arranged in 'belts' around the Earth at different **latitudes**. For example:

- tropical rainforests are found between the tropics in South America, Africa and Asia
- taiga is found in a belt stretching across Canada, northern Europe and Russia.

A biome is a large-scale ecosystem, e.g. tropical rainforest.

Latitude measures how far north or south a location on the Earth's surface is from the Equator.

Key	Vegetation
Tundra	Grasses, lichens and dwarf shrubs; no trees
Taiga (boreal forest)	Coniferous trees, e.g. pine
Temperate deciduous forest	Deciduous trees, e.g. oak
Temperate grassland	Short or tall grasses and few trees
Mediterranean	Evergreen and deciduous trees and shrubs
Hot desert	Cacti and succulents, but few of them
Rainforest	Evergreen trees growing all year round
Savanna grassland	Grass with some trees, e.g. acacia
Other biomes	Includes mountain climates, snow and ice, and other grassland and forest biomes

▼ ***Figure 2*** *The global distribution of biomes*

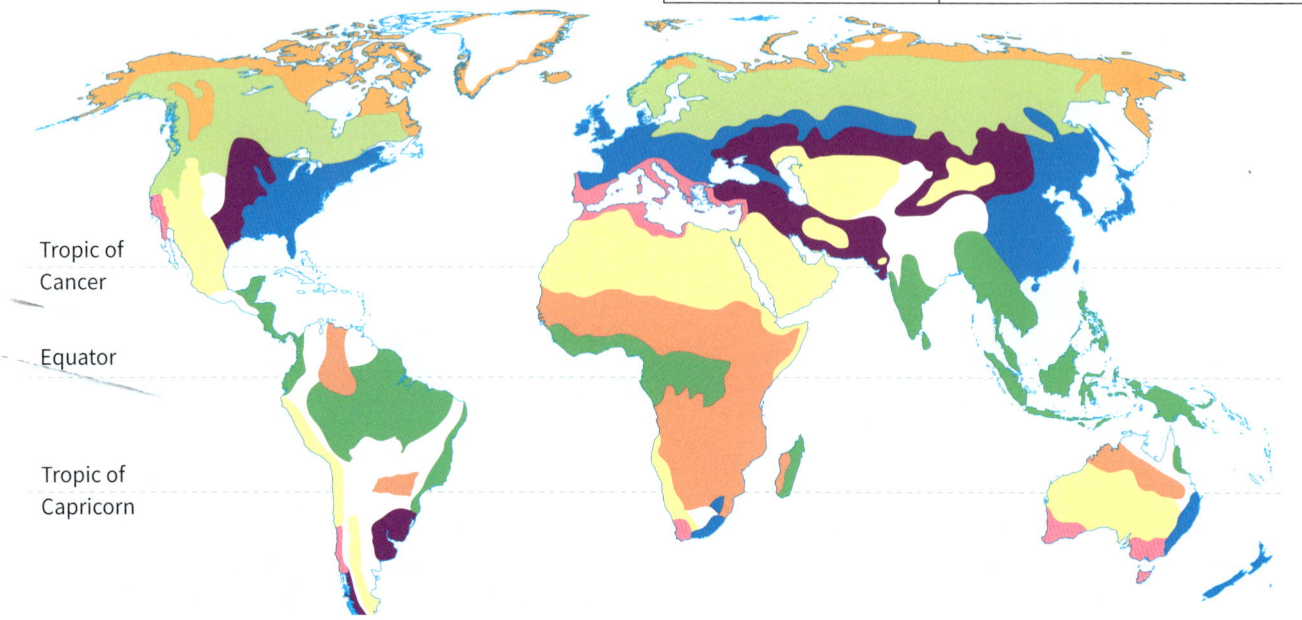

The location and characteristics of each biome are determined by climate, because it affects the growth of plants.

- Temperature is the most important factor. Most plants need temperatures of over 5°C to grow, so length of growing season varies from place to place.
- Precipitation and water availability are important, as plants need water. Plants grow if precipitation is spread across all seasons, but not if there is a dry season or water is frozen in winter.
- Sunshine hours and intensity affect photosynthesis and therefore plant growth.

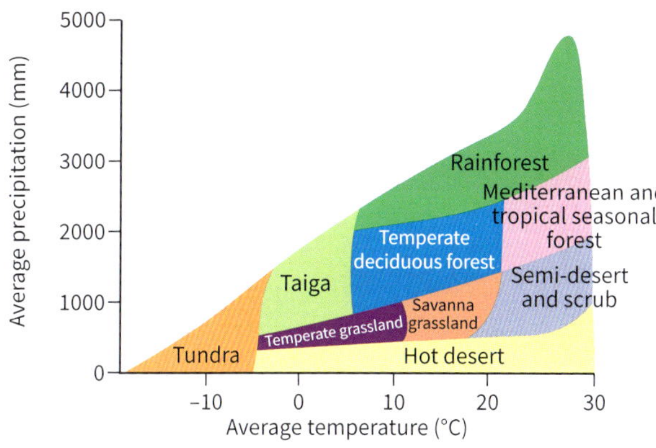

▲ ***Figure 3*** *How temperature and precipitation influence biome type*

The relationship between temperature and precipitation and biome type is shown in Figure 3. In areas with high precipitation, warm temperatures and sunshine, forest biomes are found. In areas with very dry and/or very cold seasons other types of biomes replace them, e.g. grasslands.

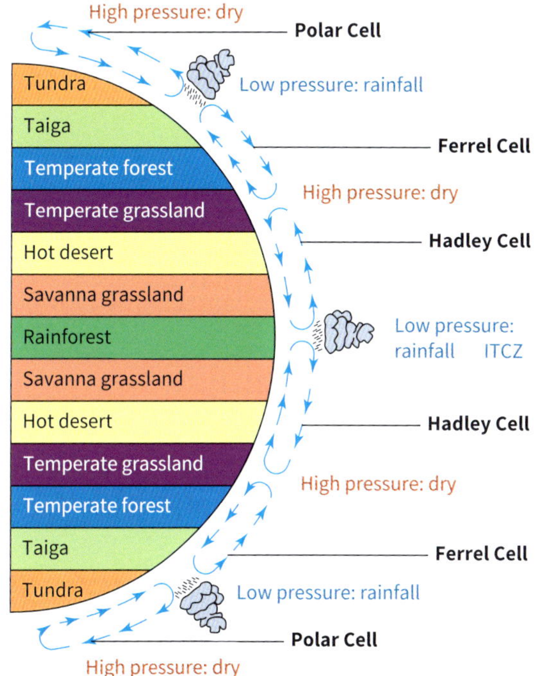

▲ ***Figure 4*** *Atmospheric circulation and biomes*

Temperature

Latitude influences biome type because temperature and sunshine intensity are each controlled by latitude.

- Locations near the Equator, where sunlight is more intense, are warmer than those near the Poles.
- Near the Equator, the sun's rays are at a high angle in the sky all year.
- As latitude increases (towards the Poles) winter becomes longer and colder, and the climate more seasonal.
- In polar areas, sunshine intensity is low; lack of heat and light limits plant growth.

Precipitation

Precipitation is also influenced by latitude. Figure 4 shows how the Earth can be divided into high and low air pressure precipitation zones.

- North and south of the Equator there are three major convection cells in the atmosphere (Hadley, Ferrel and Polar – see section 1.3).
- Precipitation is high at the rising parts of these cells because air pressure is low.
- At the descending parts of the cells, air pressure is high but precipitation is low.

Forest biomes are found in areas of low pressure and high rainfall at the boundary between cells. In high pressure areas, lack of precipitation prevents tree growth so that grasslands or, in very dry areas, deserts replace them.

Your questions

1 Explain the difference between biosphere and biome.

2 Describe the distribution of **a** rainforests and **b** taiga. 

3 Explain the difference between latitude and longitude. Which of the two affects biome distribution, and why?

4 Using Figure 3, compare **a** maximum and minimum temperatures, and **b** minimum precipitation levels needed for each of taiga, temperate deciduous forest and tropical rainforest. 

Exam-style question

5 Explain how latitude affects temperature and precipitation. (4 marks)

7.2 Local factors and biomes

In this section, you'll understand how local factors affect biome location and type.

UK ecosystems

If the UK was a totally natural environment, with no urban areas or farming, almost all of it would be covered by temperate forest. This consists of deciduous trees that lose their leaves in autumn. The forest is dominated by oak with other species, e.g. ash and hazel.

However, forests can vary because of **local factors**, which alter animal and plant species from ones we would expect. They include:

- rock and soil type
- water availability and drainage
- altitude (height of land).
- introduction of 'foreign' species by people (e.g. sycamore)

These produce different **ecosystems**, which are localised biomes. Differences in soil or drainage mean that conditions favour some plants more than others, altering the type of ecosystem (see Figure 1).

1 Rock and soil type

When rocks undergo chemical weathering (see section 4.16), they release nutrients and chemicals into soils. Soils can be neutral, acidic or alkaline, depending on rock type. The acidity/alkalinity of soil influences the plants that will grow there.

2 Water availability and drainage

Some plants can grow with their roots in waterlogged soil or boggy areas. Others prefer drier soils. How wet the soil is depends on several factors:

- the amount of precipitation
- the amount of evaporation from the soil (influenced by temperature)
- how permeable the soil is; sandy soils are dry and clay soils wet.

3 Altitude

Height affects biomes in three ways:

- Temperature drops by 6.5°C for every 1000 m increase in height.
- At high altitudes, below freezing temperatures are common, which limits the types of plants that can grow.
- Rainfall usually increases with height.

As temperature and precipitation conditions change with height, changes occur to the ecosystems. This forms a pattern called **altitudinal zonation** (see Figure 2).

▼ ***Figure 1*** *How soil type affects UK forest type and ecosystem*

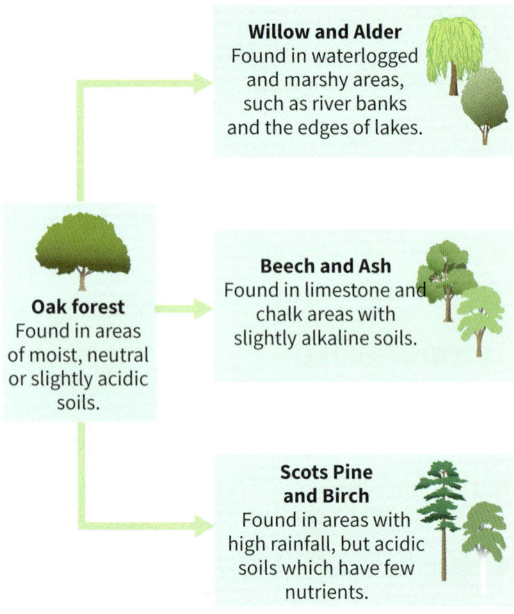

▼ ***Figure 2*** *The relationship between altitude and ecosystems in the Andes*

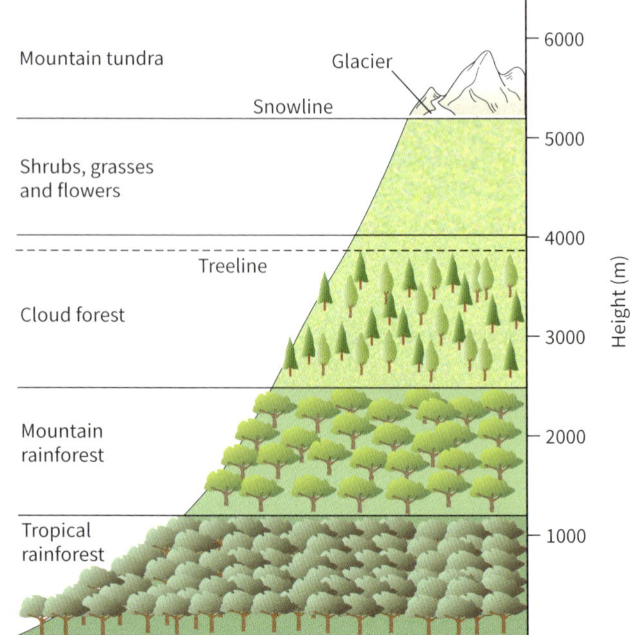

- Below 1000 m, tropical rainforest is found.
- As temperatures fall with height, this changes to mountain and cloud forest.
- Trees cannot grow above the **treeline** because temperatures are too cold.

Altitudinal zonation is the change in ecosystems at different altitudes, caused by alterations in temperature, precipitation, sunlight and soil type.

Biotic or abiotic?

Biomes consist of two parts:

- The **biotic** (living) part is made up plant (flora) and animal (fauna) life.
- The **abiotic** (non-living) part includes the atmosphere, water, rock and soil.

Dead plants and animals are biotic because they were once living.

Biotic and abiotic parts are linked as shown on Figure 3.

1. Energy is provided by photosynthesis; only plants (primary producers) can turn the sun's energy into carbohydrate.
2. Plants take in carbon dioxide from the atmosphere and release oxygen, whereas animals take in oxygen and breathe out carbon dioxide; nitrogen is also exchanged between the atmosphere, plants and soil.
3. Energy flows along the food web, from plants to herbivores (primary consumers) and carnivores (secondary consumers), then to detritivores and decomposers (worms, bacteria, fungi) that consume dead plants and animals.
4. Water (precipitation) moves through the soil, plants and animals and finally back into the atmosphere via respiration and evaporation.

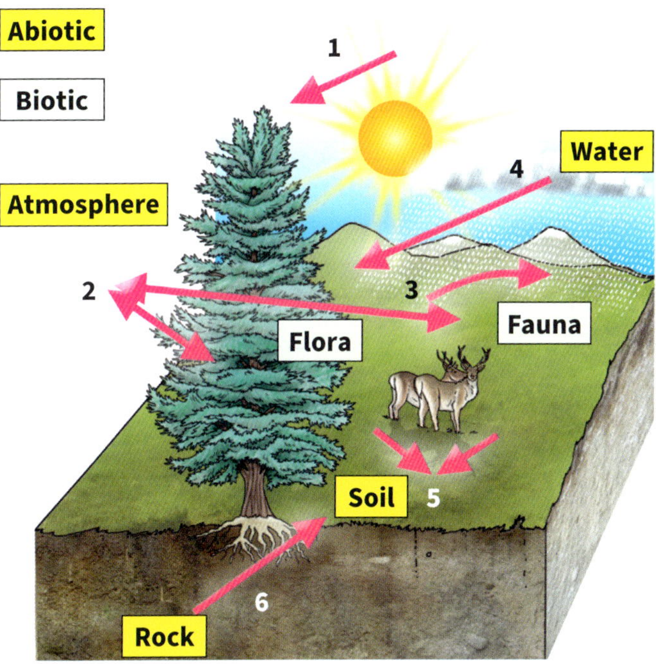

▲ *Figure 3 The biotic and abiotic parts of ecosystems*

5. When plants and animals die, decomposition returns nutrients to the soil.
6. Weathering of rock provides soil nutrients, e.g. phosphates and magnesium.

Local changes in abiotic factors (rock type, soil, local hydrological cycle and altitude) can alter ecosystem type. This changes conditions for biotic parts of the ecosystem, favouring some plants and animals rather than others.

Your questions

1 Explain the difference between a biome and an ecosystem.

2 a Study Figure 1. Complete the table to show how local factors determine forest type in the UK.

	Oak	Willow and Alder	Beech and Ash	Scots pine and Birch
Rock and soil type				
Water availability and drainage				
Altitude (height of land)				

b Using the table, explain the likely types of tree found in **i)** the Yorkshire Pennines (see section 4.1), **ii)** scarp and vale areas of south-east England (see section 4.4), **iii)** the Cairngorm mountains in Scotland.

Exam-style questions

3 Explain the difference between the biotic and abiotic components of an ecosystem. (3 marks)

4 Explain how altitude affects ecosystems. (4 marks)

5 Explain **two** reasons why all temperate deciduous forests in the UK are not exactly the same. (4 marks)

7.3 Geographical skills: learning about climate and biomes

In this section, you'll develop skills in analysing climate data and interpreting climate graphs for contrasting biomes.

1 Understanding temperate forest biomes

Nottinghamshire is famous for Sherwood Forest (of Robin Hood fame!). The legendary character of Robin Hood lived in **temperate deciduous forest** of oak, ash and beech. Figure 1 contains climate data for Nottingham in the UK. Deciduous forest needs this mild climate – no month with average temperatures below freezing, and year-round rainfall.

Months	J	F	M	A	M	J	J	A	S	O	N	D
Precipitation (mm)	61	47	50	54	52	63	58	62	59	71	66	69
Temperature (°C)	4.0	4.1	6.3	8.4	11.6	14.5	16.7	16.5	14.0	10.4	6.7	4.2
Sunshine (hours)	55	73	104	141	187	171	191	180	131	99	64	49

▲ ***Figure 1*** *Climate averages for Nottingham, UK*

Compare this to the climate data for Barrow, Alaska (Figure 2). No Robin Hoods here living outdoors. Barrow is located at latitude 71° N, inside the Arctic Circle. Figure 2 shows that no month has an average temperature above 5°C. This is far too cold for any trees to grow, so the biome here is **tundra** (cold desert).

There are no recorded sunshine data for Barrow. The sun actually disappears for two months between 18 November and 22 January! In summer, it's the reverse with 24-hour daylight between mid-May and late July!

Did you know?

Every June, Inuvik, a town of 3000 in Canada's Northwest Territories, two degrees inside the Arctic Circle, has an annual Midnight Sun Fun Run starting at 11 p.m.!

Months	J	F	M	A	M	J	J	A	S	O	N	D
Precipitation (mm)	3	3	2	4	5	8	25	27	18	10	5	3
Temperature (°C)	−25.2	−25.7	−24.7	−16.8	−6.1	2.0	4.9	3.9	0.1	−8.2	−17.4	−22.1

▲ ***Figure 2*** *Climate averages for Barrow, Alaska*

Your questions 1

Exam-style questions

1 Using Figure 1 and the guidance in Section 1.4:

- **a** Calculate the total annual precipitation for Nottingham. (1 mark)

- **b** Calculate the monthly temperature range for Nottingham. (1 mark)
- **c** Calculate the mean monthly sunshine hours for Nottingham. Show your working and answer to one decimal place. (2 marks)

2 Using Figure 2, calculate the monthly temperature range for Barrow. (1 mark)

3 State two pieces of evidence from Figure 2 that suggest Barrow has an extreme climate. (2 marks)

4 Using the data for both Nottingham and Barrow, compare their precipitation levels during an average year. (4 marks)

2 Understanding other biomes

Tropical grassland climate is different to both temperate deciduous forest and tundra. Figure 3 shows a climate graph for a tropical grassland area in Kenya, Africa.

▲ ***Figure 4*** *Photos of three biomes*

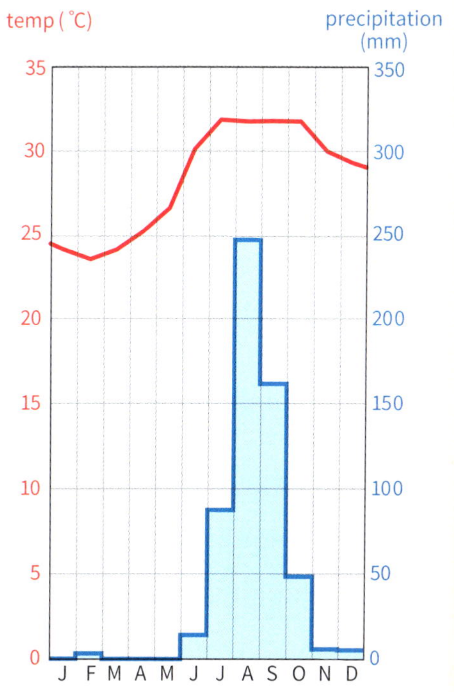

▲ ***Figure 3*** *Climate graph for a tropical grassland area in Kenya*

Your questions 2

1 a Design a table to compare the climates of Kenya, Barrow and Nottingham using the climate graph and data tables (Figures 1–3). Include columns for maximum and minimum temperatures, total annual precipitation and the number of months when rainfall is above 50 mm.

b Look at the graph of the tropical rainforest climate for Belem in Brazil on page 248 (in Section 8.1). What are the similarities and differences between the climate in Belem and the climate in Kenya?

2 Study Figure 4 showing photos of three biomes. Match the photos A, B and C to the these places: Barrow (Alaska), East Midlands (near Nottingham) and Kenya. Give your reasons.

Exam-style questions

3 Describe the differences in vegetation between photographs B and C. (2 marks)

4 State **two** factors that limit the growth of trees in tropical grassland and tundra areas. (2 marks)

5 Explain how rainfall and temperature influence the location of temperate deciduous biomes. (4 marks)

7.4 A life-support system

In this section, you'll assess the reasons why biomes are important providers of goods and services to people and the planet.

Valuing biomes

It is very difficult to put a value on biomes like tropical forests, but some scientists and economists estimate them to be worth US$4–6 trillion *each year*. They arrive at this figure because biomes provide humans with a wide range of **goods** and **services**.

- Goods are physical materials, such as timber from trees or fish caught in a lake.
- Services are functions, e.g. how forests prevent flooding or trees add oxygen to the atmosphere.

The best way of thinking about goods and services is to divide them into four categories of **ecosystem services**, as shown in Figure 1.

Provisioning services (goods)	Supporting services
These are products obtained from ecosystems: • food: nuts, berries, fish, game, crops • fuelwood • timber for buildings and other uses • genetic and chemical material.	These keep the ecosystem healthy so it can provide the other services: • nutrient cycling • photosynthesis and food webs • soil formation.
Regulating services	**Cultural services**
These services link to other physical systems and keep areas, and the whole planet, healthy: • storing carbon, and emitting oxygen, which keep the atmosphere in balance • purifying water and regulating the flow of water within the hydrological cycle.	These are benefits people get from visiting, or living in, a healthy ecosystem: • recreation and tourism • education and science • spiritual well-being and happiness.

▲ ***Figure 1*** *Four categories of ecosystem goods and services*

The importance of ecosystem services varies.

- The role of ecosystems in maintaining a healthy atmosphere is **globally** important. Trees take in carbon dioxide and store it as carbon. This helps regulate the whole climate system.
- Provisioning services are **locally** important. In some parts of the world, local people depend on a coral reef for fish, or a forest for timber for fuel, building and making furniture.

Did you know?

The gene pool found in rainforests might one day yield cures for cancer. But we won't find this out if we destroy rainforests.

Ecosystem services for indigenous people

In most of the world, people do not depend directly on ecosystems for their livelihood. However, small numbers still do. One example is the Efe people of the Ituri tropical rainforest in the Congo Basin of Africa. Only about 30 000 Efe remain today.

- Small circular houses, as shown in Figure 2, are built from wood and leaves; these are temporary as the Efe move around the rainforest hunting and gathering food.
- Wood for cooking fires comes from the forest.
- Efe hunt monkey and antelope for bushmeat and also fish in rivers.
- They gather wild yams, nuts, mushrooms and berries from the forest.
- They hunt the giant forest hog and sell its meat to other people in the region, and buy or trade items like pans or rice.
- They use plants and wild honey to make traditional medicines.

The Efe also worship the tropical rainforest itself. They call it 'father' or 'mother' and make offerings of food to the forest.

▼ ***Figure 2*** *The Efe tribe cooking bushmeat in front of their circular houses*

Exploiting ecosystem services

Slash-and-burn farming, or shifting cultivation, is a type of farming still used by up to 500 million people worldwide. It works like this:

- Farmers clear small areas of forest by cutting and burning.
- The ash from burning adds nutrients to the soil.
- The land is farmed for 5-6 years, but after that the soil becomes infertile and farmers move to a new area.

Forest grows back on the abandoned patches of farmland over time. This type of exploitation destroys small areas of rainforest, but it does regrow easily.

Ecosystem services is a collective term for all of the ways humans benefit from ecosystems.

Indigenous peoples are the original people of a region. Some indigenous groups still lead traditional lifestyles, e.g. a tribal system, hunting for food.

Destroying ecosystem services

Most exploitation of forest and grassland biomes today is not temporary like slash-and-burn. It is commercial, done for profit and often involves transnational corporations (TNCs) and governments.

Large areas of biomes are cleared for:

- commercial farming, particularly for beef cattle ranching or to grow fodder crops such as soybeans, which are fed to cattle
- commercial crops such as palm oil, cocoa beans or cereals
- mining metal ores such as copper and iron
- timber, used to make paper, furniture or construction wood
- construction of dams and reservoirs for hydroelectric power (HEP) and to supply water to cities.

Once a biome has been cleared to make way for this type of commercial exploitation, it can never grow back. Commercial exploitation of biomes provides profits for TNCs, jobs for people and income for governments. On the other hand, many of the ecosystem services are destroyed. Areas that were once forest can no longer store carbon, prevent flooding or be used for recreation. Figure 3 shows the percentage of four biomes destroyed by human activity. Most of the destruction is because biomes have been converted to farmland.

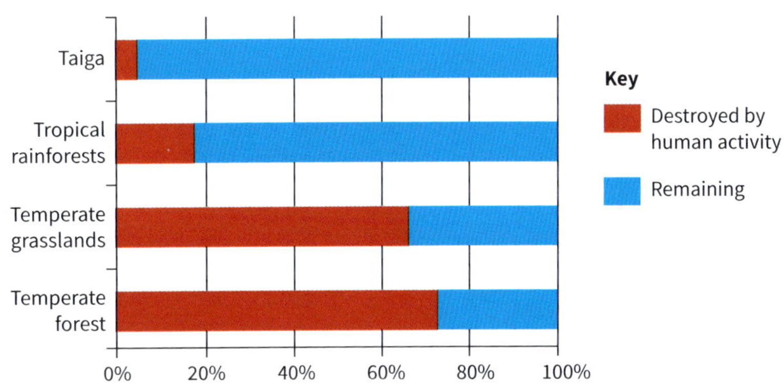

▲ ***Figure 3*** *The destroyed and remaining percentage of four biomes*

Thinking beyond

Are biomes worth more, left in their natural state, than they are if exploited?

Your questions

1. Using Figure 1, identify the category of goods and services in which the following belong: **a** hunting and gathering, **b** scientific research into new drugs from plants.
2. In pairs, write down as many ways as you can in which the lifestyle of the Efe people is 'sustainable'. To help you, use the 'sustainability stool' in Section 3.12.
3. Describe the differences in biome destruction shown in Figure 3.
4. On the basis of Figure 3, would you say that rainforest destruction can continue for the next fifty years? Explain your answer.

Exam-style questions

5. Explain the difference between provisioning and regulating services. (3 marks)
6. Study Figures 1, 2 and 3. Assess the reasons why some people are more concerned than others about the destruction of rainforest for commercial uses. (8 marks)

7.5 Biomes and global services

In this section, you'll understand the critical role biomes play in maintaining a healthy planet.

Brilliant biomes!

Without biomes, especially forests, the world's climate would be very different! Natural hazards such as flooding would be much more common. Humans depend on biomes for keeping the atmosphere in balance by regulating the level of carbon dioxide. Healthy soils are needed in which to grow food crops, and a healthy hydrological cycle to provide us with water. Ecosystems provide us with all of these key services.

Healthy air

Biomes are a very important **carbon sink**. They store carbon by removing carbon dioxide from the atmosphere and locking it up in biotic material. This is called **carbon sequestration**. The process which makes this work is **photosynthesis**.

$$\underset{\text{Carbon dioxide}}{6CO_2} + \underset{\text{Water}}{6H_2O} + \text{sunlight} \longrightarrow \underset{\text{Carbohydrate /Glucose}}{C_6H_{12}O_6} + \underset{\text{Oxygen}}{6O_2}$$

- Biomes store carbon as biomass (leaves, branches, trunks, roots and animal tissue).
- When plants and animals die the dead biomass ends up in the soil – making soil an important carbon sink.

Figure 1 shows that biomes on land absorb about 120 billion tonnes of carbon each year from the atmosphere, and the same amount is released. Biomes and soil together store 2850 billion tonnes of carbon. More carbon can end up in the atmosphere if:

- Humans destroy biomes, e.g. through deforestation, so biomes can absorb less.
- Biomass such as trees are burned, and soil destroyed, releasing their stored carbon.

These actions increase the amount of carbon dioxide in the atmosphere, which scientists link to global warming.

Carbon sinks are natural stores for carbon-containing chemical compounds, like carbon dioxide (CO_2) or methane (CH_4).

Nutrient cycle – Nutrients like nitrogen and phosphorous move between the biomass, litter and soil as part of a continuous nutrient cycle which keeps both plants and soil healthy.

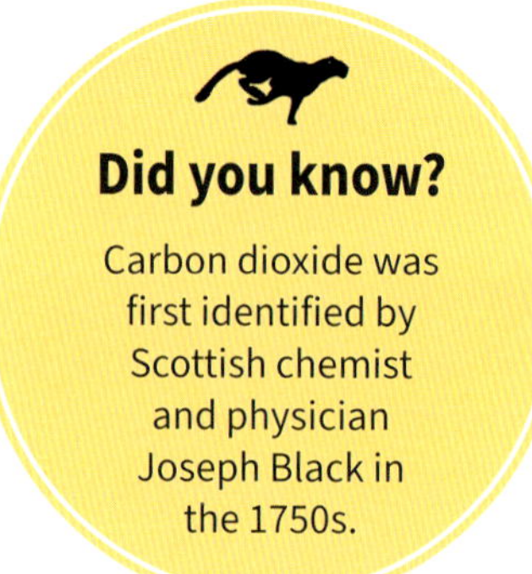

Did you know?

Carbon dioxide was first identified by Scottish chemist and physician Joseph Black in the 1750s.

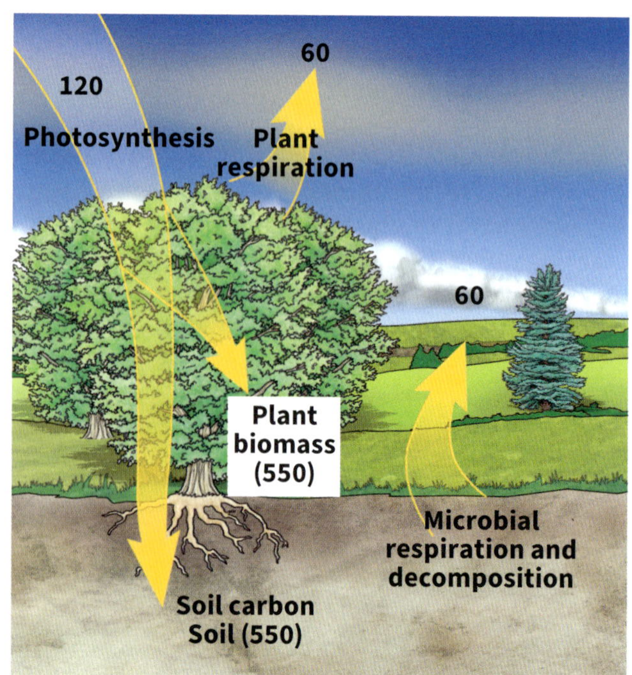

▶ ***Figure 1*** *How biomes store carbon. Figures are billion tonnes of carbon.*

Healthy soils

Biomes are very important in maintaining soil health. All farmland was once a natural biome, and the biome formed the soil that we now use for farming and food production. Soil health – or soil fertility – is maintained by the **nutrient cycle** shown in Figure 2.

Nutrient cycles can easily be disrupted, putting soils at risk:

- removing biomass, e.g. logging timber, takes away a large nutrient store
- heavy rain and surface run-off can wash away litter
- deforested areas are at risk from soil erosion, removing another store.

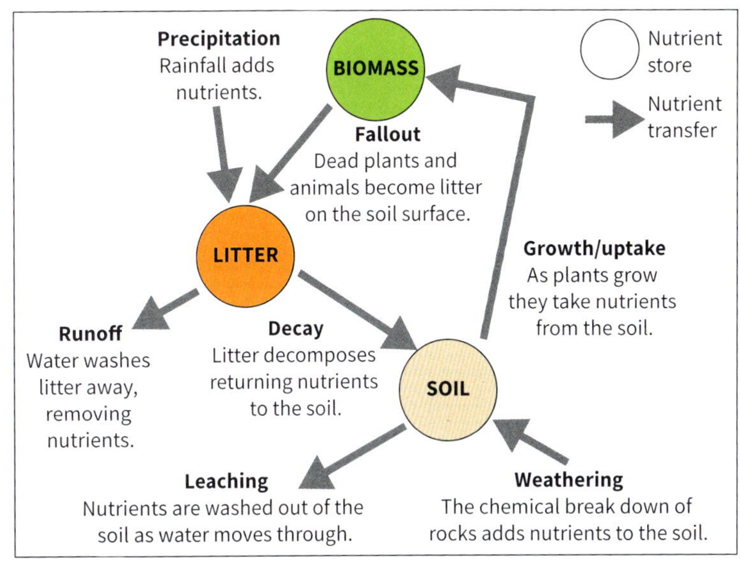

▲ ***Figure 2*** *The nutrient cycle*

Water supply and flood risk

Biomes, especially forests, are an important part of the **hydrological cycle**. In turn, the hydrological cycle provides humans with a clean, reliable water supply. Destroying a forest biome can have serious impacts on the water cycle, as Figure 3 shows.

▼ ***Figure 3*** *Ecosystems and the hydrological cycle*

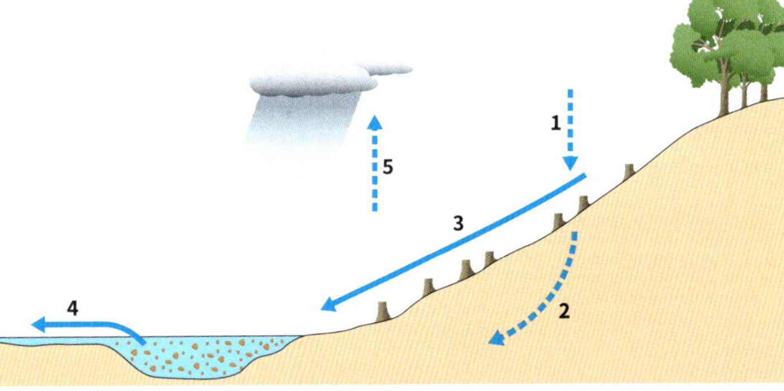

1. With no trees, interception is reduced and rainfall hits the soil surface directly, causing soil erosion.
2. Without trees, there is much less infiltration of water into the soil; this means groundwater supplies, which are an important water resource for many people, are lower.
3. Surface run-off increases, which increases soil erosion and means water gets into river channels much faster.
4. Flooding becomes more frequent, and river water is made dirty by all of the soil washed into the river.
5. With no forest, the soil dries out very quickly so overall evaporation is reduced which can mean fewer clouds, less rain and a drier climate.

Your questions

1. Explain the difference between the terms 'carbon sink' and 'carbon sequestration'.
2. Explain how the nutrient cycle works.
3. Explain the following:
   a. how soil fertility depends on leaf litter
   b. how biomass depends on soil fertility
   c. what would happen if biomass (forest) is removed.
4. Write a paragraph explaining how ecosystems help to maintain the health of the hydrological (water) cycle.

Exam-style questions

5. Explain **two** ways ecosystem services (Section 7.4) depend upon a healthy nutrient cycle. (4 marks)
6. Explain the importance of the biosphere for the water cycle. (4 marks)

7.6 More and more resources

In this section, you'll understand why there is increasing demand for resources and how demand is hard to meet.

More people, more demand

We use natural resources all of the time, without thinking about it. Even having a meal uses a huge range of resources, some of which might have directly had an effect on biomes, as Figure 1 shows.

Natural resources are materials found in the environment that are used by humans, including land, water, fossil fuels, rocks and minerals and biological resources like timber and fish.

Demand for **natural resources** such as food, energy and water is rising all the time. Human use of natural resources has implications for the biosphere.

- Biomes can be directly destroyed to create land for farming, housing, factories, mining or projects like hydroelectric power station reservoirs and dams.
- Obtaining timber, fish or hunted animals from biomes can degrade them, by removing some key species.
- Human use of natural resources can pollute the wider environment, which damages biomes. Examples include burning fossil fuels, polluting the atmosphere, and dumping waste in rivers, polluting water.

▲ ***Figure 1*** *Did your dinner do more harm than good?*

Pressure on natural resources has grown since 1975 for a number of reasons, as explained in Figure 2.

	Rising global population	Rising affluence (average global income)	Increasing urbanisation (% of people living in towns and cities)
1975	4.1 billion	US$3700 per person	38%
2020	7.8 billion	US$11 400 per person	56%
Resource demand	A larger population means a greater demand for resources, especially food and water; large areas of forest have been cleared to make new farmland.	Increasing average wealth means people use more energy resources, such as fossil fuels in the manufacture of consumer goods.	Over half of the world's population now live in towns and cities which have sprawled over biomes and increased demand for water and food.

▲ ***Figure 2*** *Pressure on natural resources*

Industrialisation

Since the 1970s, many countries have been through a process of **industrialisation**. This has been most marked in Asia. Countries such as China, South Korea, India and Thailand have seen a dramatic shift away from people living in the countryside working on farms, to people living in cities working in factories and offices.

This has meant a dramatic rise in construction and resource consumption.

- In 1998, China had no high-speed railways (with trains running at over 200 kmh), but by 2020 it had over 38 000 km of these railways.
- Between 2000 and 2020 the number of cars in India jumped from 6 million to 40 million.
- Thailand's urban population was 17 million in 1990, but by 2020 this had more than doubled to 36 million.
- In 1980, the average South Korean ate only 11 kg of meat each year but this had risen to 54 kg by 2020.

Asian growth

Figure 3 shows the dramatic rise in the population of Asia since 1975. It also shows that in other global regions like Africa, population is growing very rapidly. Growth is not happening everywhere, for instance Europe's population is barely growing at all. In developing regions, the combined impact of population growth, industrialisation, urbanisation and rising wealth has led to a dramatic increase in demand for resources, as shown in Figure 4.

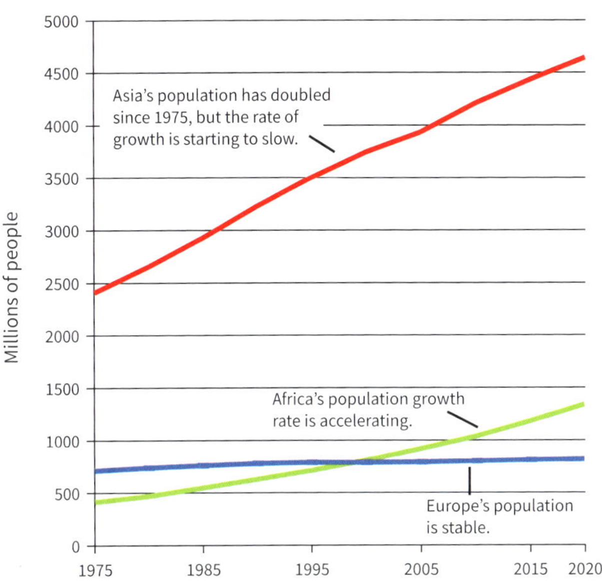

▲ ***Figure 3*** *Population growth in Africa, Europe and Asia, 1975–2020*

▼ ***Figure 4*** *The increased demand for resources in Asia*

Asia	Water consumption (cubic km)	Oil consumption (millions of tonnes)	Coal consumption (exajoules)	Meat consumption (million tonnes)
1975	1500	453	16	22
2020	2800	1660	120	145
Percentage increase 1975–2020	87%	266%	650%	559%

Figure 3 shows that Asia's population has increased by 93% since 1975. Figure 4 shows that water consumption has increased by about the same amount, but other resources have increased much more.

- Not only are there 2.2 billion more people in Asia today compared to 1975, they are on average richer, so use far more energy resources than in 1975.
- As people get richer (also called 'rising affluence'), their diets change – they eat more, but they also eat more meat, fat and dairy products.

The challenge for the 21st century is whether the Earth can provide enough resources for more people, and for increasingly wealthy people.

Your questions

1 a List all of the natural resources you are using now in your geography lesson.

b Explain how these place pressure on the world's biomes.

2 Explain how the data in Figure 2 also places increasing pressure on the world's resources.

3 Draw a spider diagram with four arms – water, oil, coal, and meat. In pairs, complete the diagram by describing all of the pressures on biomes that result from the changes shown in Figure 4.

Exam-style questions

4 Compare the population trends for Europe and Asia in Figure 3. (3 marks)

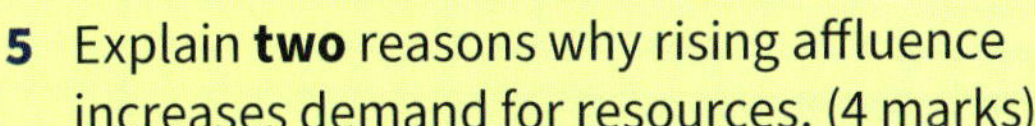

5 Explain **two** reasons why rising affluence increases demand for resources. (4 marks)

6 Study Figures 2, 3 and 4. Assess the reasons why population and industrialisation have increased demand for resources. (8 marks)

7.7 Population versus resource theories: who's right?

In this section, you'll understand the contrasting views about the relationship between population and resources.

2030: Perfect storm?

A former UK government chief scientific advisor, said:

"Our food reserves are at a 50-year low, but by 2030 we need to be producing 50% more food. At the same time, we will need 50% more energy, and 30% more fresh water."

He was describing what has been called the '2030 perfect storm' idea. It argues that by 2030 the world will be running out of resources. It is not a new idea. Geographers have worried about the number of people on the planet for at least 200 years. There are two viewpoints, outlined in Figure 1 below.

▼ ***Figure 1*** *The pessimistic and optimistic views of the future*

View	In a bit more detail...	People who held this view	Sometimes called...
Pessimistic	Population will eventually grow so large that the planet will run out of food, water, energy and other resources, leading to a crisis.	Thomas Malthus Professor John Beddington The Club of Rome	Malthusians or Neo-Malthusians
Optimistic	As population grows, humans invent new technologies to allow more food to be grown, and more resources to be supplied.	Ester Boserup	Boserupians

The Malthusian view

The Reverend Thomas Malthus wrote *An Essay on the Principle of Population* in 1798. He was writing at a very pessimistic time during the Revolutionary Wars. He argued that population would increase geometrically (1,2,4,8,16, etc.) by doubling in each generation, but that food production could only increase arithmetically (1,2,3,4,5, etc.). In this way, population would eventually outstrip food supply, leading to a 'population versus resources crisis,' as outlined in Figure 1. Population would have to fall, which would happen by:

- **Positive** checks – war, starvation and famine would reduce population.
- **Preventative** checks – people marrying later, and having fewer children.

The balance between population and food supply would therefore be restored.

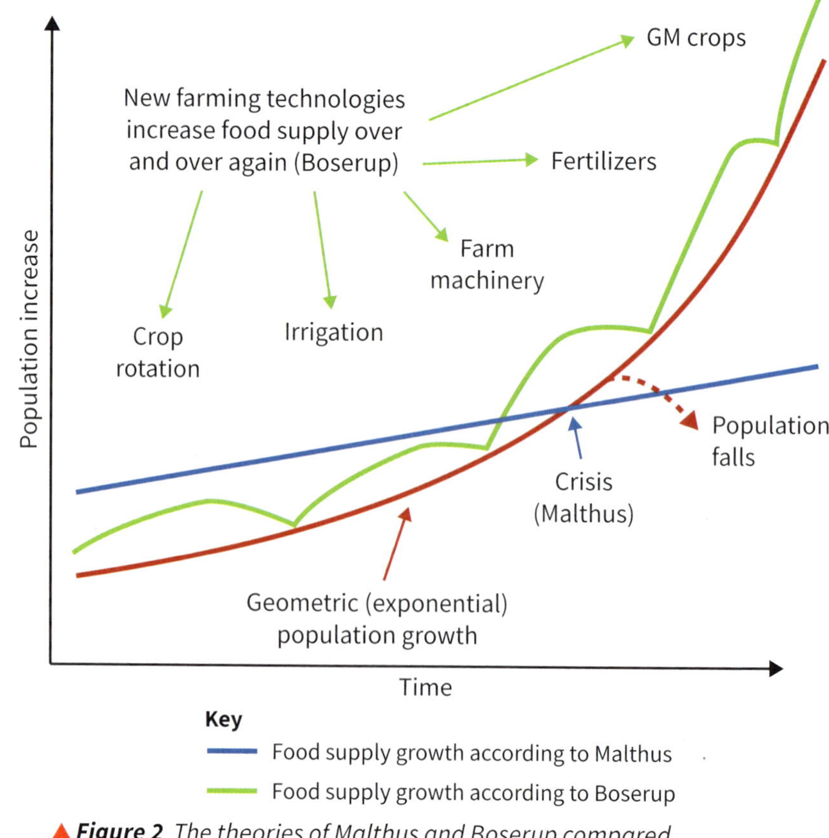

▲ ***Figure 2*** *The theories of Malthus and Boserup compared*

The Boserupian view

Ester Boserup, a Danish economist, published her view in 1965 in a book called *The Conditions of Agricultural Growth*. She argued that population growth is the driver of new inventions – as population grows, innovative humans invent new ways of producing more food. The 1960s were a time of optimism, especially in terms of technology (computers, space exploration) and Boserup's view reflects this. Technology such as farm machinery, fertilisers, genetically modified (GM) crops and irrigation are examples of this (see Figure 2).

Who's right?

Opinions are bitterly divided about the relationship between population and poverty. The Malthusian view became popular again in the 1970s because of the work of a think tank called the **Club of Rome**. 50 years later, scientists still refer to it, because:

- There has been no *global* 'crisis' since 1798, but there have been *local* ones, e.g. the Irish Potato Famine (1840s).
- The Earth provides resources for nearly 8 billion people but this doesn't mean it can do so for 12 billion people in future.

However, there are counter-arguments:

- Food production has become more technically advanced, just as Boserup said it would.
- Some resources are finite (fossil fuels), but renewable resources (wind, solar power) and sustainable ones (water, biofuels) might support many more people.

Quality or quantity?

There is a final part of the debate. A 'crisis' might occur because the **quality** of remaining resources becomes low, rather than the **quantity** becoming too small.

- Rivers and lakes could be so polluted that their water is unusable.
- Soils may become so eroded and infertile that few crops can be grown.
- When oil and gas run out, we may turn to burning very dirty, polluting coal.

Perhaps the real question is whether:

- the price of resources will become too high for some people
- the billion poor and hungry people today will ever be able to access the resources they need.

Predicting future population is important, but difficult. Since 2000, estimates of future population have increased. It was thought that world population would peak at 9 billion in 2050. This now looks unlikely (Figure 3) and population is likely to grow after 2050. However, there is a large margin of uncertainty.

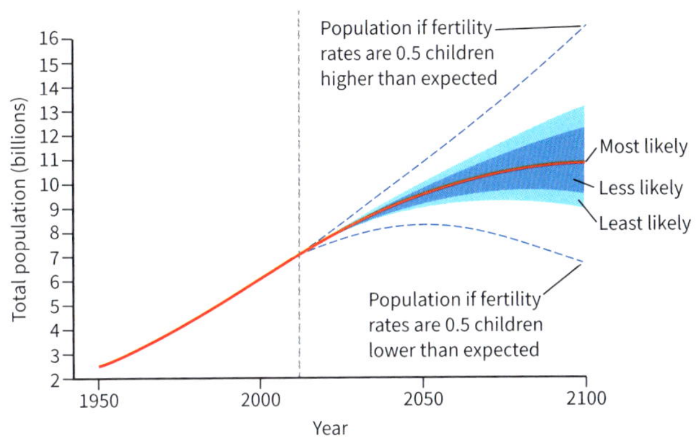

▲ ***Figure 3*** *World population projections to 2100*

Your questions

1. Look carefully at Figure 3:
   - **a** State the most likely world population by 2100.
   - **b** State the minimum and maximum population projections for 'less' and 'least' likely projections.
2. Write 200 words describing what you think might happen to the availability of food, water and energy resources by 2100 if the world's population grows to over 10 billion.

Exam-style questions

3. Explain **one** reason why the population projections in Figure 3 are uncertain. (3 marks)
4. Explain the theories of Malthus and Boserup about the relationship between population and resources. (4 marks)

8.1 What are tropical rainforests like?

In this section, you'll understand how plants and animals in tropical rainforests are adapted to their climate.

Biodiversity booms!

In Amazonia there are a staggering 16 000 different tree species, over 40 000 other plant species, 1300 bird species and six different species of sloth (see Figure 1). There are 10 000 animal species in the tropical rainforests of the Congo Basin alone! In comparison, the UK has only 30 native tree species.

Biodiversity is exceptionally high in rainforests because:

- The climate is perfect for year-round growth and reproduction.
- Rainforests are ancient and have a stable climate, so thousands of different species have evolved.
- The multiple layers in a tropical rainforest provide numerous different, specialised habitats. Plants and animals have evolved to take advantage of these.

Biodiversity means the range of different plant and animal species in an area.

▼ ***Figure 1*** *The critically endangered pygmy three-toed sloth. There are probably fewer than 100 of these animals left in the wild.*

Rainforests: climate

10% of all of the world's plant and animal species live in the Amazon rainforest. Tropical rainforests are also found in Ecuador in South America, the Congo Basin in central Africa and in many Asian countries such as Indonesia and Malaysia. One thing they have in common is an equatorial climate.

- There is no dry season, with a least 60 mm of rainfall each month; some get three metres of rain each year.
- Temperatures are high, at 26–32°C all year round, so there is no summer or winter.

Because heat, water and sunlight are available all year round, tropical rainforests grow continually. There is a huge abundance of life – both plant and animal. Figure 2 is a climate graph for Belem in Brazil. Belem's annual average temperature is 32°C, with over 2900 mm of rainfall and 2200 hours of sunshine each year – that's an average of 6 hours a day.

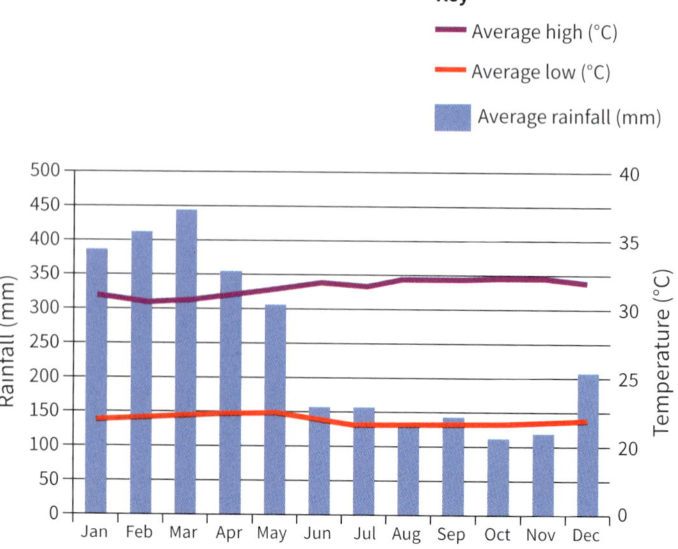

▶ ***Figure 2*** *Climate graph for Belem, Brazil*

Forest structure

All rainforests have a typical structure with layers of vegetation (Figure 3). Multiple layers of plants are found from ground level right up to 45 m above the forest floor – much taller than other forest biomes and about the height of a 12-storey building!

Adapting to a hot, wet and humid life

Rainforest plants and animals are adapted to the equatorial climate and the layered rainforest. Plants and animals have evolved over time to fit into this unique environment and thrive there, as shown in Figure 4 below.

▼ ***Figure 3*** *The structure of a tropical rainforest*

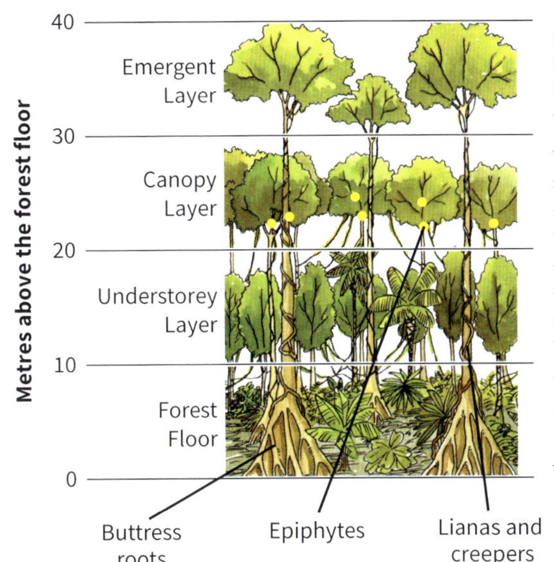

Plant adaptations	Animal adaptations
Evergreen hardwood trees Mahogany, teak and ebony trees have tall slender trunks with no branches on them, but huge triangular buttress roots. The roots support the enormous weight of the trees; leaves and branches are only at the very top, where the sunlight is.	**Sloths** Huge claws allow sloths to hang upside down in the branches, while their fur grows away from their feet to help shed rain when upside down. Green algae growing in their fur helps camouflage them from predators.
Epiphytes These plants live in the canopy on trees and have evolved to get all their nutrients from water and air rather than the soil (which is ten metres below!), so their roots dangle in mid-air.	**Primates** Lemurs and monkeys have evolved to live in the canopy where most food is; their long tails are used for balance and most have strong claws to grip trees and branches.
Lianas These climbing plants use trees as their 'trunk' – their stems cling to trees and climb up to the sunlight in the canopy, while getting water and nutrients from the soil below.	**Big cats** Jaguars, tigers and leopards all have camouflaged fur; the dark and light fur patches blend in with the shade and sunlight on the forest floor.
Drip tip leaves Most rainforest plants have thick, waxy leaves with drip tips; these shed water quickly to prevent leaves rotting.	**Birds** Rainforest birds often have very loud calls because it is easier to hear a mate than see them in the dense canopy; parrots and macaws have powerful beaks to break open nuts.

◀ ***Figure 4*** *How animals and plants have adapted to rainforests*

Did you know?

Tropical rainforests cover about 6% of the Earth's land, but contain about 50% of all plant and animal species.

Your questions

1 a Using Figure 2, describe the climate of Belem. Use the guidance points in Section 1.4 to help you describe it fully.

b What evidence is there for seasons on the climate graph of Belem?

2 a Describe the structure of a tropical rainforest.

b Would it matter if only the largest trees were removed for timber? Explain your answer.

3 In pairs, discuss whether it matters that the pygmy three-toed sloth (Figure 1) is in danger of becoming extinct. Feed back your ideas to the class.

Exam-style questions

4 Define the term biodiversity. (1 mark)

5 Explain **two** ways that plants are adapted to conditions in the tropical rainforest. (4 marks)

8.2 Soil fertility and biodiversity

In this section, you'll understand the relationships between nutrient cycling, biodiversity and food webs in tropical rainforests.

Nutrients: the stuff of life

The plants in tropical rainforests look luxuriant. You would think the soil must be fertile – but you would be wrong. If the rainforest is cut down, for example for farming, the soil quickly loses its fertility and becomes useless. So, how does the soil support such biodiverse forests? It's all to do with nutrient cycling.

Nutrient cycles

Section 7.5 introduced nutrient cycles – the circulation of nutrients between abiotic (non-living) and biotic (living) parts of ecosystems. Nutrients are critical to making biomes function.

All plants need nutrients to grow. Nutrients are tiny amounts of chemical elements and compounds like nitrogen, potassium, phosphate and magnesium that plants need. When plants are eaten by animals, it's nutrients that feed them. As animals die, those nutrients are recycled – so they move around biomes via the **nutrient cycle**.

However, it is not quite so straightforward because external factors affect the cycle, as shown in Figure 2 on page 243 (Section 7.5). For example:

- Nutrients can be **added to** an ecosystem by precipitation and weathering.
- They can also be **removed** by runoff or **leaching** (when nutrients are washed out of the soil by water moving through it).
- They are **taken up** from the soil as plants (or **biomass**) grow.
- They are **returned** to it when they die; first as **litter** (or decaying leaves and twigs) on the soil surface and then back into the soil as the litter decays.

Nutrients therefore move between biomass, litter and soil in a continuous cycle that keeps both plants and soil healthy.

The nutrient cycle in rainforests

In tropical rainforests, nutrient cycling is rapid. Large volumes of nutrients move quickly between stores via transfers. The nutrient cycle for a tropical rainforest, shown in Figure 1, is therefore different to the theoretical cycle on page 243. Key differences are:

- **Larger biomass store** – Layers of vegetation and huge trees store large amounts of nutrients.
- **Smaller litter store**, and **larger decay transfer** – In hot, wet conditions bacteria and fungi decay dead matter quickly, returning nutrients to the soil.
- **Larger growth transfer** – Plants grow all year, so draw nutrients up from the soil rapidly.
- **Larger weathering input** – Chemical weathering processes (e.g. solution) are faster in hot wet climates, so release nutrients into the soil from rocks.
- A **larger leaching output** – Heavy rainfall throughout the year brings in nutrients, but the constant flow of water through the soil removes them (leaching).

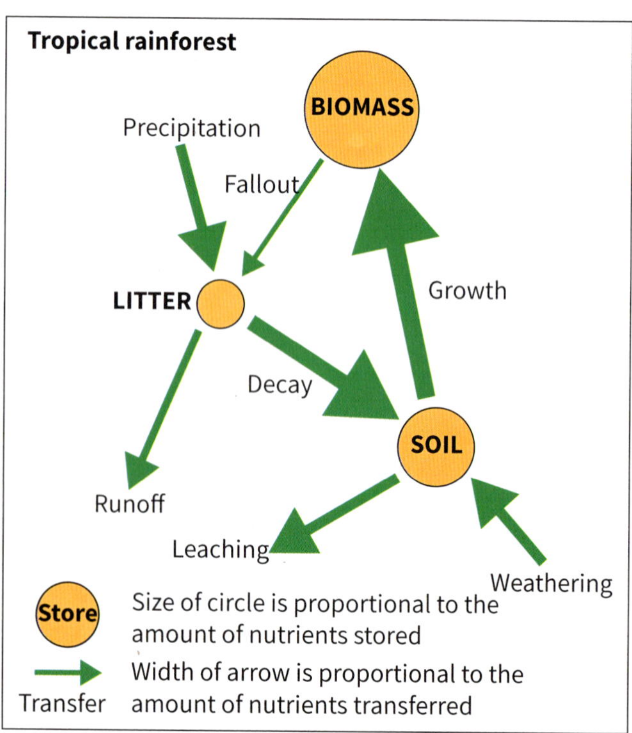

▲ ***Figure 1*** *The nutrient cycle in a tropical rainforest*

Since most nutrients in the tropical rainforest are stored in biomass, if forest is cut down (deforestation), most nutrients are destroyed. This has two effects:

- without the forest to protect it, litter and soil are easily eroded by heavy rains
- as rainforest soils contain few nutrients, land cannot be farmed for long before having to move on.

The web of life

Plants and animals in all biomes are connected through **food webs**. Sunlight is the basic energy source for all food webs. Plants convert sunlight into energy in the form of carbohydrates, through photosynthesis. As one organism feeds on another, energy passes between them (see Figure 2).

Food webs represent a delicate balance between species. If disrupted, biodiversity can be reduced. For example, if certain trees are cleared, or a species of plant dies because of disease, then primary consumers that eat it will suffer as their food source has gone.

▼ ***Figure 2*** *How energy passes through a food web in a rainforest*

Primary producers	Primary consumers	Secondary consumers	Tertiary consumers	Detritivores
Trees, ferns, flowers	Sloths, ants, butterflies	Birds, frogs	Snakes, jaguars	Fungi, bacteria
Plants ➔	Herbivores ➔	Carnivores ➔	Top carnivores ➔	Decomposers
Energy flow ➔				

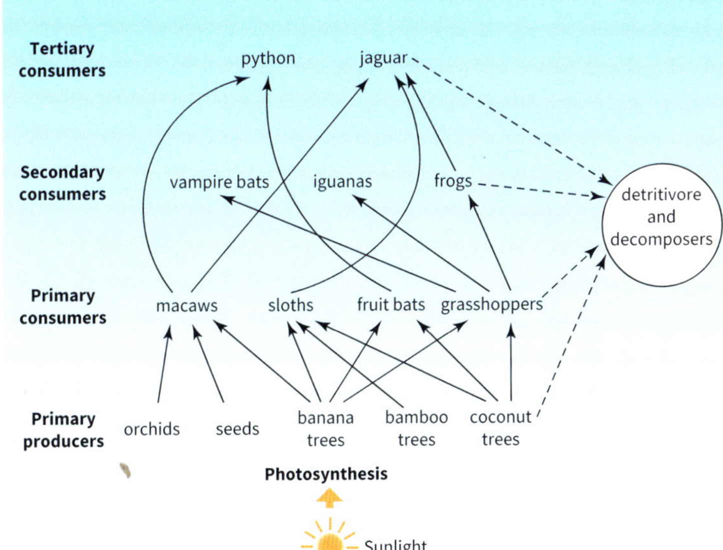

Food webs in rainforests

In tropical rainforests, there is a huge amount of biomass and thousands of different species of plants and animals. Most animals are quite selective about what they eat. This means that there are complex food webs (see Figure 3) with animals eating certain plants, and becoming the prey of other animals.

◀ ***Figure 3*** *A food web in the Amazon tropical rainforest*

Your questions

1 Identify as many differences as you can between the nutrient cycles in Figure 2 on page 243 and Figure 1 on page 250. Then complete the table below.

Differences	Reason for these differences
1	
2 etc	

2 Explain why some stores and transfers of nutrients in a tropical rainforest are so large.

Exam-style questions

3 Define the term 'biomass'. (1 mark)

4 Using Figure 3, explain what would happen to the number of pythons if macaws and fruit bats were overhunted. (2 marks)

5 Explain why tropical rainforests have complex food webs. (4 marks)

8.3 What is the taiga like?

In this section, you'll understand how plants and animals adapt to climate in the taiga, and why biodiversity is lower.

Pine trees as far as the eye can see

The photo in Figure 1 is the stuff of Christmas cards. Clear blue winter skies, and snow-covered pine trees, with trails in the snow showing where snow foxes might have hunted. Otherwise it's silent, with just the occasional thud of snow falling from a pine tree. That'll be the **taiga** in winter.

Less famous than tropical rainforests, the taiga, sometimes called **boreal forest**, is the world's largest land biome. It covers a vast 390 million square kilometres, much larger than rainforests, and makes up 30% of the world's remaining forest.

- It is found between 50° and 70° latitude, mostly in the northern hemisphere.
- Huge areas of Russia and Canada are covered by taiga.
- The trees are mostly coniferous (evergreen) and have adapted to a cold climate – their shape allows snow to fall, instead of weighing on branches, and their pine needles prevent damage by wind or snow.

▲ ***Figure 1*** *Taiga forest in winter*

The taiga climate

The taiga biome has an extreme subarctic climate, very seasonal and very different to tropical rainforests.

- Short, wet summers of three months when temperatures can rise to 20°C.
- Long cold, dry winters with several months below freezing, as low as –20°C.
- Low precipitation – below 20 mm for five months and only 350–750 mm per year.
- Snow on the ground for many months.

Figure 2 shows a typical taiga climate graph.

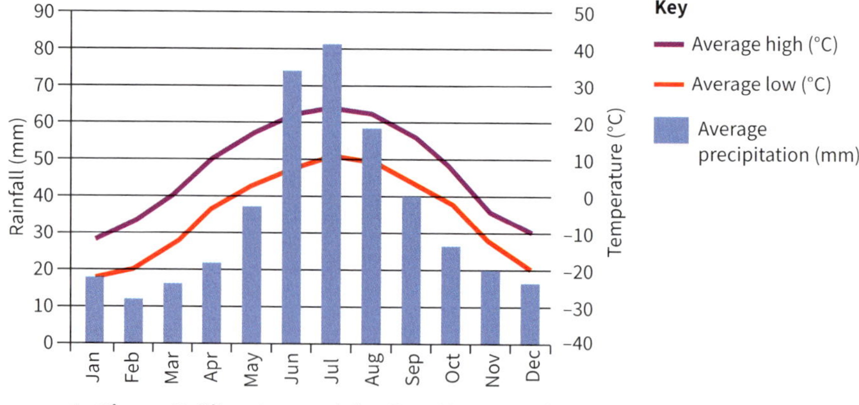

▲ ***Figure 2*** *Climate graph for Fort McMurray in Alberta, Canada, an area in a taiga biome*

Adapting to extremes

Biodiversity is quite low in the taiga because:

- Plants and animals can only survive if they are specially adapted to the cold climate; that means reptiles and amphibians are rare.
- The growing season is only 4–5 months long, meaning in winter there is little food.

Animals have to adapt to very cold winters. Plants stop growing, snow cover makes grazing even dead plants difficult, and temperatures are bitterly cold.

- Many mammals have thick, oily fur (e.g. black, brown and grizzly bears, wolves, moose and lynx) to help retain body heat and provide waterproofing.

Did you know?

When the permafrost in taiga forests melts, the ground can sag. This causes the trees to tilt in different directions, leading them to be called 'drunken forests'.

- Because food is hard to find, some taiga animals **hibernate** – e.g. bears and some species of mice, bats and squirrels.
- 300 species of birds live in the taiga in the summer, eating insects and breeding. However, another 270 species **migrate** away for winter because of the cold and lack of food.

The structure of the taiga vegetation is simpler than tropical rainforest – there is really only one layer of vegetation. However, plants have evolved to adapt to the extreme climate, as shown in Figure 3.

▼ ***Figure 3*** *Taiga forest and plant adaptations*

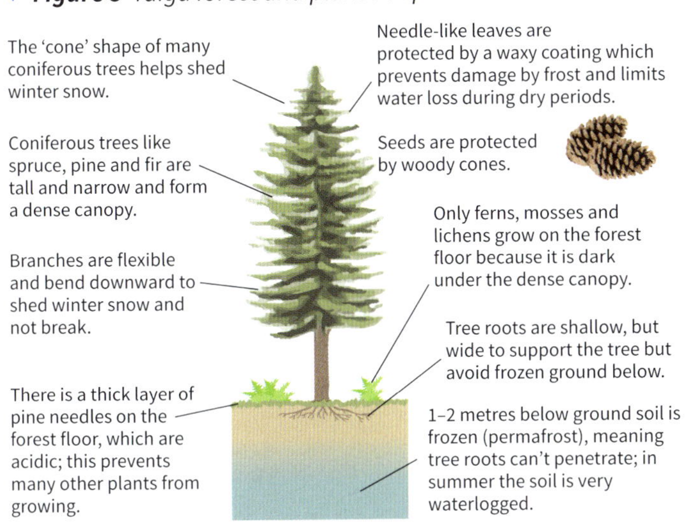

Productivity

Taiga and tropical rainforest have different levels of **net primary productivity (NPP)** (Figure 5). **NPP** is a measure of how much new plant and animal growth – or biomass – is added to a biome each year, measured in grams per square metre per year. Productivity is greatest when there is plenty of sunlight, high temperatures and precipitation. Biomes with a cold and/or dry season have lower productivity.

▶ ***Figure 5*** *Average net primary productivity (NPP) in four biomes*

Biome	Average net primary productivity (NPP) in grams/square metre/year
Tropical rainforest	2200
Temperate forest	1200
Taiga (boreal forest)	800
Tundra	140

Nutrient cycles in the taiga

In the taiga, nutrient cycling is much slower than in rainforests (see Figure 4). There are smaller flows of nutrients between stores, and stores are smaller.

- Precipitation is lower and chemical weathering is limited by cold temperatures.
- Most nutrients are in the litter – pine needles decay and release nutrients slowly.
- The biomass store is small because trees grow for only a few months each year.

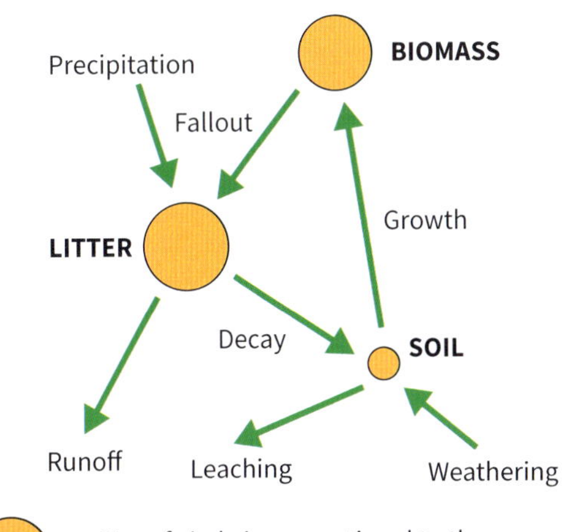

▲ ***Figure 4*** *The nutrient cycle in the taiga*

Your questions

1 Using Figure 2, explain how Fort McMurray has an extreme climate.

2 Explain how **a** plants, and **b** animals in the taiga have adapted to the extreme climate.

3 **a** Identify three differences between the taiga nutrient cycle shown in Figure 4 and that of the tropical rainforest in Section 8.2.

b Explain these differences.

Exam-style questions

4 Explain **two** reasons why litter is the largest nutrient store on Figure 4. (4 marks)

5 Explain **two** reasons why plant productivity is low in the taiga. (4 marks)

8.4 Direct threats to tropical rainforests

In this section, you'll understand why tropical rainforests are being destroyed by human activity.

Species under threat

In his book *The Singing Ark*, published in 1979, Norman Myers wrote:

'By the time you have read this chapter, one species will be extinct. We lose something in the region of 40 000 species every year – 109 a day'.

This sounds like a global catastrophe. But put another way, we will only lose 0.7% of all species over the next 50 years. This sounds more manageable, and almost as though it's acceptable to lose that much. However, we need to ask ourselves, why are we losing so many species?

People threaten forest biomes directly and indirectly.

- **Direct threats** – involve deliberate cutting down of trees for timber, to make roads or to convert the forest to farmland.
- **Indirect threats** – come from pollution, global warming or disease. These are discussed in Section 8.5.

Deforestation means the deliberate cutting down of forests to exploit forest resources (timber, land or minerals).

Did you know?

Since 1980 an area of Amazonia the size of Türkiye has been deforested.

Direct threats to rainforest biomes

The main direct threat is **deforestation**. Deforestation occurs for a number of reasons, as shown in Figure 1.

- **Poverty.** In many developing countries, local people cut down small areas of forest for land to farm because they have no other way of making a living.
- **Debt.** Countries are driven to cut down forests, export timber or grow cash crops to pay off debts.
- **Economic development.** Most tropical forests are in the developing world. In order to develop their economies, forest is sacrificed in place of roads, expanding cities, and to dam rivers and build hydroelectric power (HEP) stations (see Figure 2).
- **Demand for resources.** Tropical forests contain raw materials (see Figure 2). These include timber, but also oil, gas, iron ore and gold. To get at these, forest has to be destroyed. Land is also needed to feed growing populations.

▼ ***Figure 1*** *Direct causes of deforestation in Amazonia*

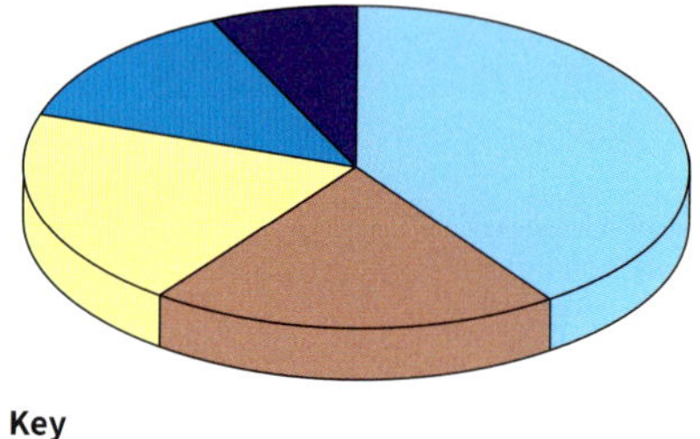

Key
- Cattle ranching (40%)
- Small-scale, subsistence farming (20%)
- Large-scale, commercial farming (20%)
- Logging, illegal and legal (13%)
- Other: mining, urbanisation, roads (7%)

▼ ***Figure 2*** *Examples of places where tropical forests are being destroyed for commercial reasons*

Cause	Example	Explanation
Mineral exploration	Coltan mining in the Democratic Republic of Congo	Coltan is a mineral used in mobile phones. It is dug from the ground from shallow mines by poor families, and sold for a few pence to transnational corporation (TNC) middlemen who sell it on. Large areas of forest are cleared to dig mines.
Hydroelectric power (HEP)	Tucurui Dam in Amazonia, Brazil	The dam opened in 1985 and cost US$6 billion. Most of the electricity is used for iron ore and bauxite mines owned by mining TNCs like Vale, especially the world's largest iron ore mine at Grand Carajas. The reservoir flooded 1750 km^2 of tropical forest, and the mines it powers still cause deforestation.
Biofuels	Palm oil in Indonesia	Indonesia has up to 12 million hectares of palm oil plantations. Palm oil is a tropical plant whose fruit contains oil used in foods, cosmetics (e.g. shampoo), and biodiesel. Huge areas of rainforest are burnt for the plantations.

Using GIS to identify rainforest loss

Until recently it was hard to know how much tropical rainforest was being cut. The advent of satellite technology and Google Earth has made it easier to find out. Figure 3 shows a GIS satellite image of part of Amazonia near the town of Porto Velho in Brazil.

- The dark green areas are tropical rainforest.
- The lighter areas are deforested places.
- Deforestation happens first along major roads, such as the one running diagonally across the top-left of the image.
- The forest is then cut down along smaller tracks which are driven into the forest from the major roads. This creates a 'herringbone pattern' of deforestation where farmers have cleared patches of forest.
- Eventually whole areas are cleared, like that in the centre of the image.

▼ ***Figure 3*** *GIS satellite image of Amazonia. The area is shown on the two maps.*

Rates of deforestation

Countries have different rates of deforestation, as shown in Figure 4. These are caused by:

- **poverty** in developing countries such as Cameroon.
- the **palm oil industry** In Indonesia.
- **protection** of forest in some emerging countries such as Brazil, which have lower rates of deforestation.
- **isolation**, for example, the Democratic Republic of Congo has a very low rate of deforestation because it is inaccessible. The rate is likely to increase as forest is opened up and poverty drives people to cut it down.

	Country	Percentage of tropical rainforest deforested 2002–19
Emerging countries	Malaysia	16.5%
	Indonesia	10.1%
	Brazil	7.1%
Developing countries	Bolivia	6.7%
	Democratic Republic of Congo	4.6%
	Cameroon	3.2%

▲ ***Figure 4*** *Annual rates of deforestation in six countries*

Your questions

1. Describe the pattern of deforestation shown in Figure 1.
2. How might the use of satellite technology and GIS help prevent deforestation?
3. **a** In pairs, list possible events in the next ten years that could cause the following changes: **i)** deforestation speeds up, **ii)** deforestation slows down, **iii)** deforestation stops.

   b Which of **i**, **ii** or **iii** do you think is most likely? Explain.

Exam-style questions

4. Name **a** one direct and **b** one indirect threat to rainforest biomes. (2 marks)
5. Explain **two** reasons why tropical rainforests are being deforested. (4 marks)
6. Study Figures 1, 2 and 3. Assess the multiple causes of deforestation in tropical rainforests. (8 marks)

8.5 Indirect threats to tropical rainforests

In this section, you'll understand how and why climate change threatens tropical forests.

Global warming: an indirect threat

Direct threats to biomes, like those in section 8.4, are manageable, and can even be reduced by countries like Brazil. Indirect threats are much harder to manage. The main indirect threat is global warming. Rising populations and resource consumption add greenhouse gases to the atmosphere, which causes the climate to change (see section 1.8). Brazil only emits about 1.5% of global carbon dioxide, so it can't prevent the global warming problem alone.

Some scientists think global warming will lead to species extinction at an unprecedented rate. Already:

- plants are flowering earlier
- bird migration patterns are changing
- the Arctic tundra is warming rapidly
- vegetation zones are shifting towards the Poles by 6 km every 10 years.

Global warming is occurring too rapidly for many species to adapt to changing climates. The table summarises expected changes as temperatures rise. A rise of 3°C could happen as soon as 2060.

Temperature rise	Impact on species	Impact on biomes
1°C	10% of land species face extinction	• Alpine, mountain and tundra biomes shrink as temperatures rise.
2°C	15–40% of land species face extinction	• Biomes begin to shift towards the Poles and animal migration patterns and breeding time change. • Extreme weather at unusual times of the year, such as heat waves and blizzards, affect pollination and migration.
3°C	20–50% of land species face extinction	• Forest biomes are stressed by drought, and fire risk increases on grassland. • Flooding causes the loss of coastal mangroves. • Pests and diseases thrive in the rising temperatures, such as bark beetles, which devastate coniferous forests.

▲ ***Figure 1*** *The impact of rising temperatures due to global warming on species and biomes*

Climate stress

The Amazon rainforest suffered severe droughts in 2005, 2010, 2016 and 2019 (see Figure 2). Droughts there are not unusual, though they do seem to be becoming more frequent and severe. At the time, the 2005 drought was described as a '1 in 100 year' drought. During these droughts, the Amazon switched from absorbing carbon dioxide to emitting it because plants stopped growing, and so stopped absorbing carbon dioxide. Forest fires broke out in the drought conditions, burning trees and litter, and releasing carbon dioxide.

Drought also puts the forest ecosystem under stress by:

- drying the leaf litter so decomposer organisms die out, threatening the nutrient cycle
- causing leaves in the canopy to die, reducing the food supply and affecting food webs.

Some scientists now argue that:

- Deforestation is making droughts more common and more severe.
- With fewer trees, there is less evaporation and transpiration. This means fewer clouds, and less rain.

Did you know?

Because of global warming, some trees in the Amazon rainforest are moving 2.5 to 3.5 metres a year upslope as they reproduce to find the cooler temperatures they need to grow!

The fear is that if drought becomes more common, rainforests will suffer permanent damage and die back. As this happens, they could become sources of carbon dioxide, not carbon dioxide sinks. This could accelerate global warming even more, making problems worse. Drought also increases the risk of forest fires. Long-term, stressed tropical forest may simply turn into tropical grasslands. Some scientists think that by 2100 between 30% and 60% of the Amazon rainforest could become a dry savanna.

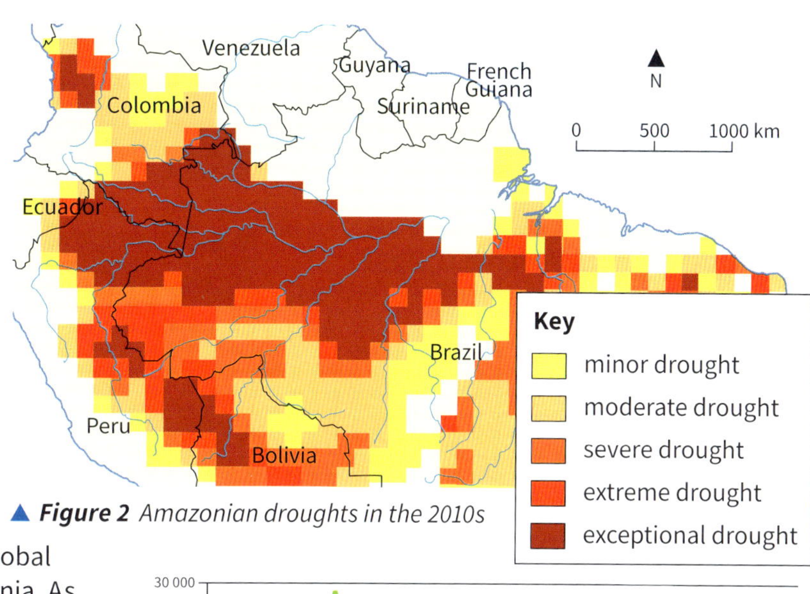

▲ ***Figure 2*** *Amazonian droughts in the 2010s*

Good news, bad news

It is worrying to think that deforestation and global warming could destroy somewhere like Amazonia. As the largest area of rainforest in the world, it produces 20% of global oxygen and is often called the 'lungs of the planet'. It contains over 40 000 plant species and over 2 million insect species. Deforestation in Brazil slowed between 2004 and 2014 (see Figure 3) because:

- From 2006 a forest area the size of France was protected by government.
- The global recession 2008–11 reduced demand for resources.
- The government cracked down on illegal deforestation by seizing land and freezing bank accounts.
- The Forest Code law, requiring landowners to preserve up to 80% of forest they own, was strictly enforced.
- Brazilians became more 'green'; 19% voted for Green Party candidates in the 2010 election.

▲ ***Figure 3*** *Annual forest loss in Amazonia, 1988–2020*

However, between 2014 and 2022 deforestation increased again. Jair Bolsonaro, president of Brazil from 2018 to 2022, strongly supported cattle ranching and commercial farming, prioritising the economy over forest protection. He reduced funding to prevent illegal deforestation. President Luiz Inácio Lula da Silva replaced him in 2022, pledging to reverse these policies and to reach net-zero deforestation by 2030.

Your questions

1. Explain the difference between direct and indirect threats to rainforests.
2. Describe the pattern of drought shown in Figure 2.
3. Explain why the speed at which temperatures are rising is such a problem for plant and animal species.
4. Create a spider diagram with five legs to show what could happen to rainforests in future if:
   - Droughts increase?
   - Further efforts are made to protect rainforests?
   - Global demand for meat continues to rise?
   - China and India's middle classes increase consumption?
   - Pests and diseases thrive in rising temperatures?

Exam-style questions

5. Describe the trends shown in Figure 3. (2 marks)
6. Suggest **one** reason for the trend in Figure 3 between 2014 and 2020. (2 marks)
7. Explain **two** ways in which tropical rainforests are threatened by climate change. (4 marks)

8.6 Direct threats to the taiga

In this section, you'll understand why the taiga is under threat from commercial exploitation

Chopping down the forest

In a major study using satellite data, the charity Global Forest Watch tracks the loss of forests worldwide, especially the loss of **intact forests**. These are:

- primary (original) forests, not secondary or replanted forests
- forests that have never been deforested
- large forest areas which allow animals to move and migrate.

Country	Type of forest	Percentage (%) of global forest deforestation 2001–20
Russia	Taiga (boreal forest)	16.9
Canada	Taiga (boreal forest)	10.7
Brazil	Tropical rainforest	6.3
Indonesia	Tropical rainforest	2.3
DRC	Tropical rainforest	1.3
Sweden	Taiga (boreal forest)	1.2

▲ ***Figure 1*** *Deforestation of intact (primary) forests between 2001 and 2020*

Its results are surprising. You might expect countries with the greatest deforestation to be those with rainforests containing endangered species, such as orangutans, but that isn't so, as Figure 1 shows.

In fact, countries which chop down the most trees are those with taiga forests. Canada and Russia account for over 25% of all deforestation between 2001 and 2020. So why is deforestation of the taiga less of an issue than rainforest destruction?

- The biome is vast. Despite Canadian and Russian deforestion, only 10% of intact taiga has been lost.
- Much of the taiga is isolated and 'out of sight' in the frozen northern latitudes.
- There are few 'cute and cuddly' species under threat that people get excited about.

Did you know?

About 40% of all wood cut down worldwide each year goes to make paper.

Write on!

Every year the world uses about 400 million tonnes of paper. Almost all comes from softwood trees – e.g. fir, pine – that grow in the world's northern forests. 80% of all trees cut down each year are softwoods. Softwood is also used for:

- Construction timber: roof beams, window and door frames.
- Board: chipboard and fibreboard used for flooring and furniture.

Paper is made by turning softwood into pulp. It is crushed and ground, sometimes using chemicals, to produce a sludge that is made into paper. A constant supply of timber is needed to run pulp- and paper-making factories. Figure 2 shows the world's largest paper producers in 2018. Many of these countries have huge areas of coniferous forest (USA, Canada, Finland and Sweden) and others import large amounts of softwood.

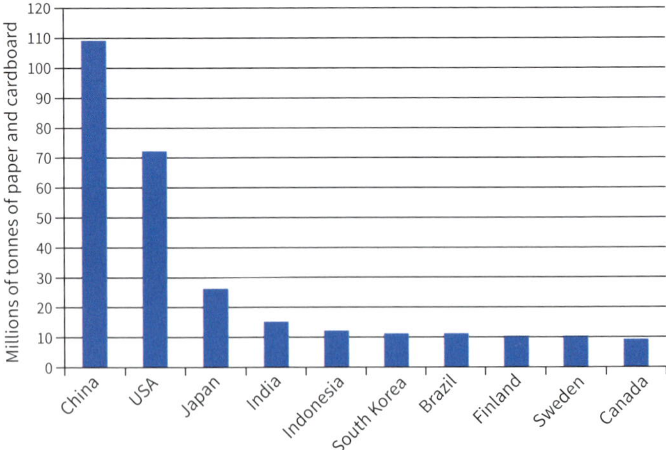

▲ ***Figure 2*** *The top paper producing countries in 2018*

Strip mining (or open-pit, opencast or surface mining) involves digging large holes in the ground to extract ores and minerals that are close to the surface.

The threat of tar sands

It's what is found underground that leads to taiga destruction. Deforestation makes way for mineral and fossil fuel exploitation, and dams (see Figure 3). Besides oil and gas extraction, there are over 4000 mines in Ontario province alone, for diamonds, gold, iron ore and other metals. Each destroys an area of taiga for exploitation, and also for roads and infrastructure.

Example of exploitation	Impact on the taiga
Athabasca tar sands, Canada Tar sands are a mixture of fossil fuel oil and sediment that can be mined and heated to separate the oil.	• Tar sands lie under an area of taiga of about 150 000 km^2. • 1000 km^2 has been mined so far. • Estimates suggest tar sands could hold 1.7 trillion barrels of oil. • Tar sands are extracted either by deforesting the taiga and **strip mining** the surface or by steaming out tar so it melts and can be collected. • Either method destroys the forest, and produces toxic waste collected in tailings ponds. • Mining uses 2–4 tonnes of water for every tonne of oil produced, plus natural gas to heat water into steam.
James Bay HEP project, Canada Water stored behind dams is used to generate hydroelectric power (HEP).	• Located close to Hudson Bay in Quebec, Canada. • One of the world's largest HEP plants generating 16 500 MW of electricity. • Built between 1974 and 2012, it has cost over US$20 billion. • 11 000 km^2 of taiga forest has been flooded during construction. • Mercury (a poison) was released as the flooded forests decayed in the reservoirs, polluting the Rupert and La Grande rivers, getting into the food web, and, via fish, eventually into the local Cree Indian population. • The roads, dams, reservoirs and electricity pylons have disrupted the caribou migrations.

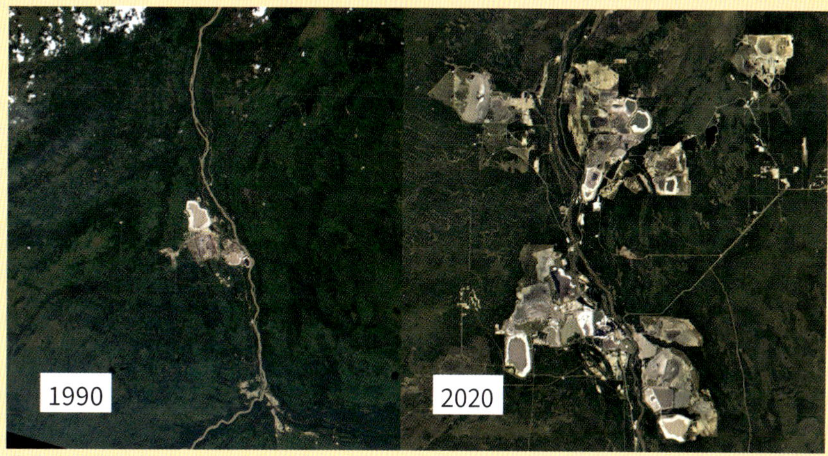

▶ ***Figure 3*** *The growth of tar sands mining near Fort McMurray, Canada between 1990 and 2020*

Figure 3 shows satellite images of the same area of taiga in 1990 and 2020. Notice these changes:

- Forest (dark green) has reduced.
- Roads and tracks have increased.
- Light grey areas are tailings ponds, containing toxic waste.
- Grey/brown areas show forest destruction to extract oil.

Your questions

1 Explain why demand for paper is such a threat to the taiga.

2 Using Google Earth, identify mining areas of Ontario, Canada (use the phrase 'mines in Ontario'). Produce a six-slide annotated photo presentation on the impact of these mines.

3 a Describe changes to the forest cover between 1990 and 2020, shown in Figure 3. 

b Describe the impact that tar sands exploitation is having in the area shown in Figure 3. 

Exam-style questions

4 Define the term 'exploitation'. (1 mark)

5 Explain **two** reasons for deforestation in the taiga. (2 marks)

6 Explain **two** ways in which economic development threatens taiga forests. (4 marks)

7 Analyse Figures 1 to 3 and 'The threat of tar sands'. Assess the ways in which mining and energy developments threaten the taiga biome. (8 marks)

8.7 Taiga under pressure

In this section, you'll understand threats facing the taiga, including threats linked to climate change.

Fire!

'Engulfed in Flames' ran the headline of the *Fort McMurray Today* newspaper on 4 May 2016. The wildfire burned 590,000 hectares of forest over two months, destroying 2200 homes and costing over £5 billion.

This seems strange. How can forest fires be a problem in cold and wet taiga? The answer is:

- Summers can occasionally be hot, and are always dry.
- The thick carpet of pine needle litter is perfect tinder to help start a fire.
- Summer storms generate lightning strikes.
- Coniferous trees contain sticky resin, which burns easily.

In fact, fire is an important, natural part of the taiga ecosystem. It allows forest to regenerate itself:

- Aspen and birch trees sprout from burned stumps.
- Black spruce, jack pine and lodgepole pine cones open when burned, releasing seeds.

Fire actually creates biodiversity; some forest areas are newly burned, whilst others have not burned for hundreds of years. Different species live in different areas.

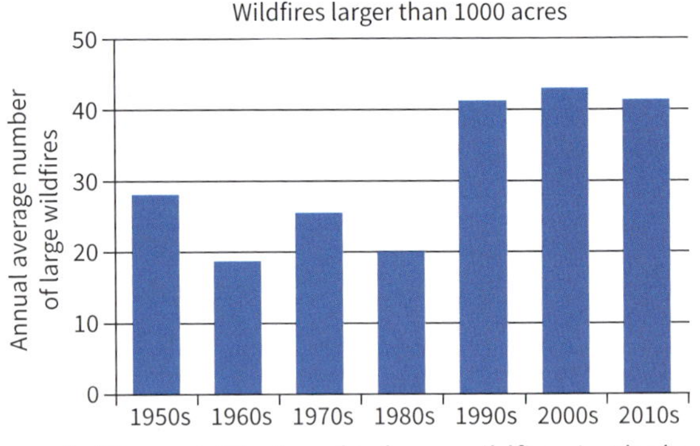

▲ **Figure 1** *The trend in large wildfires in Alaska*

Wildfire is uncontrolled burning of forest, grassland or scrub. Such fires can 'jump' roads and rivers and travel at high speed.

An **invasive species** (or alien species) is a plant, animal or disease introduced from one area to another which causes ecosystem damage.

Figure 1 shows that **wildfires** in Alaska have increased since the 1990s, and are larger and more frequent in Canada and Russia. Scientists argue that this is caused by global warming, with hotter, drier summers. The taiga needs wildfire, but too many cause long-term reductions in biodiversity:

- Forests will not regenerate properly, as trees will not mature between fires.
- Fire-tolerant species begin to dominate, reducing biodiversity.
- Trees that cannot tolerate fire decline, and so do the bird and insect species that feed on them.

▼ **Figure 2** *Aftermath of fire in the Alaskan taiga*

A plague of pests?

The taiga is home to thousands of creepy crawlies, and diseases. Normally these are kept in check by the very cold winters that kill off many insects and their larvae. In recent decades there has been an increase in insect infestation and disease, affecting coniferous trees (see Figure 3).

Pests and diseases have three consequences.

- They reduce commercial value of the forest, preventing it being sold as timber.

- They dramatically alter the ecosystem – killing tree species alters the food web.
- They change the landscape from dense forest to a more open landscape with fewer trees.

Biodiversity is reduced because only forest trees that can resist pests and diseases grow in the area.

spruce bark beetle 	• Alaska's taiga has been severely affected by spruce bark beetle. • In the Kenai Peninsula, 2.5 million hectares of spruce have been destroyed by the beetle. • The beetles bore into the spruce trees, eventually killing them. • Warmer winters, blamed on global warming, stop many beetle larvae dying in winter, creating large infestations. • Large storms, which knock down trees, seem to cause the beetles to fly off and infest new areas; an increase in storms has been linked to global warming.
mountain pine beetle 	• Mountain pine beetle has destroyed 16 million hectares of lodgepole pine forest in British Columbia, Canada. • The beetles introduce a fungus to the tree, and the fungus and feeding beetle larvae cut off the flow of water inside the tree and it dies. • Global warming, with its milder winters and warmer summers, has been widely linked by scientists to the spread of mountain pine beetles.
white pine blister rust 	• A fungal disease that attacks white pine trees. • Around 1900, it was accidentally introduced from Europe to North America so is an **invasive species**. • White pine was once an economically important tree for commercial logging in Quebec, Canada, but blister rust devastated the trees and prevented them regrowing.

▲ ***Figure 3*** *Some pests and diseases in the taiga*

Acid rain

Some parts of the taiga are affected by acid rain, particularly Scandinavia, eastern Russia, south-east Canada and eastern USA. All rain is slightly acidic, but a pH lower than 5.7 is more acid than natural. Acid rain forms when fossil fuels are burnt, releasing sulphur dioxide and nitrogen dioxide into the air. These gases react with water in clouds to form sulphuric and nitric acids. Precipitation then carries these acids down to the surface.

Acid rain makes lakes and wetlands in the taiga so acidic that fish and aquatic plants die. It also weakens trees (Figure 4).

- It damages needles (especially spruce) and their ability to photosynthesize.
- When taiga soils become too acidic, aluminum compounds are released which damage tree roots.
- Damaged soils contain less calcium and magnesium, essential plant nutrients.
- Weaker roots can't take up nutrients.
- Weak trees are more vulnerable to disease and insect attack.

Biodiversity falls when forests become stressed. Some trees die, reducing food supply. Plants that tolerate acidic soil and reduced nutrients dominate, and only insects, birds and other animals that live on these plants remain.

▲ ***Figure 4*** *Spruce trees damaged by acid rain*

Your questions

1. Draw a table to show the advantages and disadvantages of fire for the taiga.
2. Study Figure 1. What evidence is there that fire is an increasing threat to Alaska's forests?
3. Identify five threats to the taiga and draw a spider diagram to show these. Extend your diagram to show how the threats could further threaten the taiga in future.

Exam-style questions

4. Explain **one** way acid rain threatens the taiga ecosystem. (3 marks)
5. Explain **two** ways in which pests and diseases threaten biodiversity in the taiga. (4 marks)

8.8 Protecting tropical rainforests

In this section, you'll evaluate the advantages and disadvantages of global actions to protect rainforests.

Parrots in a pickle

Parrots are popular pets in North America and Europe. Some are bred legally, but there is a huge trade in illegal parrots, which means some species are threatened with extinction (see Figure 1). A breeding pair of rare parrots or macaws can sell for over US$10 000. Conservation of parrot habitats is one way of protecting them, as is banning illegal trading.

▲ ***Figure 1*** *The Red-crowned Amazon parrot*

CITES

The Convention on International Trade in Endangered Species (CITES) is an international treaty. It came into force in 1975 and is adopted by 182 countries. CITES lists 34 000 endangered species of animal (5000) and plants (29 000), including 1200 that are most threatened and in need of protection, e.g. red pandas, most rhino species, tigers, chimps and parrots. CITES bans cross-border trade in listed species. By stopping the buying and selling of endangered species, the hope is that illegal hunting and collecting will stop. However, this approach to conservation has advantages and disadvantages (see Figure 2).

Advantages	Disadvantages
Many countries have signed up; countries co-operate on trade.	Protects **species**, not **ecosystems**, so does not prevent deforestation.
Protected species include a wide variety of species geographically and by type.	Global warming could undermine its success.
Has had some key successes, e.g. reducing the ivory trade and halting the decline of African elephants.	Relies on countries setting up and funding monitoring and policing systems, which many low-income countries (LICs) can't afford.
Works well for high profile 'cute and cuddly' threatened species such as the snow leopard.	Species have to be under threat to get 'on the list' by which time the problem may be too serious to solve.

▲ ***Figure 2*** *Advantages and disadvantages of CITES*

REDD

A second global approach to conservation is REDD – Reducing Emissions from Deforestation and forest Degradation. It is a United Nations project whose purpose is to stop deforestation, which is the main cause of global warming – 20% of all carbon dioxide (CO_2) emissions come from deforestation. The aims of REDD are:

- Reduced emissions from deforestation and forest degradation.
- **Conservation** and enhancement of forest carbon stocks.
- Sustainable management of forests.

Conservation means protecting threatened biomes, e.g. setting up national parks or banning trade in endangered species.

The REDD+ programme encourages governments and transnational corporations (TNCs) in developed countries to fund forest conservation projects in developing and emerging countries. Their motive is to offset their own CO_2 emissions and thus meet their emission reduction targets. Critics argue that offsetting is an easy way for developed countries to appear to reduce their CO_2 pollution, without actually reducing it.

Did you know?

The world's largest area of protected tropical rainforest is Tumucumaque National Park in northern Brazil. At 39 000 km[2] it is larger than Belgium.

Juma Sustainable Forest Reserve (SFR)

This is an area of pristine tropical rainforest in the Amazon, 230 km south of Manaus. First protected in 2006, it is Brazil's first REDD+ project. The closest settlement is Novo Aripuana. Figure 3 shows estimates of how much deforestation would take place if Juma had not been protected.

Juma SFR is run by the Amazonas Sustainable Foundation, a non-governmental organisation (NGO). Local people are paid not to cut down the forest:

- The money to pay for conservation is donated by the TNC Marriott Hotels, Brazilian bank Bradesco, Coca-Cola Brazil, and the regional government of Amazonas State.
- Families that live in Juma SFR are paid $30 per month (given to the women) if they agree not to deforest the area. This is called the 'Bolsa Floresta' programme.
- The idea is to give people an alternative income so they no longer need to cut the forest.

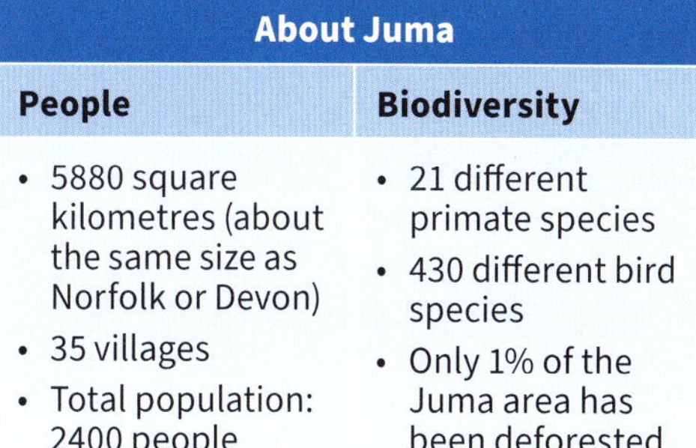

About Juma

People	Biodiversity
• 5880 square kilometres (about the same size as Norfolk or Devon) • 35 villages • Total population: 2400 people	• 21 different primate species • 430 different bird species • Only 1% of the Juma area has been deforested

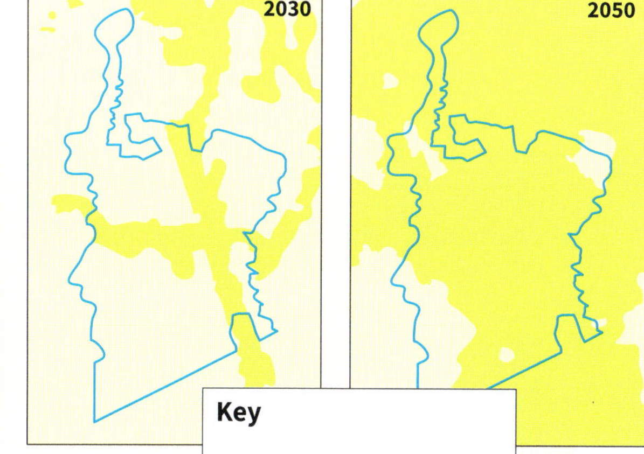

▲ ***Figure 3*** *Estimates of deforestation in Juma in 2030 and 2050 with no protection*

Has it been successful?	On the other hand ...
• Without protection, 60% of Juma's forest would have gone by 2050. • It protects a large enough area to support carnivores and larger primates. • Incomes have risen because of Bolsa Floresta payments. • The funding has built seven schools, trained people in sustainable farming, provided cleaner water and added solar panels to some roofs. • Ecotourism is being developed to provide extra income for families.	• Juma SFR relies on donations, especially from Marriott hotels. If these stop, the project may stop. • Money for families is less than $1 per day, whereas the reserve manager earns $25 000 per year. • Local people have 'signed away' their right to use products from the forest. • The huge area is hard to police, so illegal logging might continue.

◀ ***Figure 4*** *Successes and challenges ahead for Juma SFR*

Your questions

1 List the reasons for and against people keeping parrots as pets.

2 Study Figure 2.

a Take each advantage and disadvantage of CITES. Score marks out of 10 for each, and give your reasons.

b Add your total score for each. Do advantages outweigh disadvantages?

3 Compare the aims and successes of REDD with those of CITES.

4 Study Figure 4. Identify what you believe is the greatest strength of Juma SFR and its greatest weakness. Explain your reasons.

Exam-style questions

5 Study Figure 3. Describe how the pattern of deforestation is expected to have changed in the area between 2030 and 2050, if it has not been protected. (2 marks)

6 Explain **two** ways in which tropical rainforests can be successfully conserved. (4 marks)

8.9 A sustainable future for rainforests

In this section, you'll understand how tropical rainforests can be conserved by sustainable management.

Sustainable forest management

Around the world tropical rainforest deforestation rates vary from country to country. In some countries they are falling, and in others they are rising (see Figure 1). One way of conserving tropical rainforest is sustainable forest management. It conserves forests by ensuring they are not used faster than they can be renewed. Its benefits include:

- **Economic:** reducing poverty by creating income from alternative livelihoods, e.g. from ecotourism and sustainable farming.
- **Social:** may involve improving facilities to benefit the community, e.g. health clinics and schools.
- **Environmental:** protecting forest biodiversity and other resources such as rivers; using renewable energy to limit pollution.

Increasing rate of deforestation	Decreasing rate of deforestation
Democratic Republic of Congo Indonesia Nigeria	Costa Rica Colombia
• Widespread rural poverty forces people to cut forest for resources. • War and conflict make forest protection difficult. • Weak laws and little money available to police the forest. • Industrialisation drives forest destruction for timber, fossil fuels and resources.	• Strong forest protection laws; fines for illegal deforestation. • Increasing conservation in national parks and reserves, which are monitored. • Larger urban population with fewer poor people depending directly on the forest. • Public opinion supports forest conservation.

▲ ***Figure 1*** *How rainforest destruction varies between countries*

Mount Oku: Sustainable environmental management

Mount Oku forest is an area of mountain rainforest in Cameroon, Africa, home to 35 communities from three tribes (the Kom, Nso and Oku). About 250 000 people live a day's walk from the forest and it was under pressure from farming and logging for timber and fuel. In 1987 the conservation NGO BirdLife International started a project to create a sustainable forest reserve in the area (also known as Kilum-Ijim forest), shown in Figure 2.

▼ ***Figure 2*** *A sustainable forest, Mount Oku in Cameroon, Africa*

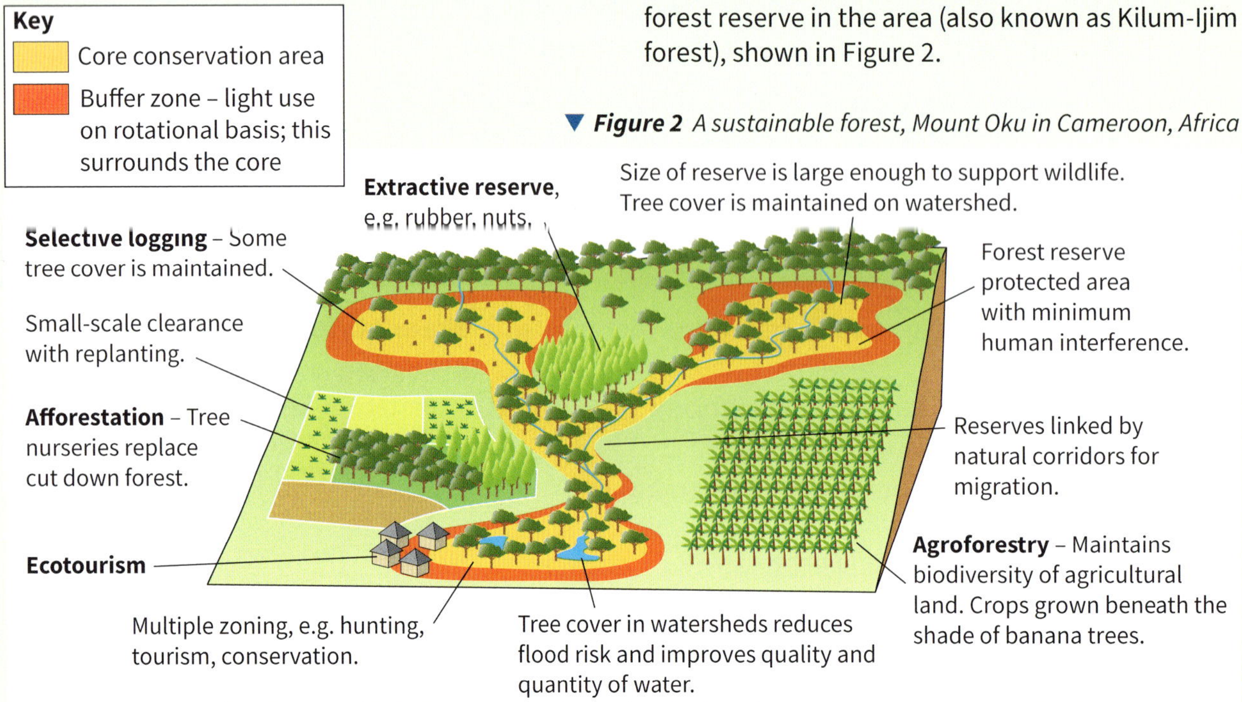

Key
- Core conservation area
- Buffer zone – light use on rotational basis; this surrounds the core

Extractive reserve, e.g. rubber, nuts.

Size of reserve is large enough to support wildlife. Tree cover is maintained on watershed.

Selective logging – Some tree cover is maintained.

Forest reserve protected area with minimum human interference.

Small-scale clearance with replanting.

Afforestation – Tree nurseries replace cut down forest.

Reserves linked by natural corridors for migration.

Ecotourism

Agroforestry – Maintains biodiversity of agricultural land. Crops grown beneath the shade of banana trees.

Multiple zoning, e.g. hunting, tourism, conservation.

Tree cover in watersheds reduces flood risk and improves quality and quantity of water.

They came together with the Cameroon Ministry of the Environment and Forestry, Kew Gardens in London and funding from the UK Department for International Development and the Dutch Ministry of Agriculture. Working with local communities they:

- marked out the forest reserve area and made lists of forests resources
- developed rules for the sustainable use of the forest
- set up a unit to manage and monitor the forest
- educated communities about replanting trees and safe levels of hunting and logging (see Figure 3).

The overall aim of the project was to conserve the forest so that future generations could continue to use it, rather than destroy it forever. The project has been a success. 50% of the Mount Oku forest was deforested between 1958 and 1988, but the forest area has increased by 8% since the project began. The key to sustainable forest management has been to provide people with alternative livelihoods such as agroforestry or ecotourism so they don't have to cut down the forest (see Figure 4).

▲ ***Figure 3*** *Local people plant new trees to replace the cut down forest*

Agroforestry	Ecotourism
• Sustainable farming – crops are grown between trees, so the trees are not cut. • Crops of different heights are grown together (bananas, maize, peppers), called inter-cropping. It helps protect the soil from erosion and reduces pest numbers.	• Small-scale, low-impact tourism. • It appeals to tourists interested in wildlife and culture. • Tourists stay with local families and eat local food. Local people act as guides. • Money from tourists goes directly to local people.

▲ ***Figure 4*** *Agroforestry or ecotourism – alternative livelihoods*

Mount Oku does face a number of challenges in the future:

- Population growth is bound to increase pressure to deforest areas.
- Urban areas, industry and even roads could encroach on the forest.
- Money and technical support from international donors could end.
- Climate change could begin to degrade the forest.

Your questions

1 In pairs, draw a table to show the strengths and weaknesses of sustainable environmental management at Mount Oku.

2 Look at Figure 2. Describe the ways in which the forest is being managed to provide sustainable economic outcomes for the local community.

Exam-style questions

3 Explain how sustainable forest management can help conserve the biosphere. (4 marks)

4 Study the information in this section. Assess how far local sustainable forest management can prevent deforestation and destruction of tropical rainforests. (8 marks)

8.10 Conserving taiga wilderness

In this section, you'll understand how taiga forests are managed and protected.

Managing taiga wilderness

Most taiga is in Russia, Canada and Scandinavia. Many pressures facing rainforests are absent in the taiga. Acute poverty isn't a major problem, nor illegal logging or deforestation. The taiga is found north of most human settlements, and is very cold. Canada's indigenous population is heavily concentrated in these regions. Surely these **wilderness** areas can just look after themselves, left to species such as the mountain bison shown in Figure 1?

However, there are pressures to develop taiga for:

- oil, gas and mineral extraction, and hydroelectric power (HEP)
- timber for paper making and construction.

Neither their isolation nor their vast size protects them.

▲ ***Figure 1*** *Mountain bison grazing in the taiga*

Wilderness – it's official!

Wilderness areas are isolated, hard-to-reach places with little human interference or settlement. In the USA and Canada, 'wilderness' is an official type of land use. In the USA, the 1964 Wilderness Act created areas of wilderness, basically untouched by human activity. All are government owned. Wilderness covers 4.5% of the USA land area. In these areas:

- motorised transport is not allowed
- recreation is allowed (e.g. camping) but people must leave no trace of their activities
- logging, mining and road building are banned.

National parks

Many taiga areas are designated as national parks in the USA, Canada and Russia. Here, conservation takes priority over exploitation of resources. National parks usually:

- exceed 1000 hectares in size
- have legal protection
- have a budget, with park rangers to protect and monitor the area
- are open to the public for recreation and leisure.

The UK has national parks too, but these are different:

- they are smaller, though their resident population is larger
- most land is privately owned.

UK national parks are managed using a system of zoning based on the UNESCO biosphere reserve model, shown in Figure 2. Different zones are identified and given different levels of protection.

▼ ***Figure 2*** *The UNESCO biosphere reserve model for conservation areas*

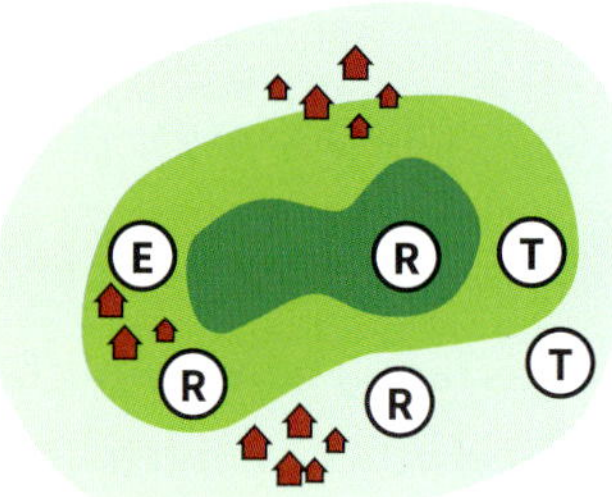

Core	No human use or settlement; protected area. Only scientific research / monitoring.
Buffer	Limited settlement and use; income generation from ecotourism. Local people educated in sustainable use.
Transition	More 'normal' economic activity but still within a protected area with scientific monitoring and education.

Wood Buffalo National Park, Canada

The world's second largest national park, Wood Buffalo, was created in 1922 to protect mountain bison (see Figure 1) from hunting. The area has grey wolves, black bears, moose and Canadian lynx populations. It is a mixture of taiga and wetland, important for migratory birds. The rarity of the mountain bison led to Wood Buffalo becoming a UNESCO **World Heritage Site** in 1983 and a **RAMSAR wetland** site in 1982.

Wood Buffalo lies north of the Athabasca tar sands mining area (section 8.6). Figure 3 shows that:

- access is from the north, with limited winter access from the south, and visitor facilities along roads to the north and east
- central and western areas are inaccessible, and are reserved for wildlife.

The area's status does not mean it is not threatened:

- Tar sands mining is proposed close by. This could pollute the park's Athabasca River, and reduce its flow when water is taken for use in mining.
- HEP dams, including one on the Peace River, could disrupt wetlands and river flow.

▲ ***Figure 3*** *Map of Wood Buffalo National Park*

In 2019 UNESCO warned the Canadian Government that it was not doing enough to protect Wood Buffalo. Canada agreed to spend $28 million to improve conservation.

RAMSAR wetland and World Heritage Sites

These are two types of conservation status reserved for areas of global importance to give them an extra level of protection.

- Countries submit sites to the global RAMSAR and World Heritage lists.
- They agree to conserve them, and provide funding to protect species from hunting, development or pollution.

However, protecting them from threats of global warming and illegal deforestation, mining and hunting is difficult.

Your questions

1 Explain the difference between 'wilderness', 'RAMSAR', and 'UNESCO World Heritage sites'.

2 In what ways can isolation be both a strength and a problem in trying to protect taiga from further development?

3 Look at Figures 2 and 3.

   a Explain the purpose of different zones in the biosphere model.

   b How far do these zones correspond to the map of Wood Buffalo National Park?

4 Explain why Wood Buffalo is still threatened, even though it is a national park.

Exam-style questions

5 Explain **two** ways in which the taiga can be conserved. (4 marks)

6 Explain **two** disadvantages of using national parks to protect the taiga. (4 marks)

7 Explain **one** advantage and **one** disadvantage of World Heritage Sites. (4 marks)

8.11 Balancing exploitation and protection in the taiga

In this section, you'll evaluate sustainable forestry and why there are conflicting views about the taiga's future.

Exploiting the taiga

Forestry is big business in the taiga, employing up to 2 million people in Russia and 500 000 in Canada. Canadian forestry generates US$27 billion every year. But not all taiga forestry is sustainable. In some areas a method called **clear-cutting**, shown in Figure 1, is used. Clear-cutting is the logging of all trees in a wide area of forest.

- It makes soil erosion likely, so the forest takes decades to regenerate naturally.
- It destroys mosses, lichen and other plants on the forest floor.
- Landslides and river bank erosion increase dramatically when all of the trees are removed.

Illegal clear-cutting is a problem in parts of Russia.

▲ ***Figure 1*** *Clear-cutting of coniferous forest*

There is an alternative to clear-cutting. **Selective logging** only removes large, valuable trees and leaves some of the forest intact.

Even if commercially-logged taiga is replanted, the regenerated forest (called secondary forest) has lower biodiversity than the original forest (called primary forest). This is because logging companies replant only the most commercially valuable tree species.

Geographical conflict means disagreement and differences of opinion linked to the use of places and resources.

Sustainable forest management in Finland

Finland has an unusual law called 'Everyman's Right'. This means anyone in Finland can use forests, including for camping, picking berries, skiing and hiking. You might think this would lead to damage to Finland's forests. In fact, the opposite is true. Because everyone has a right to use forests, people respect them and put pressure on government and industry to protect and conserve them. The features of sustainable forest management (SFM) in Finland are shown in Figure 2.

▼ ***Figure 2*** *Sustainable forest management (SFM) in Finland*

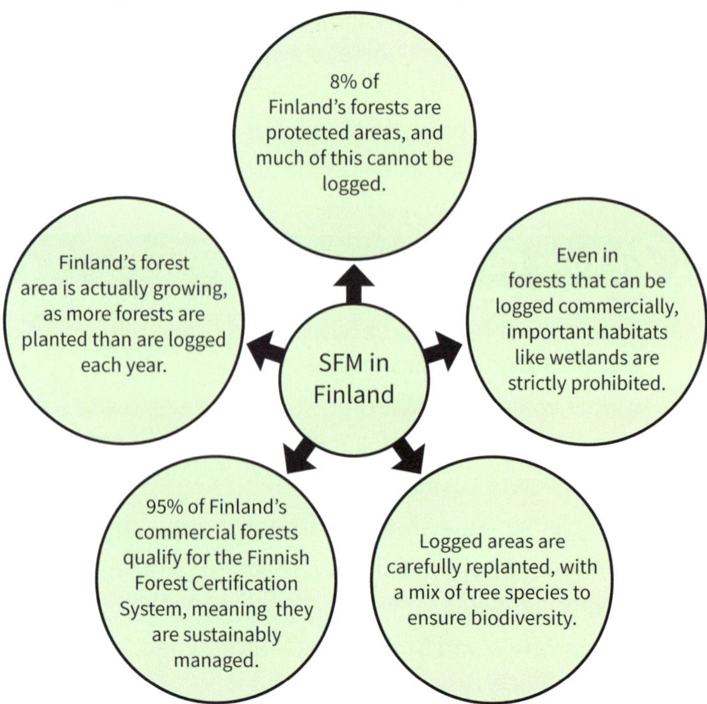

Conflict in the taiga

There are two opposing views of the taiga. Some people believe it should be conserved, whereas other believe its resources should be exploited. This is a classic argument of economic development versus environment, and is outlined in Figure 3.

Conserve the taiga's biodiversity	Exploit the taiga's resources
In favour: Environmentalists, many indigenous groups, many scientists.	**In favour:** Businesses, local government and some residents.
Reasons: • The taiga is one of the last untouched biomes. • The forests are vital global carbon sinks to help combat global warming. • They are culturally important to many indigenous peoples. • Exploitation causes deforestation, degradation and pollution.	**Reasons:** • Brings jobs and income to isolated and remote areas. • Resources can be exported, boosting national GDP. • Exploitation affects only a tiny fraction of the vast biome. • Some exploitation is renewable, e.g. hydroelectric power (HEP) and replanted forest.

▲ ***Figure 3*** *Geographical conflict in the taiga*

These opposing views cannot be easily reconciled. It often falls to national government to try and balance them by conserving some taiga, and allowing exploitation in some areas. But this makes neither side happy, as Figure 4 shows!

World Wildlife Federation spokesperson *Conservation non-governmental organisation (NGO)* 'Russia's taiga zone is fragmented by roads, rail, and infrastructure developments. Coal mining, logging, pollution, oil and gas development all pose significant threats to the region.'	**Athabasca Chipewyan First Nation spokesperson** *North American indigenous group opposed to tar sands development* 'Our Nation is not against development of our lands and territory - we want to see the respectful sharing and utilization of the land. Economic development at the expense of people and the planet makes no sense.'	**Syncrude website** *The largest oil company extracting tar sands in Canada* 'We have a long and proud history of contributing to the economic well-being of Canadians by providing rewarding employment to thousands of people, through the payment of taxes, and through the purchase of goods and services from suppliers.'
Saami spokesperson *Indigenous group from northern Norway, Finland and Sweden* 'It is our right to preserve and develop our economic activities and communities to preserve our lands and heritage for future generations.'	**Arkaim sawmill in Siberia spokesperson** 'Trees are a renewable resource, unlike coal or oil. We plant trees to replace the ones we've cut down, so there'll be plenty for the next generation.'	**International Boreal Conservation Science Panel spokesperson** 'With mounting pressures on boreal regions of Canada, it is clear that maintaining the region's globally important conservation values will require very large protected areas.'

▲ ***Figure 4*** *Contrasting views on the taiga*

Different views about how the taiga should be managed can lead to conflict between different user groups, as shown in the conflict matrix in Figure 5. Some activities – like oil and gas exploitation – can lead to major disagreement.

Indigenous people (traditional hunting, herding and use of forest resources)						
✓	Environmental groups					
✋	✓	National park and wilderness area managers				
✋	✓	✓	Climate and biodiversity scientists			
✋	✋	✋	✓	Visitors and tourists		
×	×	×	×	×	Oil, gas and mineral extraction companies	
×	×	×	✋	✋	✋	HEP companies

▶ ***Figure 5*** *Conflicting groups in the taiga* × Major conflict ✋ Minor / manageable conflict ✓ No conflict

Your questions

1 Using Figures 1 and 2, explain why logging in Finland can be seen as sustainable, whereas in Canada or Russia it is not.

2 Study the six different views expressed in Figure 4. Classify them as follows: in favour of conserving the taiga, in favour of exploiting the taiga, or a mix of conservation and exploitation.

3 Study Figure 5. Explain **a** what is meant by 'a conflict matrix', **b** how it is drawn, and **c** what it is designed to show.

4 Study Figure 5, the conflict matrix. Write 200 words explaining the conflicts between indigenous people and the other groups.

Exam-style question

5 Study Figures 2–5. Assess the reasons why different groups have contrasting views on conserving or exploiting the taiga. (8 marks)

9.1 Different types of energy resources

In this section, you'll understand how energy resources can be classified into different categories.

Using energy resources

We use energy resources every day to light our homes, to power televisions and computers, and to cook food. The energy we use comes from a variety of different sources such as oil, gas and even wind. With more people living on the planet, the global demand for energy continues to grow and more resources are being used to provide energy that we need. The International Energy Agency (IEA) has warned that current energy use is unsustainable, and that significant investment is needed throughout the world to develop renewable energy supplies. New types of energy resources are being developed to meet rising demand, such as tar sands and hydrogen fuel cells.

▲ ***Figure 1*** *Cooking the evening meal: energy use in action!*

How can we classify energy resources?

There are many different types of energy resources but they can be classified in three ways:

- **Non-renewable** – These are being used up and cannot be replaced, such as coal, oil and gas (together known as **fossil fuels**). They are sometimes known as finite resources – they're not replaceable once they've been used.
- **Renewable** – These will never run out and can be used over and over again, e.g. wind power, solar power and hydroelectric power (HEP). They are infinite resources.
- **Recyclable** – These provide energy from sources that can be recycled or reused, e.g. reprocessed uranium for nuclear power and biofuel energy.

Did you know?

The world's biggest offshore wind farm is Hornsea One, located off the Yorkshire coast. The electricity it produces is enough to power more than a million UK homes.

Non-renewable energy: natural gas supplies in Europe

Natural gas is used for electricity production, heating and cooking in Europe. In the UK, most gas comes from underneath the North Sea (see Figure 2), but much of the UK's share of the gas has been used up. Now, 44% of UK gas still comes from the North Sea and the East Irish Sea, but North Sea gas comes by pipeline from Norway which, with a smaller population, has used less of its gas. In the past, much of the EU depended on Russia and Ukraine for gas, but this changed after the Russian invasion of Ukraine in 2022. UK Government estimates expect that global gas use will peak in 2030 after which supplies are likely to diminish. Other energy sources will need to be found to meet demand.

▲ ***Figure 2*** *A North Sea gas rig providing energy for cooking and heating*

Renewable energy: wind power in the USA

Wind turbines are used to convert the power of the wind into electricity. There are now more than 16 000 large wind turbines in California, and hundreds of homes across the state are also using smaller wind turbines. In 2020 wind energy provided 7.1% of California's total energy requirements – enough to power San Francisco! However, until viable methods of electricity storage are developed, wind energy is weather-dependent, as is solar energy, and therefore unable to provide energy 100% of the time.

▲ ***Figure 3*** *Wind turbines in San Gorgonio, just outside Palm Springs, California*

Recyclable energy: biogas in India

Organic matter such as wood chips and animal dung are fed into a pit which forms part of a biogas plant and left to ferment (see section 2.14). The methane that is released is collected in a tank and can then be used to power electricity generators or used as gas for cooking. The Indian Government has encouraged communities to install biogas plants to provide energy for rural villages. There are now over 5 million biogas plants across India.

▲ ***Figure 4*** *A biogas plant at a village in Uttar Pradesh, India*

Thinking beyond

Are wind turbines the answer to our energy needs, or are they ugly blots on the landscape?

Your questions

1. Classify the following energy resources into non-renewable, renewable, recyclable: solar power, oil, wave power, coal, wind power, water, nuclear power, biogas, wood, biomass, HEP, natural gas, geothermal, tidal power.
2. Copy and complete the following table.

	Non-renewable energy	Renewable energy	Recyclable energy
Definition			
Examples			
Locations (where found)			

3. Look at the three examples of energy resources in the tinted boxes. Research one **other** type of energy resource and explain whether it is non-renewable, renewable or recyclable.

Exam-style questions

4. Define the term 'energy resource'. (1 mark)
5. Name **one** non-renewable energy resource. (1 mark)
6. Compare the meaning of the terms 'renewable' and 'recyclable' energy resources. (3 marks)
7. Explain why renewable resources cannot be relied upon for all energy production. (4 marks)

9.2 Environmental impacts of energy use and extraction

In this section, you'll understand how energy production has impacts upon the environment and landscape.

Wind turbines and solar panels – good or bad?

'For a windy country like England, wind turbines are the "foot soldiers, the pioneers" of a more intelligent energy system.' Thus speaks Piers Guy in the acclaimed film *The Age of Stupid*. In the film, he claims that wind turbines like those in Figure 1, are important for UK energy. He wants to develop three wind turbines at Airfield Wind Farm in Bedfordshire, and collect UK government payments in doing so, that will earn him a large income. He claims that three turbines would produce enough renewable energy to power 3000 homes as well as reduce **carbon emissions**. But he meets opposition from local people. Unfortunately for him, Bedford Borough Council reject the proposal. They claim that wind turbines would look out of place in a rural landscape, and have a negative impact on nearby homes, historic parkland and an ancient monument.

▲ ***Figure 1*** *Offshore wind farm off the coast of Skegness, Lincolnshire*

UK energy requirements mean that large areas of rural landscape will be used in future for wind turbines or solar panels. Solar panels allow animals such as sheep to graze the land. Landowners have been given large subsidies to erect solar panels, and they can bring farmers an income of over £15 000 each year, so are popular. But they change what people expect to see in the rural landscape.

Landscape scarring Xilinhot, China

Whichever type of energy is used, there's an impact on the landscape somewhere. China's industrialisation has led to a rapid increase in energy demand. Coal is its biggest energy resource, providing 58% of China's energy. The Chinese Government has changed policy recently, and is encouraging the development of surface opencast coal quarries instead of drilling underground mines. Quarries such as the Shengli coalfield in Inner Mongolia have become more common – Shengli is a huge opencast coal mine which is over 37 km^2 in size and has nearly two billion tonnes of coal to extract! The quarries extract 20 million tonnes of coal every year – and are expected to last 100 years! Environmental pressure group Greenpeace are concerned that such large mines create scars on the landscape and also use billions of tonnes of water to extract the coal.

▲ ***Figure 2*** *The Shengli opencast coal mine in Xilinhot, Inner Mongolia*

Oil spill – Mauritius, Indian Ocean

In July 2020, a Japanese-owned oil tanker ran aground on a coral reef on the island of Mauritius' southeast coast. When the ship's hull cracked on 6 August, over 1000 tonnes of oil leaked into the pristine waters of the Indian Ocean. The spill left a 15 km stretch of the coastline – an internationally recognised biodiversity hotspot – covered with oil, threatening marine and bird life. An international response, led by a French team, tried to prevent oil further spreading using floating booms. The United Nations sent a team of experts in oil spills and crisis management, and marine ecologists were sent from Japan and the UK. For ten days, people worked night and day, removing nearly 75% of the spilled oil.

▲ ***Figure 3*** *The Japanese tanker oil spill threatened the local environment*

Deforestation – HEP development in Pará, Brazil

The Belo Monte Dam (see page 297) is a large hydroelectric power (HEP) complex in the state of Pará in Brazil. Completed in 2019, it is one of the world's largest dams and provides electricity to support Brazil's rapidly growing economy. Currently, 45% of the energy consumed in Brazil comes from renewable energy sources, and HEP produces over 77% of all electricity used. There is, however, much opposition to the dam due to the impact on the region's people and environment. It flooded 400 km^2 of Amazon rainforest causing loss of vegetation, harming animals and changing fish migration routes.

▲ ***Figure 4*** *During its construction, the Belo Monte hydroelectric dam – the world's third largest – submerged 400 km^2 of rainforest, drowning wildlife habitats as well as people's homes*

Your questions

1 Complete this table about the environmental impacts that can arise from different energy sources, and how these might be reduced:

	Environmental Impacts	**Suggestions to reduce the environmental impact**
Wind turbines		
Opencast coal mining		
Drilling for oil at sea		
HEP development and deforestation		

2 Search online for the phrase 'UK solar panel farm'. Write a 300-word report about a UK solar farm (with photographs), giving: **i)** its name, **ii)** its location, **iii)** its size and what it looks like, **iv)** its environmental impact, **v)** any ways in which people have tried to reduce its environmental impact.

Exam-style questions

3 Define the term 'carbon emissions'. (1 mark)

4 Explain **one** way in which developing HEP could harm the Amazon rainforest. (2 marks)

5 Explain **two** ways in which opencast coal mining can have negative impacts upon the environment. (4 marks)

6 Explain why renewable energy resources do not always gain people's approval. (4 marks)

9.3 Access to energy resources

In this section, you'll understand how obtaining energy resources can be affected by accessibility and technology.

Farewell to an old friend

Maltby, South Yorkshire. A local priest reads the last rites at a service to mark the closure of the local colliery. Owners, Hargreaves Services, claim the coal is no longer economic to mine, ending an industry that has been part of the village for over a century. A piece of coal is laid next to the grave of the 'unknown miner' – the only body recovered from a pit explosion 90 years earlier, which killed 27 miners (Figure 1). The colliery brass band plays a hymn, and wreaths are laid.

Maltby lies on the Yorkshire-Derbyshire-Nottinghamshire coalfield (Figure 2). It was Britain's largest coalfield, stretching from Leeds in the north to Nottingham in the south, including cities such as Sheffield and Doncaster. It produced half of UK coal in the 1970s because of its ease of access (see Figure 2). Around Maltby, coal seams were thicker, gently tilting, and nearer the surface than on many other coalfields.

▲ ***Figure 1*** *Grave of the 'unknown miner'*

▼ ***Figure 2*** *Cross section of the northern part of the Yorks-Derby-Notts coalfield*

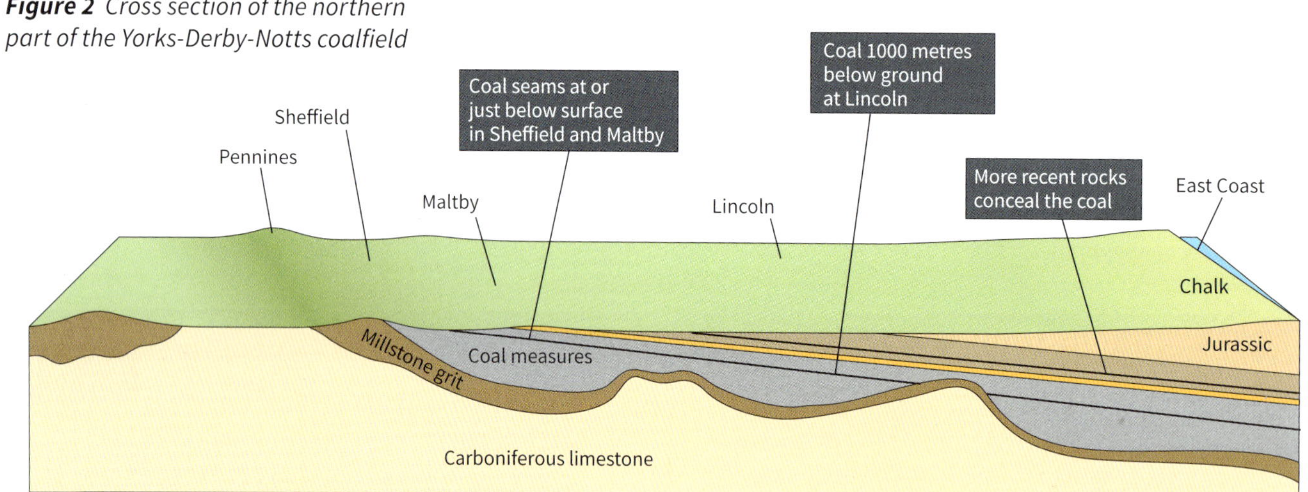

Coal – accessibility versus technology

Over two centuries, drills removed millions of tonnes of coal. Now, there is still coal there, but what's left lies further east. The seams that are left are deeper and broken up by geological faults (see section 4.1). The **technology** exists to mine the coal, but it would be too expensive compared to other energy sources.

Decline of the UK coal industry

Coal production declined sharply during the 1970s, and had almost disappeared by 2015. There were several reasons for this:

- **High cost.** Coal mining requires expensive drilling and pumping technology. Production costs rose as coal became more inaccessible – the shallowest coal had been mined.

Understanding coal

Most UK coal was formed during the Carboniferous period (section 4.1) over 300 million years ago. Coal is the remains of giant tropical plants in swamp forests. When these died, they formed layers (called **seams**) which geological pressures changed into coal. At its peak in the 1910s, coal was the UK's biggest industry, with a million jobs. It was the UK's main energy resource for industry, transport and heating.

- **Cheaper imported coal.** Cheaper imports now come from countries including, Colombia and the USA.
- **Declining demand.** Trains switched from steam to diesel and electric, and homes moved from coal heating to oil and gas. Gas is a cleaner, cheaper way of producing electricity.
- **Other energy sources** replaced coal (e.g. oil, gas, nuclear and renewables).
- **Greenhouse gas emissions.** Coal produces more greenhouse gases than any other fuel. Pressure by the EU and environmental pressure groups such as Greenpeace forced the UK to reduce coal usage.

Meeting UK energy needs

Coal is still important, but the UK now has an energy mix of different types:

- **Fossil fuels.** In the 1960s, natural gas and oil were discovered beneath the North Sea. New drilling technology and deep sea oil rigs were developed to access these. The most accessible reserves have been extracted and production is declining.
- **Renewable energy.** These depend on landscape, climate and developments in technology. Renewables provide 20% of UK energy, from wind, solar, biomass and HEP. Most wind energy potential is out at sea (see Figure 3) but the cost of building turbines is much higher there.
- **Recyclable energy,** e.g. nuclear energy which provides 20% of UK energy. Nuclear technology, using uranium and plutonium, was developed in the 1950s and the UK developed several power stations. These may now be replaced by a new generation of reactors, like the one at Hinkley Point in Somerset.

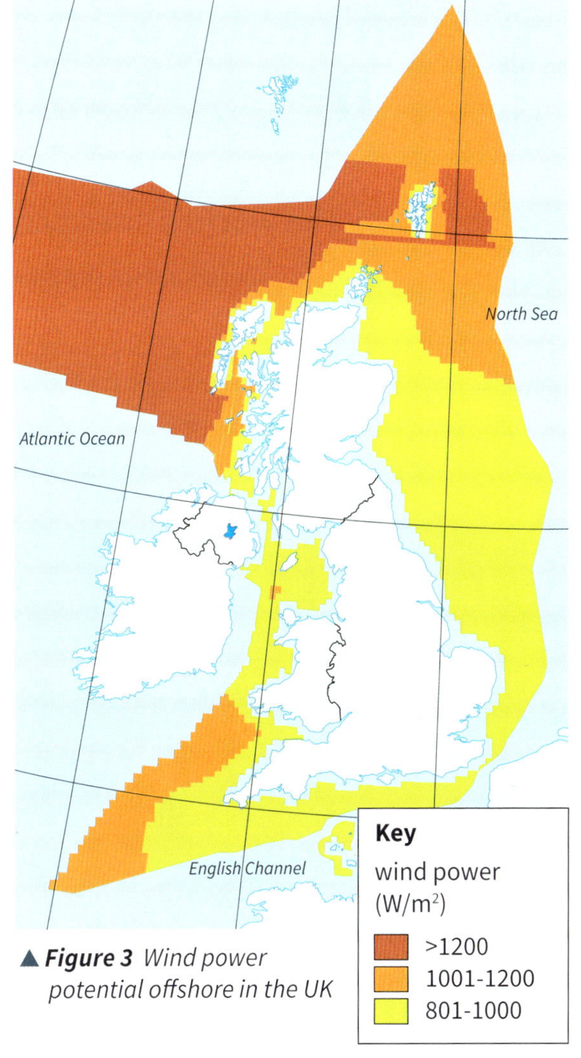

▲ *Figure 3* *Wind power potential offshore in the UK*

Your questions

1. Using Figure 2, outline the likely disadvantages of exploiting coal between Lincoln and the East Coast.
2. **a** Explain how much the UK still depends on fossil fuels for its energy.
   b In pairs, discuss and decide how much the UK is likely to depend on fossil fuels by the time you are 50.
3. Copy and complete the following table to show how different energy types depend on technology and accessibility.

Energy type	Factors that affect its accessibility	How technology can help access it
Coal		
North Sea oil and gas		
Wind energy		

Exam-style questions

4. State **one** cause for the end of coal mining in Britain. (1 mark)
5. Describe the location of coal seams in Figure 2. (2 marks)
6. Explain **two** reasons why it is harder to obtain coal in some locations on the cross section in Figure 2 than others. (4 marks)
7. Analyse the material in this section. Assess the factors that affect accessibility to energy resources in the UK. (8 marks)

9.4 Geographical skills: investigating global energy resources

In this section, you'll use a variety of skills and techniques to investigate global energy resources, and how the UK could develop its renewable energy potential.

1 Where does the world's energy come from?

Fossil fuels are not evenly spread across the planet. Some places have an abundance of them, while others have few. The map on this page shows the location of coal, oil and natural gas deposits globally. Coal is most abundant – there are probably over 900 billion tonnes of coal left – enough for 150 years, at current rates of usage!

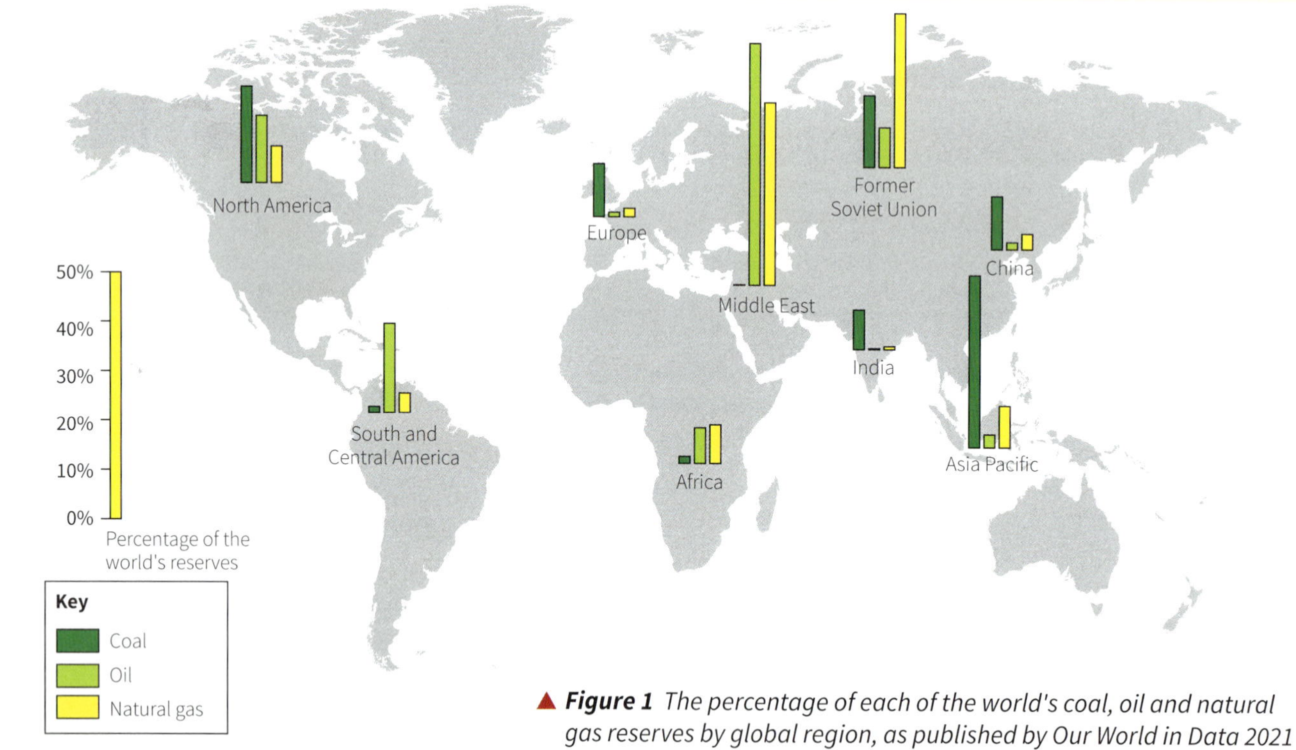

▲ ***Figure 1*** *The percentage of each of the world's coal, oil and natural gas reserves by global region, as published by Our World in Data 2021*

Your questions 1

1 Study Figure 1.
   - **a** Describe the global distribution of coal reserves.
   - **b** Rank the continents in terms of coal reserves.
   - **c** Which continent has the least potential for coal extraction?

2 Using Figure 1,
   - **a** Describe the global distribution of the known oil deposits.
   - **b** Identify which continent has the **i)** most, and **ii)** least potential for oil extraction.
   - **c** Identify which continent has the **i)** most, and **ii)** least potential for gas extraction.

3 **a** Within the region with the largest oil reserves, use an atlas to name four individual countries which are important oil exporting countries. 
   - **b** What potential conflicts can arise when one region of the world has so much of the world's oil and gas?

Exam-style questions

4 Explain **two** benefits for a country of having coal or oil deposits. (4 marks)

5 Suggest **two** ways in which potential problems can arise for countries with few or no deposits of coal or oil. (4 marks) 

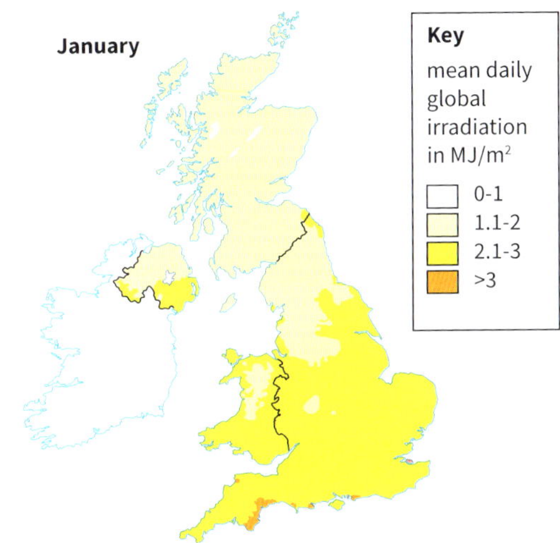

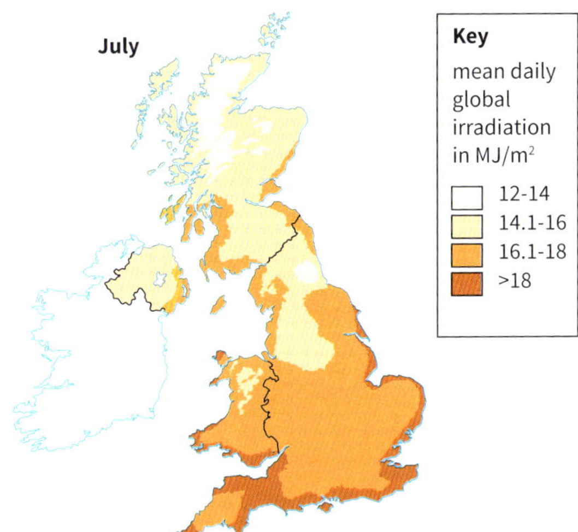

▲ ***Figure 2*** *Solar radiation maps of the UK for January and July*

2 Identifying renewable energy potential in the UK

The UK has great potential for renewable energy. It is the windiest country in Europe, has 11 000 miles of coastline and enough daylight sunshine to generate ample solar power! The maps on this page show the potential for developing renewable energy resources.

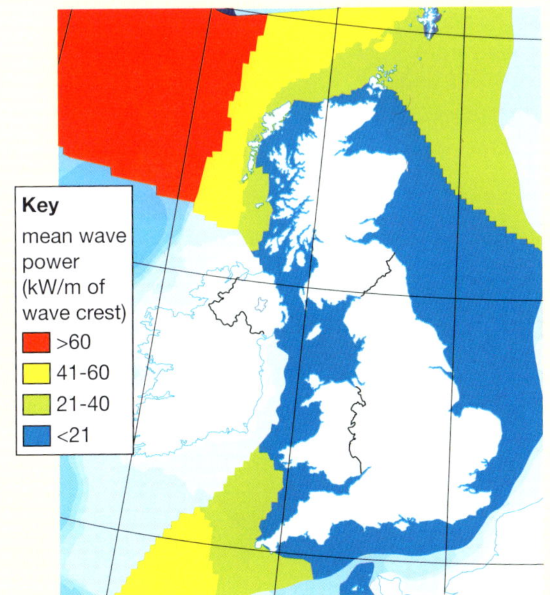

▲ ***Figure 3*** *Mean wave power in UK coastal waters*

Your questions 2

1 Study Figure 2, showing solar radiation in the UK.
  - **a** Describe the distribution of solar radiation in the UK for **i)** January, and **ii)** July.
  - **b** Identify the regions of the UK that would be most suitable for developing solar power. Explain why these are good locations.

2 Study Figure 3, showing mean wave power for UK coastal waters.
  - **a** Describe the distribution of wave power around the UK.
  - **b** Identify the regions of the UK that would be most suitable for developing wave power. Explain why these are good locations.

3 Study Figure 3 in section 9.3, showing wind potential around the UK.
  - **a** Describe the distribution of wind power around the UK.
  - **b** Identify the regions of the UK that would be most suitable for developing wind power.
  - **c** Wind farms have recently been built in large numbers offshore in the UK. Suggest the advantages and disadvantages of locating wind turbines offshore.

4 Using the information in spreads 9.1–9.4, work in pairs to produce a spider diagram to show how access to energy resources is affected by **a** technology, **b** geology, **c** accessibility, **d** climate influences on renewables, and **e** landscape influences on renewables.

Exam-style question

5 Assess the influence of physical and human factors on the development of renewable energy resources. (8 marks)

9.5 Global energy use

In this section, you'll understand the global pattern of energy use, and why it varies.

2019 – A global turning point?

Will 2019 be the year that changed everything? In that year, global energy consumption grew only very slightly and looked as though it might level off completely, as Figure 1 shows! This happened despite a growing global economy. Cars use less fuel, power stations waste less energy, and houses are better insulated. The experience from many countries, the UK included, is that the world is finally becoming more energy efficient.

But the pattern in Figure 1 is still clear – global energy consumption has continued to grow since the year 2000. The increase in energy consumption actually began in the 1700s with Britain's Industrial Revolution! But demand from a rapidly-growing global economy since the 1980s has increased global energy consumption by a half since 2000, much of the increase coming from Asia, particularly China and India.

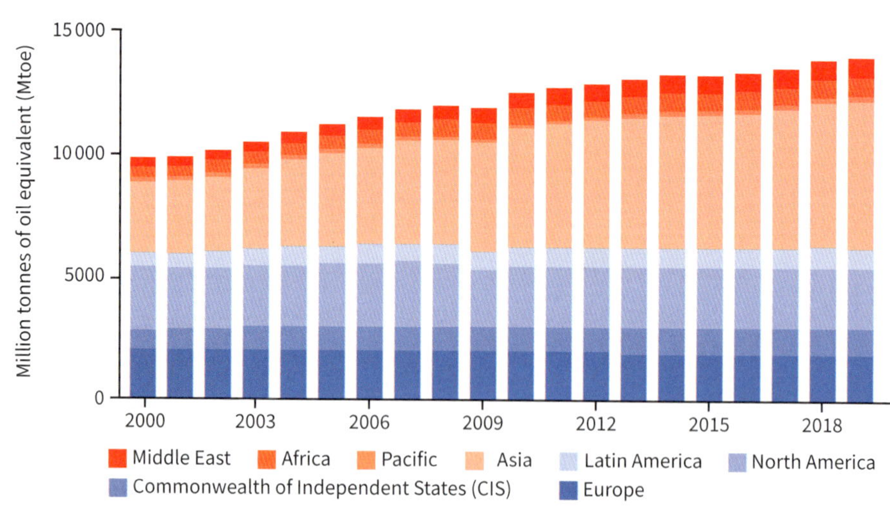

▲ ***Figure 1*** *Total energy consumption by region, 2000-2019*

Economic development and energy use

There are huge variations in the use of energy. Figure 2 shows the amount of energy consumed by different countries. The most is consumed by the USA, Russia and China, and most High Income Countries (HICs) have high consumption of energy. The USA is the world's largest user of energy – with just 4.25% of the world's population, it consumes 17% of global energy each year! As countries develop, energy use increases. Most increase in the world's HICs comes from increased ownership of domestic appliances (e.g. washing machines, TVs) and cars.

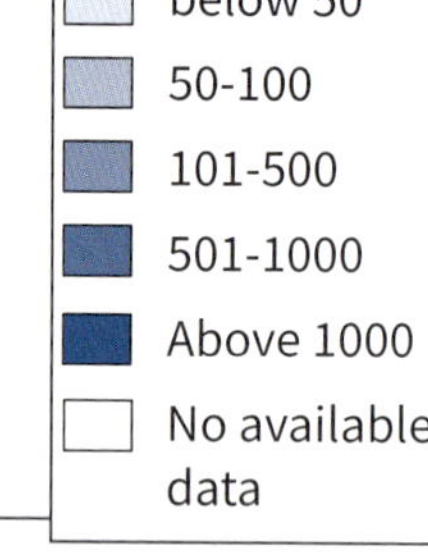

▼ ***Figure 2*** *Total energy consumption by country in million tonnes of oil equilvalent, 2019*

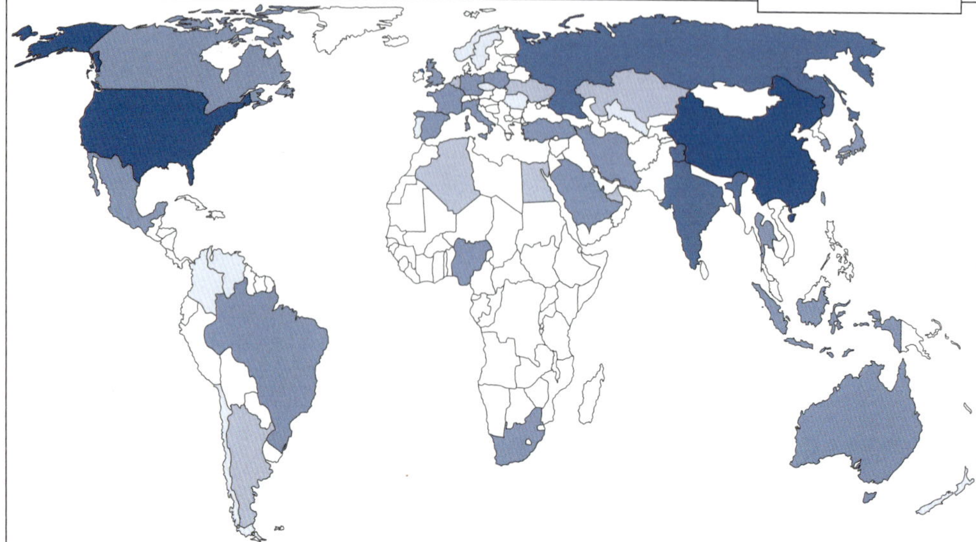

Demand from different economic sectors

Energy consumption increases with economic growth. As countries develop, their energy use changes:

- Low Income Countries (e.g. Malawi, section 2.1) have primary economies with low energy consumption. Most energy use is domestic – see below.
- Industrialisation increases energy consumption. Newly industrialising or emerging economies (e.g. India, section 2.8 onwards) focus on manufacturing, using high volumes of energy (e.g. steel-making). Large infrastructure projects (e.g. the Sardar Sarovar Dam, section 2.13) increase energy production, allowing economic growth.
- HICs consume more energy. But many have reduced energy use by 'exporting' their manufacturing overseas to countries such as China. It reduces their own energy consumption, but increases China's! Service industries and the knowledge economy demand less energy than manufacturing.

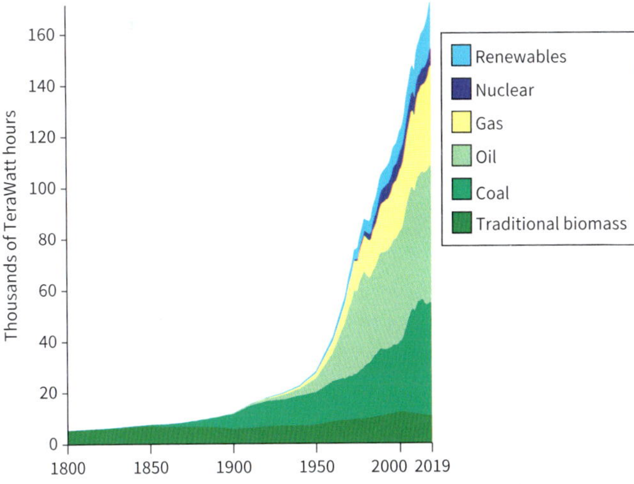

▲ ***Figure 3*** *Global primary energy consumption by source, 1800–2019*

Relying on traditional fuels

Every day in rural sub-Saharan Africa, many women carry home an average of 20 kg of fuelwood (Figure 4). They are **energy-poor**, lacking electricity or income to pay for it. What they can't pay for, they must collect. Energy poverty keeps women poor by limiting their chances to attend school, and earn an independent income. Globally, 1.1 billion people have no access to electricity. 3 billion rely on biomass for cooking and heating. Their health suffers – 4 million women and children die each year from lung conditions caused by smoke from traditional cooking stoves.

▲ ***Figure 4*** *Women in Ethiopia carrying dried animal dung and wood to use as fuel*

Your questions

1 Study Figure 1.
   - **a** Identify parts of the world in which energy consumption has **i)** grown, **ii)** remained the same, and **iii)** declined.
   - **b** Suggest what might happen to global energy consumption by 2050. Give reasons for your suggestions.

2 Study Figure 2.
   - **a** Name the continents with the highest and lowest energy consumption.
   - **b** Suggest reasons for each of these.

3 Study Figure 3.
   - **a** Describe the trends shown on the graph.
   - **b** Explain why 'traditional biomass' has not grown in use like the rest.

Exam-style questions

4 Define the term 'energy poor'. (1 mark)

5 Describe the trend in total energy consumption in Figure 1 since 2000. (2 marks)

6 Study Figure 3. Explain **two** reasons for the increase in global primary energy consumption since 1950. (4 marks)

7 Explain **two** reasons why energy consumption varies globally. (4 marks)

9.6 How much oil is there?

In this section, you'll understand how oil reserves and production are unevenly distributed, and where and why oil consumption is increasing.

Black gold!

Oil is used in many ways in modern society. It fuels cars, heats buildings, provides electricity and is used to make plastics for everything from milk containers to computers. Daily oil consumption soared from under a million barrels in 1900 to 99 million barrels in 2019. The International Energy Agency predicts that demand will rise to around 100 million barrels by 2030. But is there enough? The world may face a problem – is there enough oil to go round?

Where does the world's oil come from?

Oil production is unevenly distributed as Figure 2 shows. In 2020, 72% of the world's oil came from just ten countries! The USA produced most (18.6 million barrels a day), followed by Saudi Arabia (11 million). New reserves of shale oil in North Dakota allowed the USA to increase production to become the largest global supplier while Canada and Brazil are also producing record levels.

▲ ***Figure 1*** *Building pumps that will tap into an oil well several hundred metres below*

Drilled out?

It is difficult to tell either how much oil exists underground, or how much may yet be found. It makes sense to say that the more oil that is used, the more difficult and expensive it will become to extract what is left. Oil pessimists believe that the world has already reached, or is close to reaching **peak oil** – so oil can only get more expensive. Others believe that this point may be decades away because there is so much oil yet to be discovered.

But things change:

- It was estimated in 2016 that 1.65 trillion barrels of oil were still to be extracted. This would mean that at current rate of use there would be enough oil for another 46.6 years.
- Estimates in 2020 now show that the world has an **extra** 300 billion barrels more in reserve than it thought in 2004! As fast as oil was being used, more reserves were

Black gold – A term used for oil, as it is regarded as such a valuable commodity.

Peak oil – The theoretical point at which half of the known reserves of oil in the world have been used.

▼ ***Figure 2*** *Oil production by region in 2019, in thousands of barrels daily*

Key	
World	94 961
North America	24 363
South & Central America	6206
Europe & Eurasia	18 151
Middle East	30 162
Africa	8452
Asia Pacific	7628

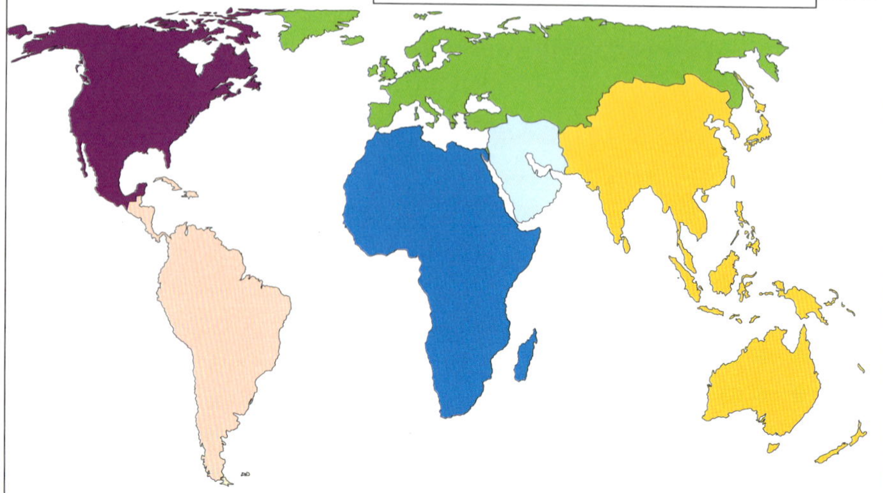

being discovered, including huge reserves of tar for refining in Canada (see section 10.3). The chief economist at BP predicts *'People will run out of demand before they run out of oil'*.

Growing demand for oil

In 2019, global consumption of oil grew by about 0.4%, a much slower rate than in 2018. Will growth will be slower than in the past? Oil consumption in developed countries has peaked, but significant increase in global consumption is likely to come from emerging economies. Increasing wealth, particularly in Asia, has led to higher demand for oil as more people use energy and car ownership increases.

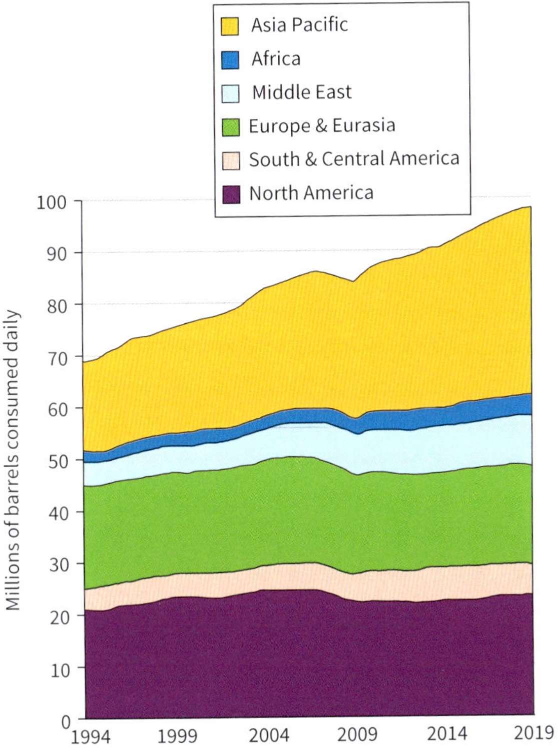

▲ ***Figure 4*** *Global consumption of oil by region, 1994–2019*

Rapid industrialisation in China

Since 1990, China has undergone rapid industrialisation with huge numbers of factories producing goods for export. Continued growth has resulted in a massive increase in energy demand, especially for oil. China's oil consumption almost doubled between 2009 and 2019, to 14.3% of global consumption. It is now second to the USA as a consumer, but its population is four times larger than that of the USA! How long before China's oil consumption overtakes that of the USA?

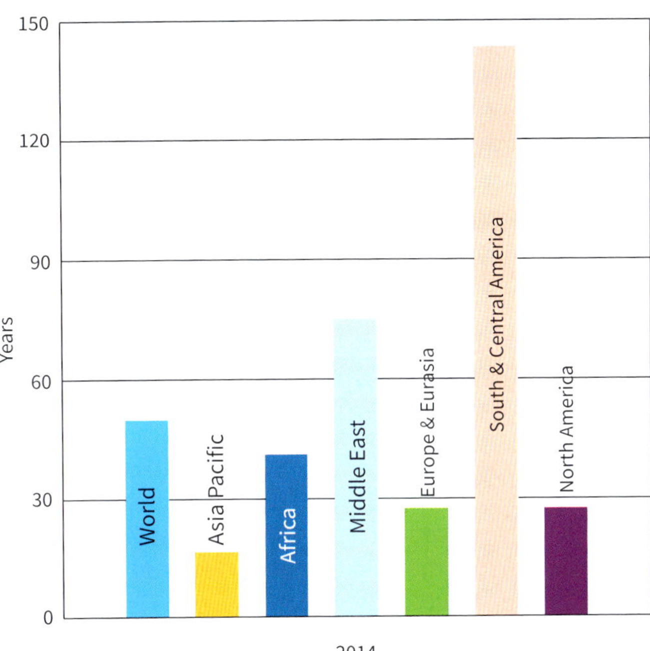

▲ ***Figure 3*** *World oil reserves by number of years at current production rates, 2019*

Your questions

1. Study Figure 2. Describe the distribution of regions producing the highest amounts of oil. 
2. Explain potential problems that countries may face as follows:
   - **a** Countries that rely on imported oil for energy and transport.
   - **b** Countries that are major exporters of oil.
3. In pairs, produce a six-slide PowerPoint about the Bakken oilfield in North Dakota, USA. Research how oil is obtained, how much is produced, and its impacts on North Dakota and on oil prices in the USA.

Exam-style questions

4. Study Figure 4. Explain **one** reason why world oil consumption has continued to rise. (2 marks)
5. Suggest **two** reasons why higher income countries are using a smaller percentage of the world's oil as time goes on. (4 marks)
6. Explain **two** reasons why oil reserves and production are unevenly distributed. (4 marks)

9.7 The changing price of oil

In this section, you'll understand how oil supply and oil prices are affected by changing international relations and economic factors.

Oil – it's a volatile market!

April 2020. The phones are busy at heating oil companies because the price of oil has collapsed to 24p per litre, from over 55p a month earlier. Those with oil-fired heating systems are keen to fill their storage tanks while the price is so low – news headlines say that the global oil price might increase, so they are keen to get supplies now.

Oil prices vary frequently – from month to month, day to day, even hour to hour. The 2020 price collapse was caused by lack of demand for oil, as the Covid-19 pandemic took hold (see Figure 2). Commuters worked from home, and shoppers shopped online, reducing demand for petrol and public transport. 70% of oil is used for transport, so the fall in demand was serious.

But oil prices can rise, as happened in 2021–22, largely because of supply restrictions. These include:

- **Security of supply.** Nigeria has suffered occasional security problems which have shut down 20% of its production.
- **Conflict in gas-producing countries.** Gas and oil prices work together – as the price of one rises, so does the price of the other. Conflicts in gas-producing countries (e.g. Russia) can mean that gas supplies are reduced, and so the price of gas – and therefore oil – goes up.
- **Deliberate reductions** to oil production by OPEC countries (such as Saudi Arabia) in order to keep prices high.

▲ ***Figure 2*** *A London street during the Covid-19 lockdown in spring 2020, when office workers were working from home. Oil prices fell by nearly half due to lack of demand.*

The Iraq war 2003–2011

In 2003, following more than a decade of intermittent conflicts involving Iraq, US and Allied forces invaded Iraq. It has the world's fourth largest oil reserves, and its then leader, Saddam Hussein,

Year	Average price per barrel (US$)
1975	12.21
1980	36.83
1985	27.56
1990	23.73
1995	17.02
2000	28.5
2005	54.52
2010	79.5
2015	55.0
2020	41.96

▲ ***Figure 1*** *The average global price of one barrel of crude oil, 1975–2020*

OPEC (Organisation of Petroleum Exporting Countries) – established to regulate the global oil market, stabilise prices and ensure a fair return for its 12 member states who supply 40% of the world's oil.

Why do oil prices change?

The global price of oil is decided by traders on commodity markets (see section 2.5), based on:

- **demand.** High demand causes prices to rise, and falling demand causes lower prices.
- **supply.** Too much oil and the price falls, too little and it rises.

Countries sometimes deliberately increase supply to increase income, even if they know increasing supplies will drive the price down. The price of oil fell from US$115 per barrel in 2014 to US$50 per barrel in 2015. It was caused by oversupply of oil from Saudi Arabia, Iraq, USA fracking sources, and Iran.

was thought to pose a threat to global oil supplies. The conflict led to shortages of oil and an increase in prices. Other Allies (e.g. Saudi Arabia) then increased production in order to stabilise prices.

▲ ***Figure 3*** *Burning oil pipelines during the Iraq war (2003–2011) which was largely about oil*

The era of fracking

Oil exists in many rocks; one of the most common is shale (see section 4.3). In recent years, it has become commercially viable to extract oil from shale. The key change has been caused by a revolutionized form of oil drilling technology which makes it possible to drill horizontally (see Figure 4). Water is blasted into rock fractures under pressure, a process known as hydraulic fracturing, or **fracking**. The impact of fracking is revolutionary for the oil industry, and explains why the USA now exports more oil than it imports.

However, fracking has its opponents. It involves huge water consumption in blasting shale to obtain oil, and the water is often polluted as a result. Loosening the rock is also said to cause earthquakes locally.

▼ ***Figure 4*** *The fracking process*

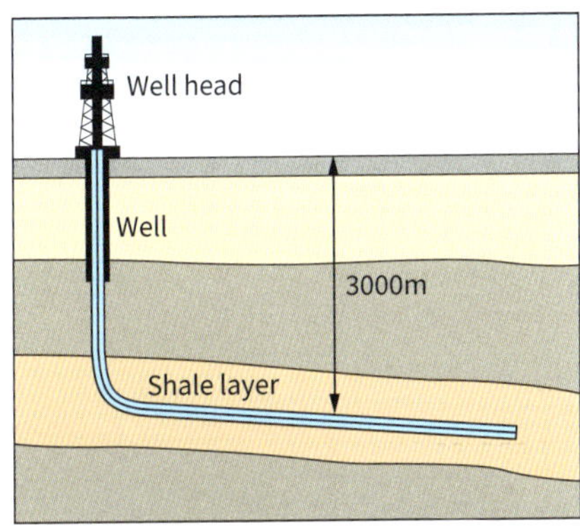

1. Wells are bored using directional drilling, a method that allows drilling in vertical and horizontal directions to depths of over 3000m.

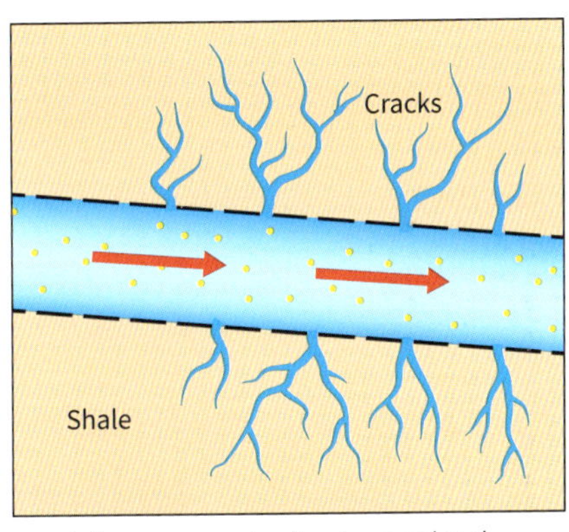

2. Large amounts of water, sand and chemicals are injected into the well at high pressure, causing cracks in the shale.

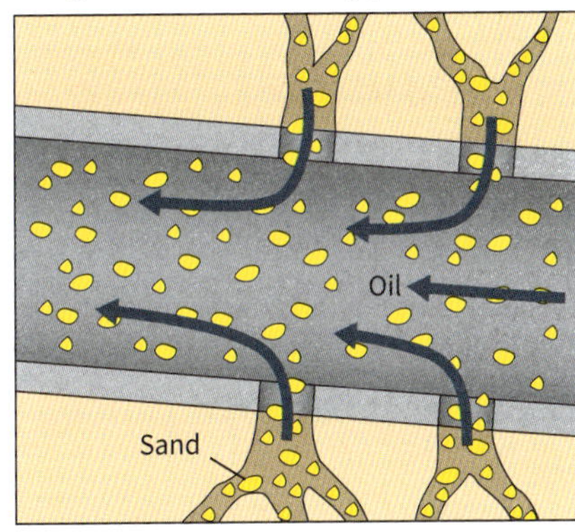

3. Sand flows into the cracks keeping them open so that the oil from the shale can flow up and out of the well.

Your questions

1 Study Figure 1. Draw a line graph for the data.

2 Using Figure 2 to guide you, suggest why city streets were so empty during the Covid-19 pandemic.

3 Copy and complete the following table. Take each event and explain its likely impact on the global oil price.

Event	Likely impact on the global oil price, and why
2003 Invasion of Iraq	
2008 Global financial crisis	
2011 'Arab Spring'	
2020 Covid-19 pandemic	

4 Research how much oil is meant by the term 'one barrel'.

Exam-style questions

5 Suggest why global oil prices might be likely to rise. (4 marks)

6 Suggest **two** reasons why global oil prices might be likely to fall. (4 marks)

7 Analyse the information from this section. Evaluate the statement that 'changing oil prices can be affected by international relations'. (8 marks)

9.8 The costs of developing fossil fuels 1

In this section, you'll assess the economic benefits and environmental costs of developing oil and gas in environmentally sensitive areas.

Gazprom – the polar pioneers

The vast, white, barren Arctic Circle northwest of Russia. A speck of red punctures the view – it is the *Prirazlomnaya*, a giant oil rig owned by Gazprom, one of the world's largest oil and gas companies. Gazprom has spent US$4 billion developing technology to search for oil and gas beneath the sea in a region that holds over 70 million tons of oil. The platform was built specifically for the Arctic region and is used for all production operations, including oil drilling, extraction, storage, treatment, and offloading. Drilling in the Arctic presents challenges. It's cold, fragile, and so remote that it took a fleet of ships and aircraft to take everything needed for the operation.

As the *Prirazlomnaya* prepared to depart from Murmansk in September 2013, it was surrounded by 30 activists and crew (shown in Figure 2) from the Greenpeace International ship, *Arctic Sunrise*. They were protesting against drilling for oil in the Arctic, and impacts of further use of fossil fuels on global climate.

▲ ***Figure 1*** *The* ***Prirazlomnaya*** *in the Pechora Sea*

▲ ***Figure 2*** *Protestors against Arctic drilling*

Exploring remote places

Why drill in such difficult environments? It offers huge potential revenue for oil companies – and a guarantee for customers that petrol won't dry up yet. Profits from high oil and gas prices throughout the 2000s (section 9.7) have helped. They allow oil companies to drill in regions previously thought too expensive (e.g. in deep seas), or too difficult to work (in intense cold). Costs are high – so profits need to be guaranteed.

New technologies enable companies to exploit oil and gas in extreme environments:

- **drilling technology** allows deep water reserves of oil and gas to be accessed. Estimates suggest that over 5% of the world's oil and gas lie under the seabed.
- **seismic imaging** has led to the discovery of oil off the coasts of Brazil and West Africa, and in the Gulf of Mexico. It uses sound waves bounced underground to detect rock structures containing oil and gas.
- **liquefaction** of natural gas (converting gas into liquid) means it is economically possible to transport gas by ship. This allows development of remote gas fields such as Australia's Ichthys Project (see text box).

Thinking beyond

Is the Arctic worth exploring, even if we decide not to exploit anything that is found?

The Ichthys LNG project

The Ichthys gas field lies 220 kilometres off the coast of north-western Australia. It's one of the world's biggest offshore gas fields, below 260 metres of water. It produces 10 million tonnes of liquefied natural gas (LNG) *per day* to meet demand in Asia. At that rate, there's enough gas there for 120 years! The benefits are huge too – 900 jobs were created in pipeline construction alone.

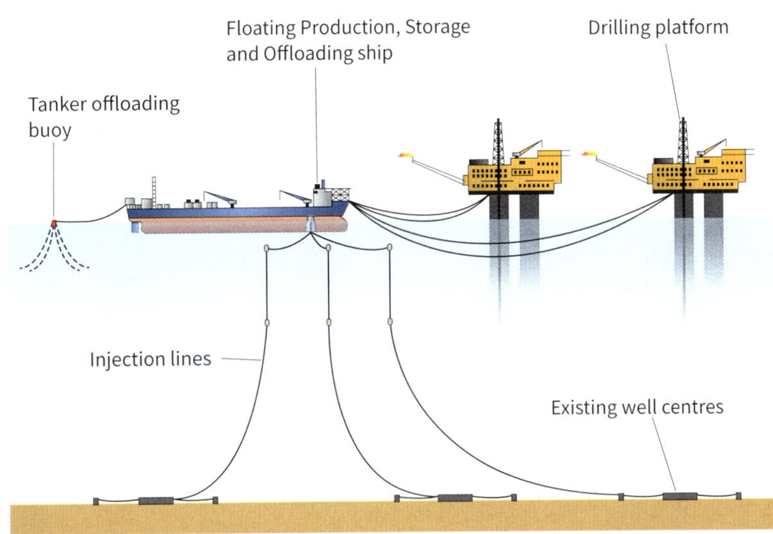

▶ ***Figure 3*** *How liquid natural gas is obtained*

The Arctic at risk?

Oil and gas companies have long known that the Arctic region (see Figure 4) has up to 25% of the world's remaining oil and gas.

- The region – which includes the Arctic Ocean, northern Russia and northern Canada – is huge, nearly four times larger than Australia!
- It is a vast wilderness, with a quarter of the world's taiga forests (see Chapter 8). Only 4 million people live there, from over 40 different ethnic groups.
- Several countries lay claim to the Arctic.

Environmental groups are concerned that oil companies have already damaged many parts of Alaska and Siberia. They think that the world should be using renewable energy instead of oil and gas.

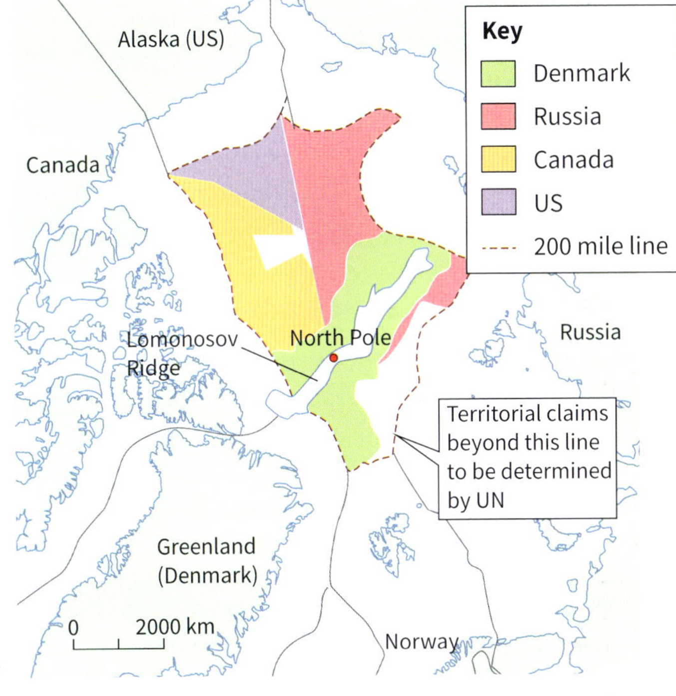

▶ ***Figure 4*** *Territorial claims in the Arctic*

Your questions

1 In pairs, produce a PowerPoint showing photographs of the Arctic region, using locations in Figure 4. Include a map of the region, and explain the difficulties of drilling there.

2 Draw a table listing **a** the economic benefits, **b** the environmental costs associated with developing oil and gas from the Arctic.

3 Now list the wider environmental pressures in developing a wilderness region such as the Arctic. Use sections 8.3, 8.6 and 8.7 on the taiga to help.

4 Explain in 200 words why it may not be easy to decide whose oil or gas lies beneath the Arctic.

Exam-style questions

5 Explain **one** cost and **one** benefit of searching for energy resources in environmentally sensitive areas. (4 marks)

6 Analyse the contents of this section. Assess the arguments for and against exploiting the Arctic region. (8 marks)

9.9 The costs of developing fossil fuels 2

In this section, you'll assess the environmental costs of developing unconventional oil and gas resources in ecologically sensitive and isolated areas.

Unconventional fossil fuels

February 2021. Joe Biden, the newly elected President of the USA, holds his first international virtual meeting with Justin Trudeau, Prime Minister of Canada. It's not good news – Biden is cancelling the permit for the Keystone pipeline, which was to transport 830 000 barrels a day of highly polluting carbon-intensive tar sands oil, from Alberta (Canada) to Nebraska (USA).

Tar sands oil extraction is part of a new wave of unconventional fossil fuel extraction that is now possible because of a combination of improved technology and high oil and gas prices. Vast reserves of oil and gas are stored in deposits often in ecologically sensitive and isolated areas such as Athabasca in the central wetlands of Canada (see section 10.3). The oil cannot be drilled by conventional technology, and needs new unconventional technology. The problem is the huge environmental cost.

▲ ***Figure 1*** *Canada's tar sands oil extraction in Athabasca*

Extracting tar sands and shale gas

The extraction of tar sands to produce oil and shale gas is unconventional (or unusual).

- **Tar sands** occur naturally – they are a mixture of sand, clay, and water, and a very dense sticky form of petroleum called **bitumen**. Bitumen is extracted from tar sands by injecting hot steam underground. This heats the sand and makes the bitumen far less sticky so that it can then be pumped out. The process requires enormous amounts of energy and water. Unusually, much of this is obtained by opencast mining (see Figure 1) which strips away surface rock to get at the shale beneath. After this, the land has no further use.
- **Shale gas** is natural gas that is trapped underground in shale rock (see section 4.3). Shale is impermeable which means that any gas trapped inside cannot be reached or pumped out using conventional vertical drilling. Instead, **fracking** is used to extract the gas (see section 9.7).

▶ ***Figure 2*** *Map showing the location of the Athabasca region of Canada*

▶ *Figure 3 Digging at the Muskeg River Mine, near Athabasca*

Tar sands oil extraction in Athabasca, Canada

Most of Canada's tar sands are found in and around Athabasca, in the western province of Alberta. It is a challenging environment to work in – hot in summer and –30°C in winter. The area covers 140 000 km^2 (half the size of the UK!) that is mainly home to taiga and peat bogs. The extraction process of tar sands oil in the region often involves large-scale open pit mining requiring the destruction of forests and peat bogs, and therefore the loss of ecosystems.

The other environmental challenge to the region is the huge amounts of water required for the extraction process. It can take between two and five barrels of water to produce one barrel of oil and it is estimated that in Athabasca, 359 million tonnes of water is used annually in the extraction process (twice the consumption of a large city). There have also been several reported leaks of water polluted with oil into the nearby Athabasca River.

Greenpeace are concerned that the development of the tar sands industry has led to a decline in many animal species in the region including caribou, lynx and wolverines (see section 10.3). It is also thought that more than six million birds will be lost by 2050.

Your questions

1 Explain the difference between tar sands and shale gas.

2 List the environmental problems associated with developing tar sands oil and shale gas resources from remote regions.

3 **a** Compile a list of costs and benefits of extracting shale oil and gas.

b Score each cost and each benefit out of 10 in terms of its degree of cost or benefit.

c Add a total score, and decide which is greater – costs or benefits.

Exam-style question

4 Analyse the information in this section. Assess the problems that result from exploiting Canada's unconventional oil and gas resources. (8 marks)

Shale gas extraction in Taranaki, New Zealand

Taranaki is located on the west coast of New Zealand's North Island and is famed for Mount Taranaki, an active volcano in the nearby Egmont National Park. It is an exceptionally fertile rural area with a rich volcanic soil. There are 48 hydraulic fracturing activities taking place across the region and the main environmental concerns are the risk of water pollution and the possibility of seismic activity.

▲ *Figure 4 Drilling site in Taranaki*

9.10 Reducing reliance on fossil fuels

In this section, you'll understand how energy efficiency and energy conservation can cut demand, help finite energy supplies last longer and reduce carbon emissions.

How much longer can it last?

By 2050 there are likely to be more than 9.7 billion people on Earth. Two thirds will be living in urban areas. Each person will demand energy for everyday life. As living standards improve, more people will want refrigerators, computers and cars. Global energy demand could be 80% higher by 2050 than in 2000!

Carbon footprint is a calculation of the total greenhouse gas emissions caused by a person, a country, an organisation, event or product. These emissions can be:

- **direct** – created by burning fossil fuels for energy use at home or on transport.
- **indirect** – those that come from owning a product, from its manufacture to its final disposal.

Every use of fossil fuels contributes to the world's **carbon footprint**. Fossil fuels release greenhouse gases when they are burnt, which contribute to climate change (see section 1.8). The more fossil fuels that are used, the greater the carbon footprint and its impacts. Fossil fuels are finite; reducing their use means they'll last longer and also cuts carbon emissions.

Energy efficiency and conservation at home

Reducing energy consumption at home can significantly cut energy use. Several home improvements can reduce energy consumption, e.g. installing energy efficient boilers and double glazing (see Figure 1). To help achieve this, the UK government offers a 'Green Deal':

- **Loans** – repaid in installments which are added to electricity bills over several years. These pay for home improvements to cut energy consumption e.g. wall and loft insulation.
- **Grants** – The government will provide a voucher that covers up to two thirds of the cost of installing two energy-saving home improvements.

In June 2019, the UK government set a goal to reach net-zero carbon emissions by 2050. Following this, it introduced the Smart Export Guarantee (SEG) for small-scale low-carbon electricity suppliers to receive payment for electricity supplied to the National Grid:

- Householders are encouraged to install small-scale, low-carbon energy generators such as PV solar panels.
- Energy companies pay for any renewable energy exported to the National Grid.

Now the government is banning the sale and installation of new oil and gas heating boilers in new homes from 2025, and in all houses from 2030.

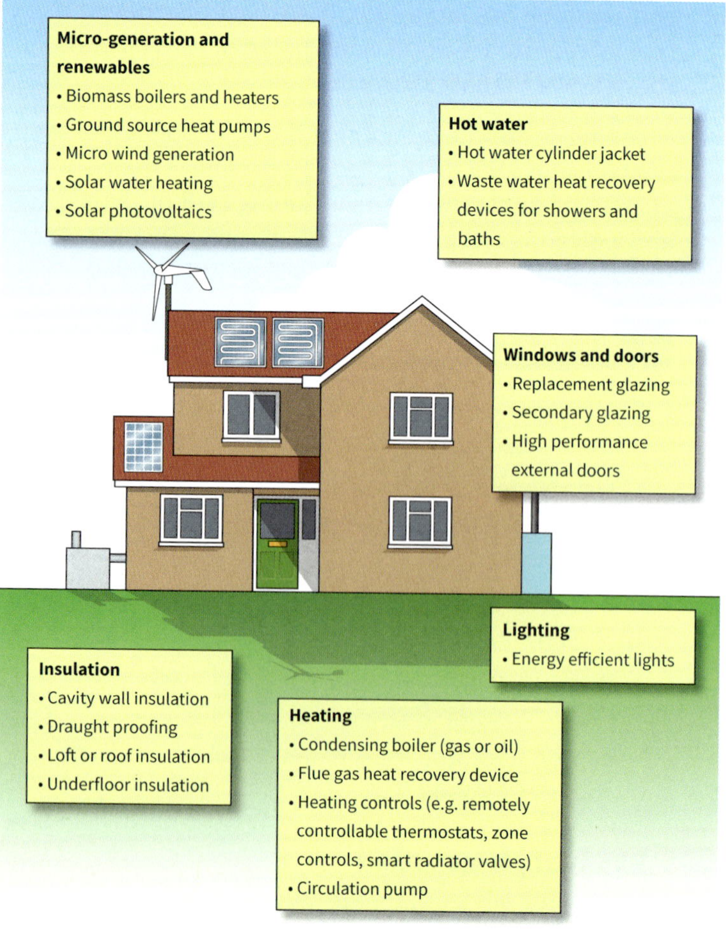

▲ **Figure 1** *Improving energy efficiency in the home*

Improving transport energy efficiency

In the UK, Vehicle Tax Rates – often called road tax – depend on levels of vehicle pollution. Vehicles with low CO_2 emissions (<100 g/km) pay no road tax. They are designed to encourage drivers to buy vehicles with lower CO_2 emissions. High polluting vehicles that are less fuel-efficient are taxed at higher rates. In 2020 global sales of electric cars rose by 43% to more than 3 million. Tesla were the most popular brand of electric cars, with the Tesla Model 3 achieving a range of over 300 miles before recharging is needed.

▲ ***Figure 2*** *The Tesla Model 3, the world's best-selling electric car*

Reducing London's carbon footprint

- Since 2019 all of central London's double-decker buses have been hybrid, electric or hydrogen, and since 2020 all single-decker buses emit zero exhaust emissions. By 2037 all 9200 buses across London will be zero emission.
- London's **cycle hire scheme** was introduced in 2010. Funded by Santander, it provides 11 500 bikes for hire at 750 docking stations. In 2021 there were 450 km of Cycleways across London. These are routes that link communities, businesses and destinations in one cycle network. More Cycleways are planned for the future.

▲ ***Figure 3*** *London's new 21st century Routemaster buses which use hybrid technology*

Your questions

1 a In pairs, research 'carbon footprint' online and calculate your own footprint. Compile class results.

b Write a paragraph describing **i)** how your footprint was calculated, **ii)** which parts contributed most to your score.

c Outline three ways in which your footprint could be reduced.

2 Draw a table to compare the strengths and weaknesses of UK initiatives in homes and transport to reduce people's carbon footprints.

Exam-style questions

3 Define the term 'carbon footprint'. (1 mark)

4 Explain **two** pieces of evidence which indicate that the UK is trying to reduce its carbon footprint. (4 marks)

5 Explain **two** ways in which efficiency and conservation measures can reduce energy consumption. (4 marks)

9.11 What are the alternatives?

In this section, you'll assess the costs and benefits of alternatives to fossil fuels.

Will the lights go out?

What if there's a world ahead without enough energy? India already cannot meet its energy demand – a huge power cut in 2020 affected 20 million people living in India's financial capital Mumbai! What if there wasn't enough energy to go round all those who want it? One of the worst future scenarios is a world which does not have **energy security.** As Figure 1 shows, fossil fuels generate 84% of global energy supplies. The world relies on a future which depends on limited supplies of fossil fuels being available – forever!

The solution is **energy diversification** – that is, spreading energy sources around more types. As global energy demand grows, renewable energy is more important. Solar and wind energy help to reduce carbon footprints and improve energy security, and they won't run out!

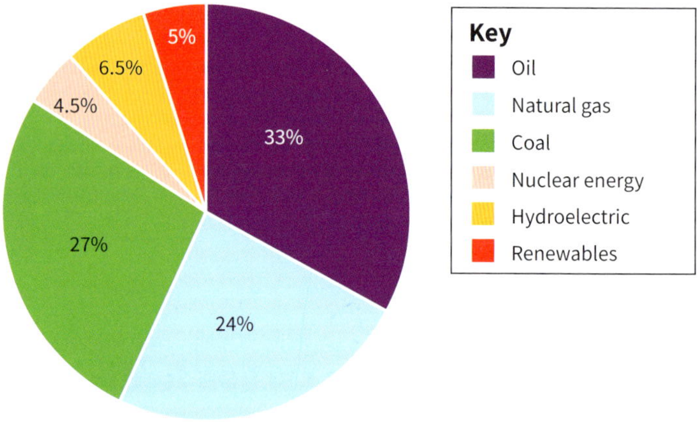

33%
24%
27%
4.5%
6.5%
5%

Key
Oil
Natural gas
Coal
Nuclear energy
Hydroelectric
Renewables

▲ ***Figure 1*** *The current sources of global energy supplies*

Energy security means having access to reliable and affordable sources of energy. Countries with access to enough energy are **energy secure**, whilst those without enough are **energy insecure**.

Diversifying the energy mix

Hydroelectric power (HEP) in California

HEP uses water energy to drive a turbine that generates electricity. A dam is built to hold back water that is then fed by gravity through pipes to the turbines. HEP is a major source of power in California, with 271 HEP plants providing 19% of the state's electricity in 2019. Dam construction is enormously expensive.

▲ ***Figure 2*** *The Shasta Dam in northern California*

Biofuels

Fuels extracted or burned using plants and crops are known as **biofuels**, or biomass. The most common is Bioethanol which is blended with petrol for cars. These fuels are renewable, can be grown anywhere, have lower carbon emissions than fossil fuels, and cost the same as oil to produce. Biofuel includes straw and wood chippings which are burnt with lower emissions to produce heat, e.g. as in the power station in Figure 3, which provides electricity for London's Olympic Park and local housing.

▲ ***Figure 3*** *Providing electricity from biomass*

Solar energy in California

Solar energy is of two types:

- Concentrating solar power (CSP) units generate electricity by concentrating solar energy to heat fluid, producing steam to power a generator.
- Solar photovoltaic (PV) directly converts solar energy into electricity, using PV cells grouped into panels.

The warm climate and long hours of sunlight make California ideal for solar generation. It has 400 000 small-scale solar projects and the world's largest solar power plant, the Ivanpah Solar Electric Generating System (Figure 4). It cost over US$2 billion to build and requires government subsidies to keep electricity costs down.

▲ ***Figure 4*** *Ivanpah Solar Electric Generating System*

Hydrogen technology and Toyota

Hydrogen is the most abundant element in the universe, but is usually combined with other elements e.g. carbon (oil, natural gas). Once separated, it can provide an alternative to oil – it powers vehicles whose exhaust is water that's so pure that you can drink it! Separating hydrogen requires energy, but this can be provided from a renewable resource such as solar or wind power.

Toyota produced the world's first mass-produced dedicated hydrogen fuel cell electric vehicle. The Mirai (Japanese for 'future') uses hydrogen in a 'fuel cell' which converts hydrogen into electricity to power the car's four motors (one for each wheel). The Mirai has a range of up to 312 miles before refuelling with hydrogen is required. The cars are expensive but are popular in the USA where government subsidies help to keep the cost competitive with petrol-driven cars.

▲ ***Figure 5*** *The Toyota Mirai*

Your questions

1. Explain the difference between 'energy security' and 'energy diversification'.
2. Explain how continued reliance on fossil fuels makes energy security unlikely.
3. Copy and complete the following table to compare the costs and benefits of the different renewable schemes shown in this section.

Renewable energy	Costs	Benefits
HEP schemes		
Solar power for hot water		
Solar power for electricity		
Hydrogen fuel for cars		

4. In pairs, decide which of the schemes has greatest potential for:
   - **a** reducing carbon footprint
   - **b** improving energy security
   - **c** diversifying the energy mix.

   Explain your ideas to the class.

Exam-style questions

5. Define the term 'energy security'. (1 mark)
6. Explain how renewable energies could alter the world's dependency on fossil fuels. (4 marks)
7. Explain why developing alternatives to fossil fuels is challenging. (4 marks)

9.12 What does the future look like? 1

In this section, you'll assess the different views that different groups have about energy futures.

World energy outlook

As bedtime reading, the International Energy Agency's (IEA) *World Energy Outlook 2020* does not sound like a gripping story! But it contains startling projections about the future. As global energy use continues to slowly rise, it predicts that the following will happen to fossil fuels:

- the world will become less dependent on them for energy needs
- coal will fall to below 20% of global energy supply by 2040
- they will continue to contribute to climate change.

The IEA puts forward two possibilities to show how global use of energy could be in future:

- **Business as usual.** This assumes that the world will continue to rely on fossil fuels as the main source of energy, with oil production and natural gas increasing just to meet extra demand.
- **A sustainable future.** More countries would adopt renewables and rely on mixed energy supplies in order to reduce CO_2 emissions as a way of combating impacts of climate change.

Whichever turns out to be true, decisions will be driven by different economic and political factors and by key players. This will include:

- those with vested interests in energy production
- those who are concerned about impacts of climate change, as shown in the text box.

Study the table on the next page and think about why the people and organisations shown have different viewpoints.

▲ ***Figure 1*** *Energy will come from a number of different sources in the future*

The 450 Scenario

The IEA has set out a '450 Scenario' – that is, it wants to limit greenhouse gases in the atmosphere to 450 parts per million of CO_2. Its motive is to reduce the increase in global temperatures from 6°C to 2°C. The scenario assumes that different countries will adopt CO_2 emissions targets. These countries will remove any fossil fuel subsidies and impose taxes on fossil fuels – called **carbon taxes**, which would make fossil fuels much more expensive – to discourage their use.

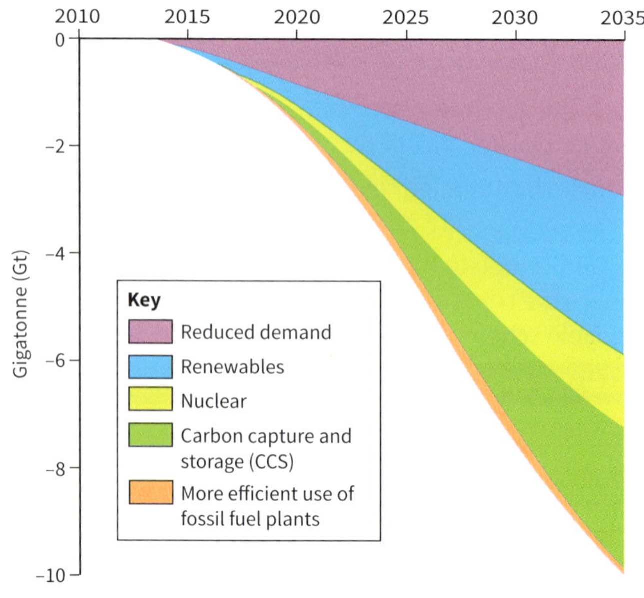

▲ ***Figure 2*** *Possible reasons why CO_2 emissions might fall by 2035*

<table>
<tr><th>Name</th><th>Who they are</th><th>What they say</th></tr>
<tr><td>Matt Ridley</td><td>Journalist</td><td>• Describes himself as a climate 'lukewarmer'.
• Thinks that 'global warming is real, mostly man-made and will continue' but isn't as dangerous as is often suggested.
Believes that:
• fossil fuels have a place in the modern world
• phasing out coal is a good long-term solution.</td></tr>
<tr><td>Shell</td><td>TNC</td><td>'A range of sources will be needed to supply this vital energy over the coming decades. Up to 30% of the world's energy mix could come from renewables in 2050, with fossil fuels and nuclear providing the rest.'
Shell say they are committed to:
• finding ways to provide energy from cleaner sources
• helping customers to use energy more efficiently.</td></tr>
<tr><td>Department for Business, Energy and Industrial Strategy</td><td>UK Government</td><td>Committed to:
• working to secure global emissions reductions
• reducing UK emissions
• adapting to climate change in the UK.
Believes that if we take action now:
• we will avoid burdening future generations with greater impacts and costs of climate change
• economies will be able to cope better by mitigating environmental risks and improving energy efficiency
• there will be wider benefits to health, energy security and biodiversity.</td></tr>
<tr><td>Professor Susan Solomon</td><td>Professor of Atmospheric Chemistry and Climate Science, Massachusetts Institute of Technology</td><td>• Has studied atmospheric chemistry and climate change since the 1980s.
• Was among the first to show that, if the world got rid of CFC pollutants, then the Antarctic ozone hole would heal – and it has.
• Believes that
  • almost all atmospheric warming in the past century is 'due to human activity'
  • understanding the chemistry of the atmosphere helps us understand the causes of climate change, and might also offer some solutions
  • climate change poses huge political challenges because fossil fuels are so important to the world economy
  • countries must collaborate on climate change.</td></tr>
<tr><td>Greenpeace</td><td>Environmental campaign group</td><td>'It's about getting the world from where we are now to where we need to be … cutting CO_2 emissions while ensuring energy security.'
Proposes an 'Energy Revolution' which would protect the climate by:
• phasing out fossil fuels
• investing in renewable energy.</td></tr>
</table>

▲ ***Figure 3*** *Views of different players (or decision-makers) in the debate about energy futures*

Your questions

1 Explain the difference between 'business as usual' and 'sustainable future'.

2 Explain why carbon taxes are:

a good in theory,

b difficult to put into practice.

3 In pairs, discuss the five viewpoints presented in the table on this page. Explain which would call themselves:

a Boserupian (see section 7.7)

b a believer in 'business as usual'

c a believer in 'a sustainable future'

d a supporter of the 450 Scenario

e a supporter of carbon taxes.

Exam-style questions

4 Suggest **one** reason why agreement about a sustainable energy future is likely to be very difficult. (3 marks)

5 Analyse Figure 3. Assess the reasons why people's views differ about energy futures. (8 marks)

9.13 What does the future look like? 2

In this section, you'll understand how attitudes are changing towards energy consumption, and ecological footprints.

Rising affluence and the environment

Rising wealth – called **affluence** – places greater pressure on the planet. As people become wealthier, their tastes widen (think strawberries at Christmas!) and they develop exotic tastes for foods that must travel tens or hundreds of miles to where it's bought. That's called **food miles** (i.e. how far food has to travel to where it is consumed). Fossil fuels used in travel took millions of years to form, yet are gone in an instant.

But there is another side. Rising affluence also gives developed countries the means to 'care' more about the environment and resources. In recent years, the world's most affluent countries have actually started to reduce energy consumption per capita, as shown in Figure 1. They have cut energy use per capita by:

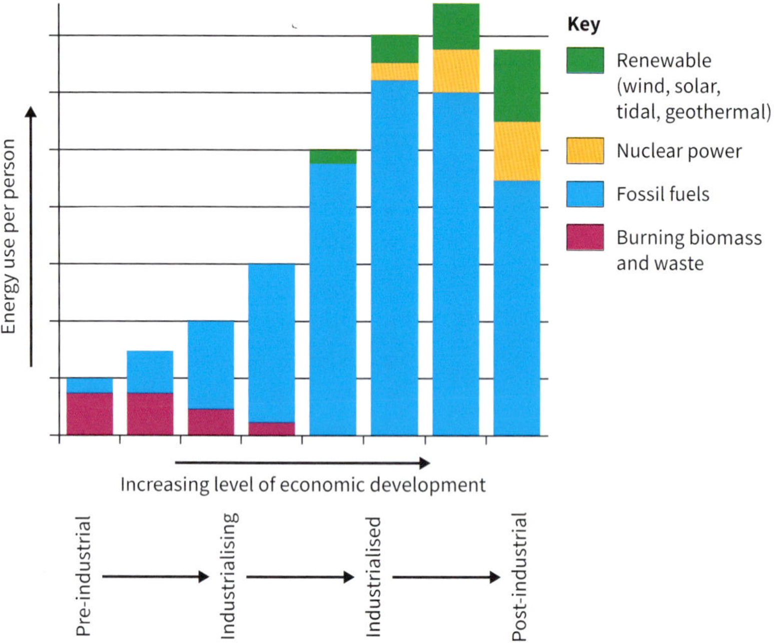

▲ ***Figure 1*** *How energy use per capita changes as countries become more affluent*

- using new technology such as LED bulbs to cut domestic energy use at home
- investing in more efficient cars which travel more miles per litre of fuel
- 'exporting' most dirty polluting industries, thus reducing their own energy consumption.

In addition, the development of renewable energies has reduced carbon emissions and has cut demand for fossil fuels.

Did you know?

There are over 650 Farmers' Markets in the UK where local farmers sell food directly to the public, reducing food miles.

Changing attitudes to the environment

Since the 1970s, political ideas have gained ground which place environmental concerns higher up the agenda in developed countries. Across the EU, for example, green political parties have emerged, together with pressure groups such as Friends of the Earth or Greenpeace. In countries where human development, water and food supply are no longer priorities because they have been achieved, many people realise that greater care about polluting and damaging the Earth is important. This is a concept known as the **Kuznets Curve**, which is shown in Figure 2.

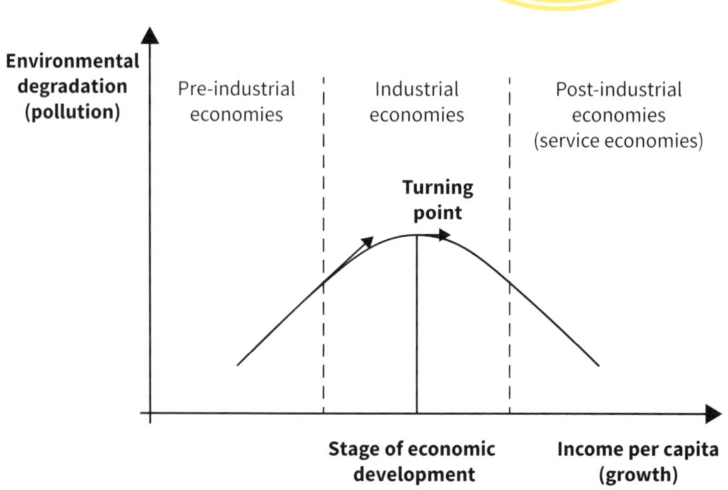

▲ ***Figure 2*** *The Kuznets Curve*

Ecological footprints

Environmental concern is about changing how people lead their lives – e.g. the foods we buy and how we consume energy. The sum total of energy used is measured by a concept known as **carbon footprint** (see section 9.10). It calculates the energy resources used in producing food and consumer goods. From this, we can calculate our **ecological footprint**, shown in Figure 3. Knowing your own ecological footprint is a first step in reducing it, and begins to help towards **sustainable development**.

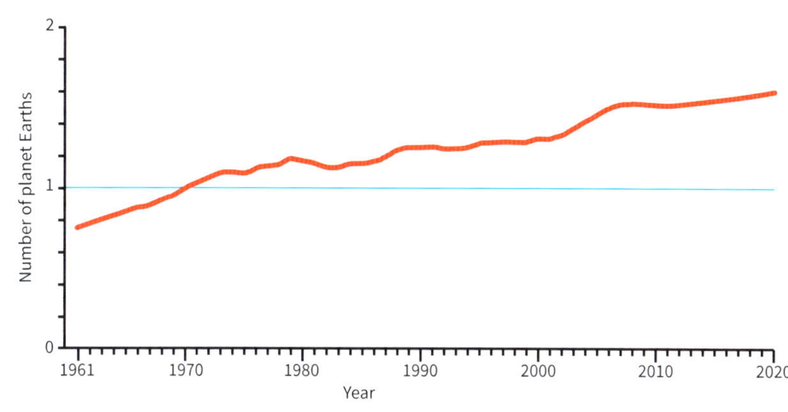

▲ ***Figure 3*** *A line graph to show humanity's ecological footprint*

Ecological footprint is measured in global hectares (gha) and shows the amount of land and water required to produce resources for each country. If people live within the Earth's ability to supply resources, their ecological footprint is 1. The higher the figure, the more space is needed – and the greater the inability of the Earth to supply their wants. In 2021, the Global Footprint Network calculated that humanity is currently using the resources of 1.6 planets to provide the goods and services we use, when we only have one Earth. Figure 3 shows that the global trend in ecological footprint is rising.

> **Sustainable development** – defined by the Brundtland Commission as that which 'meets the needs of the present without compromising the ability of future generations to meet their own needs'.

Educating for a changing world

Education and action is essential to the future of the planet. Increasingly, schools in the UK are teaching about sustainability in the curriculum (like the photo in Figure 4). It's local in focus rather than global, so that schools are able to change attitudes and take whatever actions they can. Actions taken by schools include recycling campaigns, using local foods that reduce the number of food miles and encouraging more students to travel to school by public transport.

▶ ***Figure 4*** *Primary school students learning about sustainability*

Your questions

1. Explain the differences between 'carbon footprint', 'ecological footprint', and 'sustainable development'.
2. Describe the relationship between economic development and energy use per capita over time in Figure 1.
3. Investigate and name one country for each stage of the Kuznets Curve shown in Figure 2.
4. Calculate your own ecological footprint – go online and visit the WWF-UK's footprint calculator.
5. In which parts of your life is your ecological footprint greatest? List how you could reduce your ecological footprint.
6. Describe actions being taken by your school **a** to increase recycling, **b** to reduce food miles, **c** regarding how people travel to school, and **d** any other actions.

Exam-style questions

7. Explain **one** way that ecological footprints can be reduced. (2 marks)
8. Explain how rising global affluence could lead to a more sustainable future. (4 marks)

10.1 Decision-making examination (DME)

In this section, you'll prepare for Component Three, the decision-making examination (DME).

Getting to know the exam

The decision-making exam lasts for 1 hour 30 minutes. It has 64 marks, of which 4 are for spelling, punctuation, grammar (known as SPaG) and use of specialist terminology. The exam is based on an unseen Resource Booklet, which will be about 10 pages long. You should spend some time at the start of the exam looking through the booklet. This will help you get familiar with the resources in it. Most questions will test your understanding of information in the booklet, but some will test your wider understanding of topics taught in Component Three.

What topics are covered?

The topics covered in the Component Three exam are:

- Topic 7: People and the biosphere
- Topic 8: Forests under threat
- Topic 9: Consuming energy resources.

The Resource Booklet in the exam will blend these three themes together through the study of an issue which affects a particular place. For example, section 10.3 explores tar sands in Canada, which affects the taiga (Topic 7), forests under threat (Topic 8) and the energy resources obtained from the tar sands (Topic 9). To prepare for the exam, you will need to revise all three topics because some of the shorter questions will test your knowledge and understanding of them. You will also need to apply your knowledge and understanding to the issue in the Resource Booklet.

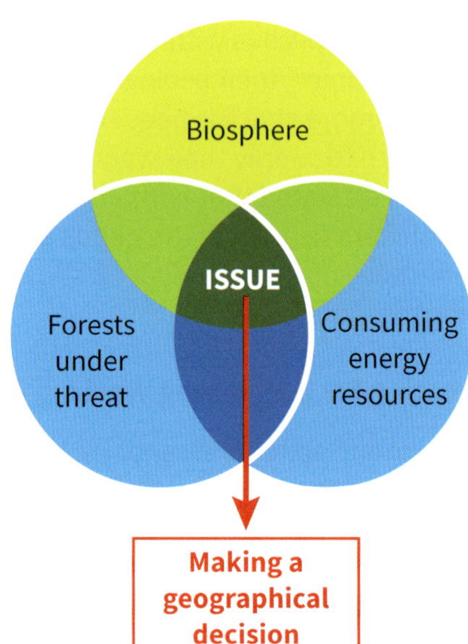

▲ ***Figure 1*** *How Topics 7, 8 and 9 link together*

The Resource Booklet

The Resource Booklet will contain geographical resources that you won't have seen before. It will be a mix of:

- Short passages of text
- Maps
- Photographs
- Graphs (e.g. pie, line and bar charts)
- Tables of data
- Views and opinions.

You will recognise the types of resources, even if you don't know anything about the place. You don't need to know anything about the place – the exam is testing your ability to examine resources, to understand them, and then to make a decision about the place.

Both the Resource Booklet and exam paper are organised into a number of sections:

- Section A: People and the biosphere
- Section B: Forests under threat
- Section C: Consuming energy resources
- Section D: Making a geographical decision.

Handy hint

You can get particular advice on how to answer the exam questions in Paper 3 in sections 11.3 (1–2 mark questions), 11.4 (3–4 mark questions), 11.9 (8 mark questions) and 11.10 (12 mark questions).

People–environment issues

The issue or problem about which you will have to make a decision will always involve **people**, the **natural environment**, and **issues** (see Figure 2).

The **people** could be:

- Local residents or indigenous groups
- Companies and their employees
- Environmental groups, charities, national and local government.

The **natural environment** will always involve each of the following:

- Forest biomes (either the tropical rainforest or taiga), and
- The development of energy resources set in a forest context (renewable, non-renewable or both).

Examples of energy-related people–environment **issues** might include:

- Whether to develop oil and gas extraction in the Amazon rainforest.
- Whether to develop biofuel crops in the rainforest of Indonesia.
- How to exploit tar sands in Canada's taiga sustainably.
- Whether to build an HEP dam in Russia's boreal forest.

An example of an issue

Figure 2 shows indigenous people from Amazonia protesting about the construction of the Belo Monte HEP dam in the Amazon. Opened in 2016, the dam provides electricity to Manaus, a major city, but as the dam filled, it destroyed an area of rainforest which is home to these people. Is HEP the best energy source, or is there an alternative? This is the sort of people–environment issue you might have to make a decision about.

▲ ***Figure 2*** *Indigenous people from Amazonia protesting about the dam*

Exam skills

Exam questions will generally fall into four types, as shown in the table in Figure 3.

It's very important to understand that the last question is about making a decision. You'll have a number of options to choose from, but there is no 'right' or 'wrong' answer. You get marks for:

- the quality of your explanation
- your justification – why you have decided to choose one option
- your reasons for rejecting the other options.

The next two sections are decision-making exam-style practice exercises about oil extraction in the Niger Delta (section 10.2) and Canada's tar sands (section 10.3). This will help you understand better how the exam works.

Knowledge and understanding	Skills	Applying skills and understanding	Making a decision
Worth 1–4 marks	Worth 1–4 marks	Worth 8 marks	Worth 12 marks
Command words State, Describe, Explain, Define	**Command words** Identify, Calculate, Describe	**Command words** Assess, Evaluate	**Command words** Select ... and justify
Types of question Short questions either linked to the component content, or a figure in the Resource Booklet.	**Types of question** Short questions linked directly to some data in the Resource Booklet, testing how you handle data and information.	**Types of question** Extended writing questions where you need to weigh up more than one view, opinion or impact.	**Types of question** The last question, where you will be given between two and four options. You'll need to choose one, and then justify your choice.

▲ ***Figure 3*** *Types of exam questions that will likely occur in the Component Three exam*

10.2 Resource Booklet 1 – Oil exploitation in the Niger Delta region

In this section, you'll investigate the development of oil and gas in the Niger Delta region, as a trial exercise to prepare you for Paper 3.

This is an example of what to expect from a Paper 3 Decision Making Exercise.

- Pages 298 to 300 contain the resources you will need to use to investigate the issues.
- Page 301 shows you exam-style questions to help you understand the Decision Making exam.

Section A: People and the biosphere

The issue: the real cost of Nigeria's oil?

- Oil and gas make up 85% of Nigeria's exports and 60% of government revenue (through taxes on oil and gas drilling).
- Oil and gas wealth only seems to have benefitted some Nigerians: 40% live below the income poverty level of $380 per year, but there are at least 9 billionaires and 15,000 millionaires in the country.
- Oil exploitation has been linked to serious pollution and deforestation of tropical and mangrove forests in the Niger Delta region.
- Nigeria could continue to exploit oil and gas, or move towards less oil dependency and greater forest conservation.

Getting to know Nigeria

- Nigeria is a developing country in Africa with a low HDI of 0.54. Its population is growing rapidly, by 2.6% per year, and it is the largest country in Africa by population.
- 30% of Nigerians are farmers. In rural areas, many people rely on fuelwood and biomass for their energy needs, which make up 80% of Nigeria's energy resources.
- Nigeria was a British colony until 1960. Commercial quantities of crude oil were discovered there in 1956. Oil was first exploited by British (Shell, BP) and American (Texaco, Mobil, Gulf) TNCs. It is estimated that Nigeria has 37 billion barrels of oil reserves.
- Most oil is found in the Niger Delta region, where the River Niger enters the Atlantic Ocean. The delta has biodiverse mangroves, swamps and tropical rainforests.
- Oil, and some natural gas, are extracted from oil wells on land and from oil rigs in the shallow sea of the Niger Delta region.
- Despite being the world's eighth largest oil exporting country, Nigeria imports most of its petrol and diesel because it has limited oil refining capacity.

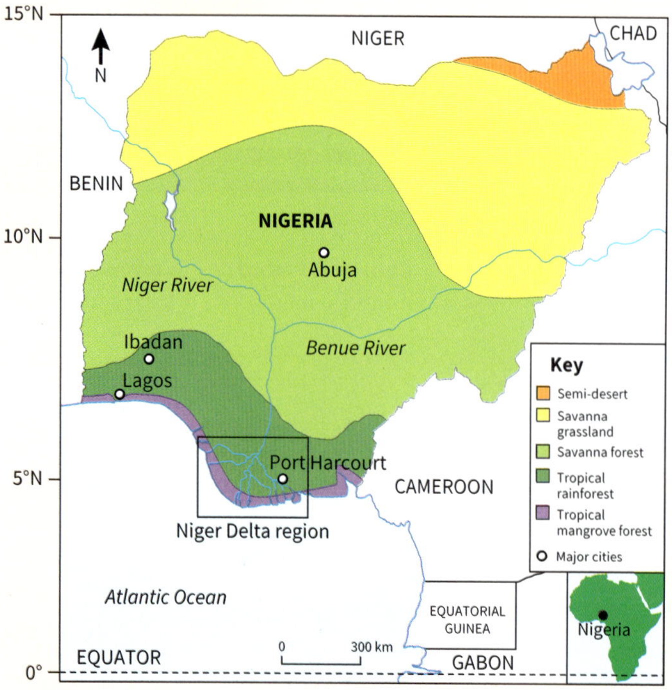

***Figure 1** Geography and ecosystems of Nigeria*

Section B: Forests under threat

Year	Total population (millions)	GDP per capita (US $)
1960	45	93
1980	73	875
2000	122	568
2020	205	2230

▲ ***Figure 2a*** *Population and economy of Nigeria, 1960–2020*

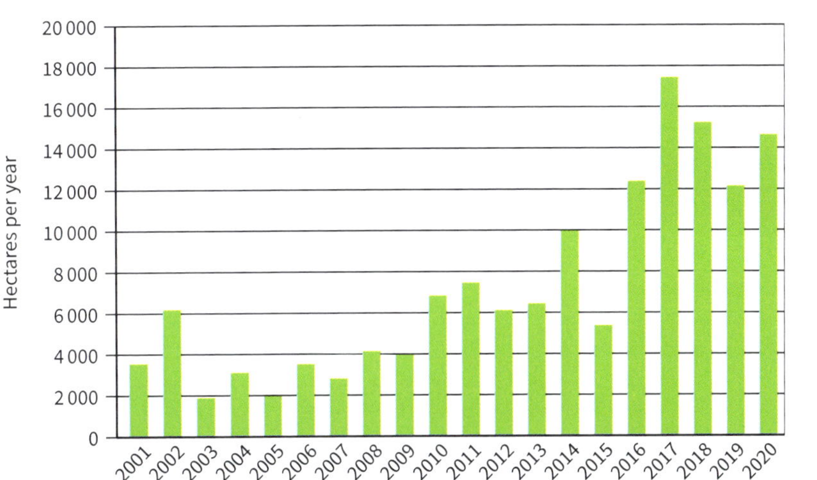

▲ ***Figure 2b*** *Primary forest deforestation in Nigeria, 2001–2020*

Section C: Consuming energy resources

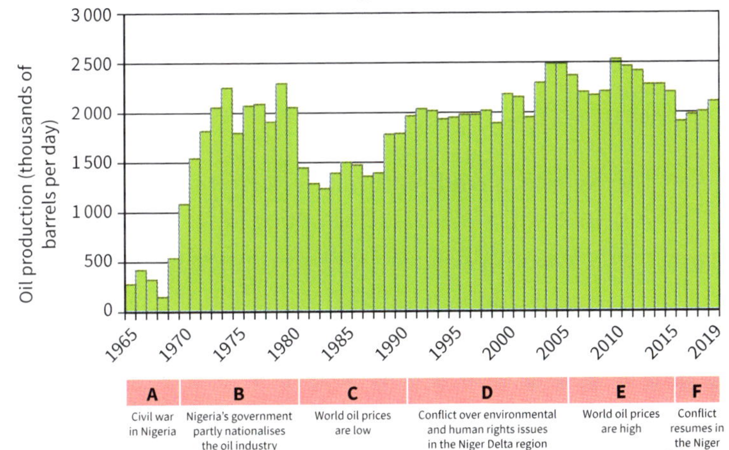

▲ ***Figure 3a*** *Oil production in Nigeria, 1965–2019*

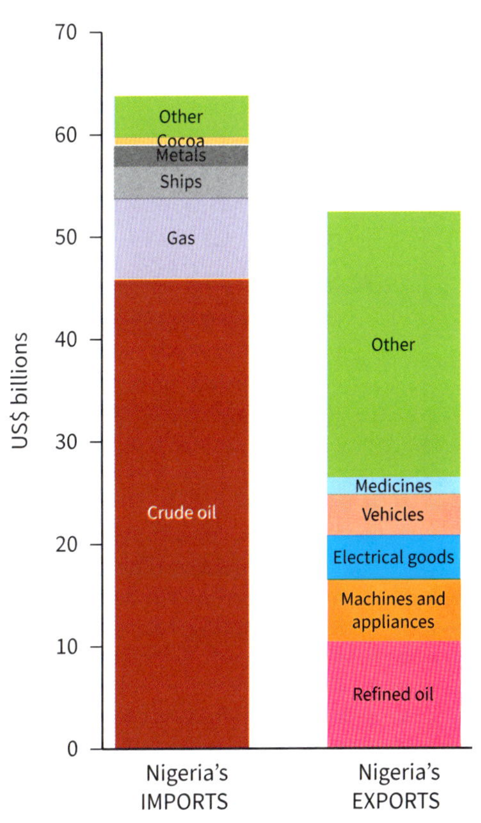

▲ ***Figure 3b*** *Nigeria's imports and exports, 2019*

◀ ***Figure 3c*** *Oil pollution on the water and deforested mangroves near Port Harcourt in the Niger Delta region*

Environmental Issue	Explanation	Impact score: Nigeria's environment	Impact score: Nigeria's people	Overall impact score
Carbon emissions	Over 60% of Nigeria's carbon emissions come from deforestation. The oil and gas industry is a major contributor to Nigeria's carbon emissions. It is ranked 36th by carbon emissions per country. Greenhouse gas emissions of 0.7 tonnes per person per year are below the world average of 4.7 tonnes. Carbon emissions contribute to global warming.	–2	0	–2
Oil spills	There are estimated to be 300 oil spills in the Niger Delta region each year from oil rigs (out at sea) and oil wells (on land). Since the 1950s, 5–20 million barrels may have been spilled. Oil pollutes rivers and groundwater water supplies, and kills mangroves and rainforest. Oil pollutes farmland, reducing food supply. Oil can also cause skin disorders and can contaminate fish.	–3	–3	–6
Natural gas flaring	About 70% of the natural gas found alongside crude oil is 'flared' or burnt, rather than being collected and used as an energy source – due to the high cost of collection. Gas flaring releases methane, a greenhouse gas. Flaring creates air pollution which affects human health in places close to flare sites.	–1	–2	–3
Impact scores: 0 = minimal impact; –1 = small, local impact; –2 = large impact, wider area; –3 = widespread, severe impact				

▲ ***Figure 3d*** *Oil and environmental issues in Nigeria*

- Conflict has been a feature of the Niger Delta region since oil was discovered.
- Since the 1990s, the indigenous Ogoni and Ijaw people have fought against oil exploitation, pollution and loss of their traditional lands to oil companies. Non-violent protests turned violent by the mid-2000s, leading to armed conflict, kidnappings and destruction of oil company pipelines and infrastructure.
- The violence has reduced foreign investment and turned the Niger Delta region into a region dominated by the military, police and private security firms.
- Nigeria ranked 149th out of 180 in Transparency International's 2020 Corruption Perception Index.
- It is widely believed that corruption has helped a small number of businesspeople and political leaders in Nigeria benefit from the country's oil wealth.
- Many of the workers in Nigeria's oil industry are foreign high-skilled workers employed by global TNCs, not local Nigerians.
- About 14% of Nigeria is protected (forest reserves, national parks, conservation areas) but corruption, illegal deforestation and lack of monitoring mean that having protected status rarely prevents ecosystem destruction.

"The production of oil and gas in Nigeria is linked to deep social inequality and environmental disasters. Nigeria has one of the highest rates of energy poverty in the world and suffers from chronic power cuts." [Carbonbrief.org]	"Nigeria has overtaken India as home to the largest population of people living in extreme poverty, with 87 million citizens living on less than $1.90 a day." [AfricaPortal.org]	"Nigeria is Oil Dependent, not Oil Rich." [Council for Foreign Relations]
	"The continued failure of oil companies and government to clean up oil spills have left hundreds of thousands of Ogoni people facing serious health risks, struggling to access safe drinking water, and unable to earn a living." [Amnesty.org]	"Oil major Chevron plans to cut 25% of its Nigeria workforce (1000 jobs). Already, the oil companies do not provide enough jobs to satisfy employment needs in the region, which has few other industries." [Reuters]

▲ ***Figure 3e*** *The human cost of oil*

Exam-style questions 1 – Oil exploitation in the Niger Delta region

Section A: People and the biosphere (9 marks in total)

Use Section A of the Resource Booklet to answer this question. Study Figure 1.

1 a Define the term 'biodiversity'. (1 mark)

b Describe the distribution of tropical rainforest in Nigeria. (2 marks)

1 c State **two** features of the climate of tropical rainforests. (2 marks)

1 d Explain **one** reason why the north of Nigeria has a drier climate than the south. (2 marks)

1 e Explain **one** way tropical rainforest plants are adapted to their climate. (2 marks)

Section B: Forests under threat (7 marks in total)

Use Section B of the Resource Booklet to answer this question. Study Figures 2a and 2b.

2 a i Calculate the percentage increase in Nigeria's population between 2000 and 2020. Show your working out. (2 marks)

2 a ii Explain **one** way rising affluence can threaten tropical rainforests. (2 marks)

2 b Explain **one** way that poverty could have contributed to the increase in deforestation shown in Figure 2b. (3 marks)

Section C: Consuming energy resources (32 marks in total)

Use Section C of the Resource Booklet to answer this question.

3 a i Study Figure 3a. Identify the year when Nigeria's oil production was highest. (1 mark)

3 a ii Identify which time period, A to F, on Figure 3a shows when oil production increased most rapidly. (1 mark)

3 a iii Suggest **one** political reason why oil production varies in Figure 3a. (2 marks)

3 a iv Suggest **one** economic reason why oil production varies in Figure 3a. (1 mark)

3 b i Study Figure 3b. Compare Nigeria's imports and exports. (3 marks)

3 b ii Study Figure 3c. Explain **one** negative environmental impact of oil exploitation in the Niger Delta. (4 marks)

3 c Study Figure 3d. Using evidence from Figure 3d, assess the impact scores given to the three environmental issues. (8 marks)

3 d Explain what is meant by **i** renewable energy resources and **ii** recyclable energy resources. (4 marks)

3 e Study Figure 3e. Using evidence from Figure 3e, assess how far people in Nigeria have benefitted from the oil industry. (8 marks)

Section D: Making a geographical decision (12 marks in total)

Use all of the resources in Sections A–C to help you answer this question.

4 Study the three options to the right. They show what the Nigerian government could choose for the future of the oil and gas industry.

Select the option that you think is the best plan for Nigeria's people and environment. Justify your choice.

Use information from the Resource Booklet and knowledge and understanding from the rest of your Geography course to support your answer. (12 marks)

Option 1	Focus on developing the remaining oil and gas reserves.
Option 2	Slow down oil and gas development and increase forest conservation efforts.
Option 3	Increase taxes on TNC oil companies to pay for pollution to be cleaned up and forests replanted, and prevent any new oil and gas drilling.

10.3 Resource Booklet 2 – Tar sands in Canada's taiga

In this section, you'll investigate the development of tar sands in Alberta province, Canada, as a trial exercise to prepare you for Paper 3.

This is an example of what to expect from a Paper 3 Decision Making Exercise.

- Pages 302 to 304 contain the resources you will need to use to investigate the issues.
- Page 305 shows you exam-style questions to help you understand the Decision Making exam.

Section A: People and the biosphere

The issue: exploiting tar sands in Canada's taiga

- Global demand for oil is high. To meet demand, unconventional resources like tar sands and oil shale may need to be exploited.
- Canada has large reserves of tar sands in Alberta, found beneath taiga forest.
- Employment opportunities outside of the oil industry are limited in Alberta, and northern Alberta is isolated and undeveloped.
- Oil from tar sands is a valuable export. Much is exported to the USA.

Where tar sands are extracted

- Tar sands are found in three areas of Alberta (Figure 1a). Only Athabasca is mined.

▲ ***Figure 1a*** *Location map of the Canadian tar sands*

- Alberta's tar sands have known reserves of 173 billion barrels, the world's third largest.
- 500 000 jobs in Canada depend on the tar sands industry. This could grow to 800 000 by 2030 as Canadian oil production reaches a peak (see Figure 1b).
- Taiga forest covers 75% of Alberta, including the tar sands area (about the size of Scotland). But only 3% of tar sands can be mined using current methods.
- The area is inhabited by many indigenous groups, called First Nation people in Canada.
- Tar sands are extracted by Canadian companies, as well as global oil and gas TNCs such as Shell, Cenovus, Nexen, Suncor and Imperial Oil.

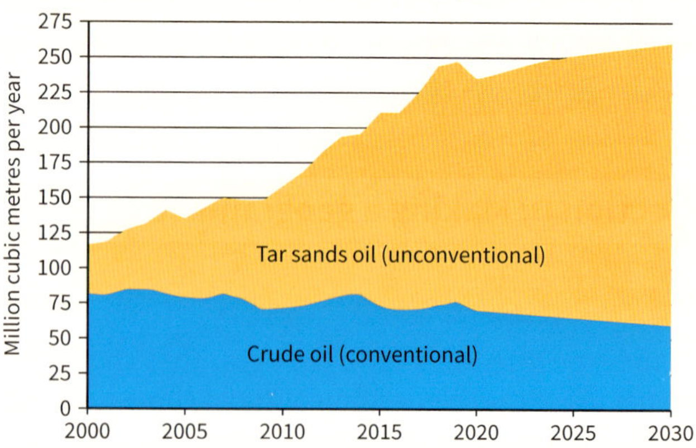

▲ ***Figure 1b*** *Canadian oil production, 2000 to 2030 (estimated)*

Section B: Forests under threat

▲ ***Figure 2a*** *Photo showing the tar sands mining and processing complex near Fort McMurray in the Athabasca tar sands area*

Tar sands mining is said to have a range of impacts on people and resources in the area.

- Each barrel of oil produced uses two to six barrels of water. 80% of this is taken from the Athabasca River.
- About 1.8 million tonnes of this water *each day* becomes highly toxic waste because it is used in processing the tar sands.
- Toxic waste water is stored in ponds, but these leak into the Athabasca River and into groundwater. The water then flows into First Nation territories.
- In 2006, unexpectedly high rates of rare cancers were found in the First Nation Chipewyan community. In 2008, Alberta Health confirmed a 30% rise in the number of cancers between 1995–2006. Oil companies question these figures.
- Other energy resources (mostly fossil fuel natural gas) have to be used to heat and process the tar sands to make crude oil. This emits greenhouse gases.

▼ ***Figure 2b*** *Some ecological impacts of tar sands mining*

CARIBOU

Caribou populations have declined. The Cold Lake herd has declined by 74% since 1998 and the Athabasca River herd 71% since 1996. In 2014, 175–275 caribou remained. By 2025, the total population is expected to be less than 50, and locally extinct by 2040.

WOLVES

The few remaining caribou have become easy prey for grey wolves. In order or preserve caribou, over 1000 wolves were shot between 2005 and 2015 in a deliberate cull to reduce wolf numbers.

BIRDS

Large numbers of migrating duck and geese land on tailings ponds each year, and almost all die quickly in the toxic water. Numbers are not known with certainty, but 500–5000 each year seems likely.

FORESTS

Only about 0.2% of Alberta's forest has been destroyed by mining, and some has been regenerated by the oil companies. One company, Syncrude, claims to have replanted 20% of its mined land but environmentalists argue the real figure is less than 1%.

Section C: Consuming energy resources

Extracting oil from Canada's tar sands is expensive.

- Extreme mining conditions, with winter temperatures plunging to –30°C.
- Northern Alberta is isolated, so equipment and resources have to be transported long distances. Oil must be transported by train or pipeline to customers.
- The extraction process uses natural gas and water, increasing costs.

As the world oil price changes, so does the profitability of tar sands mining (Figure 3c).

- Competition from renewable resources like wind power, and the switch to electric vehicles, means future demand is uncertain.
- Canada is one of the world's top ten oil producers.
- 19% of Canada's oil is used in Canada. The remaining 81% is exported, mostly to the USA.
- The USA's energy security is improved by having a friendly major oil exporting country very close by.

	Country	Oil production in barrels per day
1	USA	17000000
2	Saudi Arabia	11800000
3	Russia	11500000
4	Canada	5700000
5	Iraq	4800000
6	United Arab Emirates	4000000
7	China	3800000
8	Iran	3500000
9	Kuwait	3000000
10	Brazil	2900000
	Top 10 producing countries total:	68000000

▲ ***Figure 3a*** *The top ten oil producing countries in 2019*

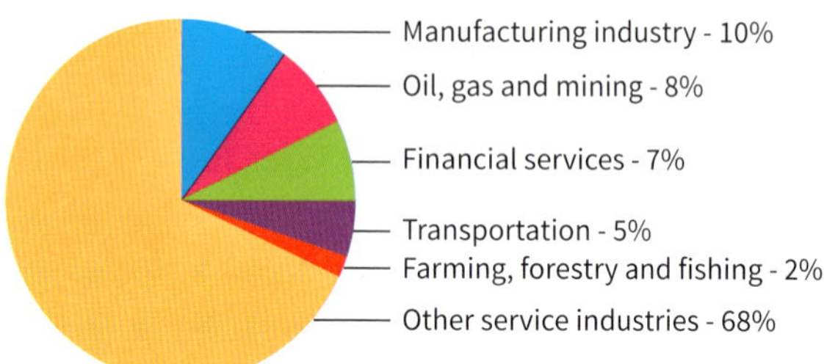

▲ ***Figure 3b*** *Economic sectors as a share of Canada's total GDP, 2018*

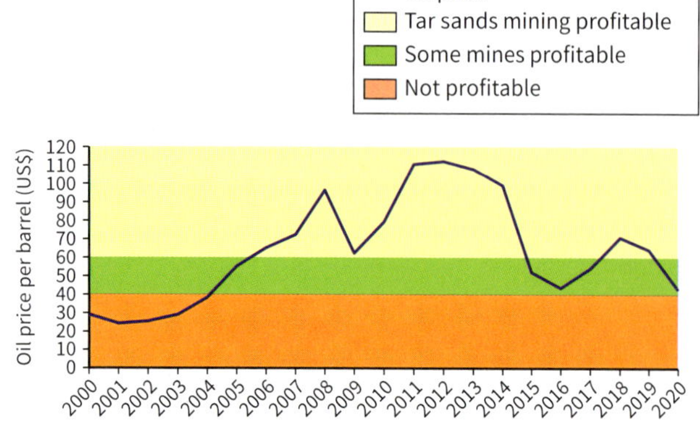

▲ ***Figure 3c*** *Global oil price 2000–2020 and the profitability of tar sands mines*

FOR	Syncrude (one of the largest tar sands companies)	'As a large producer of crude oil from the oil sands, Syncrude has a tremendous and positive impact on the economies of Alberta and Canada. We are one of the largest employers of First Nation people in Canada that comprise about 9% of our workforce.' – *From the Syncrude website at www.syncrude.ca*
FOR	American Petroleum Institute (which represents oil companies in the USA)	'Canadian oil can help meet our growing energy needs and make the United States more energy secure. Canada sends more than 99% of its oil exports to the United States.' – *From the API report 'Canadian Oil Sands'*
FOR	Government of Alberta province	'Experts say the world will be dependent on fossil fuels for the foreseeable future. With the 3rd largest proven reserves of oil in the world, Alberta must help meet the global demand while ensuring sustainable and responsible extraction.' – *From the Alberta government website at www.oilsands.alberta.ca*
AGAINST	Sierra Club (an environmental organisation from the USA)	'There is no limit to the amount of tar sands development the Alberta and Canadian governments are willing to allow – despite the serious social, economic and environmental problems this growth will unleash on the world. Tar sands mining decimates Canada's Boreal Forest.' – *From the Sierra Club website at http://content.sierraclub.org/beyondoil/tar-sands*
AGAINST	ACFN (An organization of the Chipewyan First Nation People)	'Tar sands have been widely recognized as the most destructive project on Earth because of the serious impacts on treaty and rights, ecological destruction and global greenhouse gas emissions (GHG).' – *From a press release of the ACFN*
AGAINST	A Canadian journalist writing about the tar sands	'Canadians don't universally support construction of the pipeline. A poll in 2012 found that nearly 42% of Canadians were opposed. Many of us, in fact, want to see the tar sands industry wound down and eventually stopped, even though it pumps tens of billions of dollars annually into our economy.' – *From 'The Tar Sands Disaster', www.homerdixon.com*

▲ ***Figure 3d*** *Views on tar sands development in Alberta*

Exam-style questions 2 – Tar Sands in Canada

Section A: People and the biosphere (10 marks in total)

Use Section A of the Resource Booklet to answer this question.

1 a Other than Canada, name **one** country that has large areas of taiga. (1 mark)

1 b Name **two** resources that indigenous people (First Nation people) are likely to get from the taiga. (2 marks)

1 c Study Figure 1b. Estimate the production of tar sands oil in Canada forecast for 2030. (1 mark)

1 d Explain **one** reason why conventional crude oil production in Canada is forecast to decline by 2030. (2 marks)

1 e Explain **two** ways forests help regulate the hydrological cycle. (4 marks)

Section B: Forests under threat (10 marks in total)

Use Section B of the Resource Booklet to answer this question.

2 a Describe the impact of tar sands mining on the taiga landscape shown in Figure 2a. (2 marks)

2 b Explain **one** way that plants in the taiga are adapted to the climate. (2 marks)

2 c Identify **two** resources that are used during the extraction and processing of tar sands. (2 marks)

2 d Using evidence from Figure 2b, explain **two** ways that biodiversity in the taiga has been affected by tar sands mining. (4 marks)

Section C: Consuming energy resources (28 marks in total)

Use Section C of the Resource Booklet to answer this question.

3 a Study Figure 3a. Calculate the percentage of oil production from the top 10 countries that is produced in Canada. (1 mark)

3 b Explain **one** way that conflict could affect oil supply from some countries. (3 marks)

3 c Study Figure 3c. State the oil price at which mining tar sands becomes unprofitable. (1 mark)

3 d Describe the trend in oil prices between 2000 and 2020 shown on Figure 3c. (3 marks)

3 e Explain how changes in the oil price since 2015 might affect tar sands oil production in the future. (4 marks)

3 f Using evidence from the resources in Section C, assess the importance of the tar sands industry to the Canadian economy. (8 marks)

3 g Study Figure 3d. Using evidence from Figure 3d, assess the reasons why there are many different views about Canada's tar sands industry. (8 marks)

Section D: Making a geographical decision (12 marks in total)

Use all of the resources in Sections A–C to help you answer this question.

4 Study the three options to the right about how Canada should use the taiga forest region in Alberta in future.

Select the option that you think would best balance the needs of people and the forest environment in Canada. Justify your choice.

Use information from the Resource Booklet and knowledge and understanding from the rest of your Geography course to support your answer. (12 marks)

Option 1	Focus on the expansion of tar sands and oil exports to the USA to help develop the economy.
Option 2	Allow existing mines to continue, but only allow new expansion when the oil companies agree to replant mined areas and stop water pollution.
Option 3	Stop the mining of tar sands and convert the remaining forests into conservation areas managed by First Nation peoples.

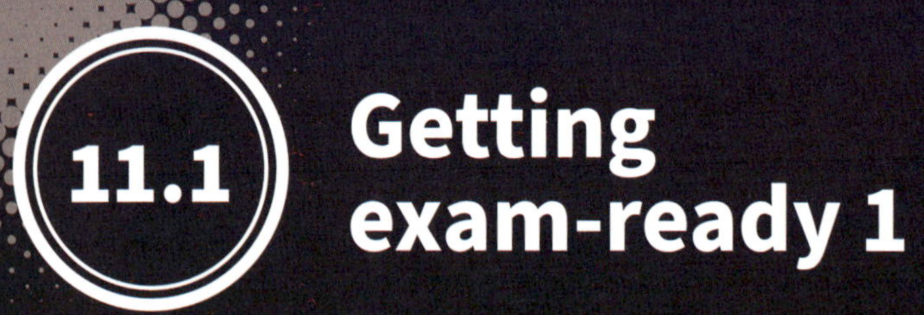

11.1 Getting exam-ready 1

In this section, you'll get to know details of the three GCSE exam papers.

The exam papers

There are three parts to the specification, called Components. Each Component is assessed by an exam paper. Each Component has three topics, with different sections in the exam.

Within each Topic, you'll be taught different places, examples, and case studies which you need to learn for the exam. To help you, complete 'Keeping a record of examples for Paper 1/2!'.

Paper 1 – Global Geographical Issues

Paper 1 assesses all the topics in Component One. It has three sections (A, B and C), each section on a different Topic.

- **Section A** tests **Topic 1 Hazardous Earth**
- **Section B** tests **Topic 2 Development dynamics**
- **Section C** tests **Topic 3 Challenges of an urbanising world**

Each section carries 30 marks, split into 22 marks of short (1- to 4-mark) questions, and an 8-mark question. All questions are compulsory.

- You should therefore split your time equally between sections – 30 minutes per section.
- That means you have a mark per minute – so take no more than 1 minute on 1-mark questions, 2 on 2-mark questions, and so on.
- Section B's 8 -mark question also has 4 marks for Spelling, Punctuation and Grammar (SPaG).

Keeping a record of examples for Paper 1!

In Topic 1 Hazardous Earth:

- I studied earthquakes* / volcanoes* in tectonic hazards.
- My examples of earthquakes* / volcanoes* were ________________ in an emerging* / developing* country and ________________ in a developed country.
- The tropical cyclones I studied were ________________ in an emerging* / developing* country and ________________ in a developed country.

In Topic 2 Development dynamics:

- My case study of an emerging country was ________________.

In Topic 3 Challenges of an urbanising world:

- My case study of a megacity in a developing or an emerging country was of ________________, which is in a developing* / emerging* country.

**means delete one of these*

Paper 2 – UK Geographical Issues

Paper 2 assesses all topics in Component Two. It has three sections (A, B and C), each on a different Topic. Section C is split into Section C1 and C2.

- **Section A** tests **Topic 4 The UK's evolving physical landscape**
- **Section B** tests **Topic 5 The UK's evolving human landscape**
- **Section C** tests **Topic 6 Geographical investigations**. You have to make a choice:
  - **Section C1** contains questions on physical fieldwork (choose **either** Coastal change and conflict **or** River processes and pressures)
  - **Section C2** contains questions on human fieldwork (choose **either** Dynamic urban areas **or** Changing rural areas)

Like Paper 1, allow about a mark per minute, splitting your time in proportion to the marks in each section. This time it's a little more complex:

- Sections A and B are each worth 27 marks, split into 19 marks of shorter (1- to 4-mark) questions, and one 8-mark question. These questions are compulsory. Allow 25 minutes per section.
- The 8-mark question in Section B will also have 4 additional marks for Spelling, Punctuation and Grammar (SPaG).
- Sections C1 and C2 each carry 18 marks, split into 10 marks of shorter (1- to 4-mark) questions, and one 8-mark question. Allow 20 minutes per section.

Keeping a record of examples for Paper 2!

In **Topic 4 The UK's evolving physical landscape**:

- My two in-depth examples of physical landscapes were ________________ (name of stretch of coast) for coastal change and conflict, and ________________ (name of river) for river flooding.

In **Topic 5 The UK's evolving human landscape**:

- My case study of a major UK city was ________________.

In **Topic 6 Geographical investigations** (fieldwork), I carried out two investigations (one physical, one human) in two places:

- **For my physical fieldwork** I studied Coastal change and conflict* / River processes and pressures*. We visited ________________ to study this.
- **For my human fieldwork** I studied a Dynamic urban area* / a Changing rural area*. We visited ________________ to study this.

**means delete one of these*

Paper 3 – People and Environment Issues: Making Geographical decisions

Paper 3 is different from the others, as it's based on a Resource Booklet about a geographical issue concerning energy and forests, based on Topics 7–9. It's sometimes called the decision-making paper, or DME (Decision-Making Exercise). The exam has four sections. All questions are compulsory.

- **Section A** tests **Topic 7 People and the biosphere**, a short section, worth 7–8 marks.
- **Section B** tests **Topic 8 Forests under threat** , also a short section, worth 7–8 marks.
- **Section C** tests **Topic 9 Consuming energy resources**, worth 32–33 marks.
- **Section D** tests your ability to make decisions about options for the future, based on the issue in the Resource Booklet, worth 12 marks.
- Most questions are based on your skills in interpreting geographical information (e.g. maps, diagrams, statistics, or photos).

Because you'll need time to read the Resource Booklet, there are 90 minutes for this exam but only 64 marks. This means you have 30 minutes in which to read the booklet, and an hour to answer the questions. Most candidates read a section then do the questions on that section, working through the paper. Allow the following time for each section:

- Sections A and B are worth about 15 marks in total, all on shorter (1- to 4-mark) questions. Allow 20 minutes in total.
- Section C's 32–33 marks are split, with 16–17 marks on shorter (1- to 4-mark) questions, and 16 marks on two 8-mark questions. Allow 45–50 minutes for this section.
- Section D consists of a single 12-mark essay question with an additional 4 marks for Spelling, Punctuation and Grammar (SPaG). Allow 20 minutes for this section.

There is more detail about Paper 3 in section 10.1 (pages 296–297).

11.2 Getting exam-ready 2

In this section, you'll get to know what kinds of questions to expect, and to understand how your exam questions will be marked.

What are the examiners assessing?

Examiners use Assessment Objectives (called AOs) to decide what they'll look for in setting questions and marking answers. There are four in GCSE Geography:

AO1 – Knowledge recall: These questions test your knowledge, e.g. of definitions of key words, or locations of places, or of landforms. These are usually very short questions worth 1 or 2 marks.

AO2 – Understanding of concepts, places and environments: These questions test your understanding of processes (e.g. how a river erodes), or of places and change (e.g. the impacts of rural-urban migration), or of environments (e.g. the impacts of deforestation of rainforests). These are usually short questions worth 2–4 marks.

AO3 – Applying ideas, analysing, and making informed judgments: These questions test your ability to interpret and apply information in order to develop an argument (e.g. whether top-down or bottom-up changes in megacities are more effective in improving places). Because of what they ask you to do, many questions of this type are 8-mark questions where an argument is given space to develop.

AO4 – Geographical skills (including statistics and maths skills): These questions are about your ability to interpret resources (e.g. maps, photos, and data) in order to get information, or to present information. They also include mathematical skills questions (e.g. calculate).

Answering short questions

In each section, the first questions are short, and worth between 1 and 4 marks.

- These questions are point marked.
- These include multiple-choice, short and paragraph answers, and skills questions, including geographical and mathematical skills (e.g. calculations).
- You can use a calculator to help with any part of the exam.
- There are resource materials (e.g. data, photos) on which you'll be asked questions. These will be based on Topics you've learned.
- Detailed knowledge is not normally required in shorter questions, except brief references to examples you've studied, like the ones in the records that you've completed in Section 11.1.
- Usually, you'll have two lines of answer space per 1 mark – so 2-mark questions have four lines, 3-mark questions have six lines, and so on.

Answering longer questions

The last question on each section in Papers 1 and 2 is worth 8 marks, and requires extended writing.

- These questions are level marked.
- You'll need to learn detail to answer these, because questions are likely to be on those parts where you have been taught examples (e.g. in Paper 1 examples of tectonic hazards, tropical cyclones) or case studies (e.g. of an emerging country or a megacity).
- 8-mark questions are marked differently in Paper 1 (see section 11.5) from how they are marked in Paper 2 (see section 11.6) or Paper 3 (see section 11.9).
- There is particular guidance on answering the 12-mark question in Paper 3 in section 11.10.
- You'll have a minimum of 16 lines of space for an 8-mark question, and 24 for a 12-mark question. However, examiners sometimes give more space, particularly if the question also assesses SPaG, where candidates might want more space to allow for paragraphing, etc.

How will my exam answers be marked?

It's important to know how questions are marked:

- Questions carrying 1–4 marks are **point marked**. This means that you earn a mark for every correct point you make. See sections 11.3 and 11.4 for more about shorter questions on each of the exam papers, and sections 11.7 and 11.8 for specific guidance on point-marked fieldwork questions on Paper 2.
- Questions carrying 8 marks are **level marked** using level-based mark schemes. Here, examiners read the whole answer, and judge it against a set of qualities, called levels.
- There are three levels; Level 1 is lowest and 3 is highest.
- On 8-mark questions, Level 1 answers earn 1–3 marks, Level 2 earn 4–6 marks, and Level 3 earn 7–8 marks.
- Your ability to explain and develop arguments is what counts.
- Level 1 answers have little explanation, detail, or judgement.
- By contrast, Level 3 answers have full explanations, detail about particular places, and make evidenced judgements that lead to a conclusion.

Know the command words

Answering questions properly is the key to success. When you read an exam question, look at the **command word** – that's the word that examiners use to tell you what to do. Figure 1 gives you the command words to expect in Papers 1 and 2, the number of marks you can expect for questions using each command word, and which Assessment Objective each is testing.

Point marking

All questions worth 1–4 marks are point marked on all three exam papers.

- This means that you earn a mark for every correct point you make.
- Learn how to develop answers so that you turn 1 mark into 2 on a 2-mark question, or add even more points for 3- and 4-mark questions.

Command word	No. of marks	What it means	The AO that it's testing
Identify / State / Name	1	Find (e.g. on a graph), or give a simple word or statement	AO1
Define	1	Give a clear meaning of a word or term	AO1
Calculate	1 or 2	Work out an answer (1 mark) including workings (2 marks)	AO4
Label	1 or 2	Write a name on a map or diagram	AO1
Draw	2 or 3	As in sketching or completing a graph	AO4
Compare	3	Identify similarities or differences	AO3
Describe	2 or 3	Say what something is like (AO1); identify trends or patterns (e.g. on a graph) (AO4)	AO1 or 4
Explain	2, 3 or 4	Give reasons why something happens	AO1 or 2
Suggest	2, 3 or 4	For an unfamiliar situation, explain why something might occur, with a reason	AO3

▲ ***Figure 1*** *Command words used for 1- to 4-mark questions in Papers 1–3*

Know the difference between 'describe' and 'explain'

Students often mix up 'describe' and 'explain'.

- **Describe** means you must say what the main features of something are like (e.g. coastal features) or what the steps in a process are (e.g. how the Milankovitch cycle operates). You don't have to give any reasons.
- **Explain** means that you must give reasons why something occurs (e.g. rural-urban migration).

11.3 Tackling 1- and 2-mark questions in Papers 1–3

In this section, you'll understand how to answer questions carrying 1–2 marks in Papers 1, 2 and 3. These kinds of questions are similar across all three exam papers.

Tackling 1-mark questions

1-mark questions are straightforward – although not always easy! Examples are shown in Figure 1. They tend to be of three types:

- Questions testing what you know (AO1), e.g. defining key words.
- Questions testing skills, such as completing a graph (e.g. of urbanisation trends), or maths questions (e.g. calculating an increase in GDP) (AO4).
- Multiple-choice questions (AO1 or AO4 depending on the question).

1-mark questions usually have two lines of writing space. Answering these questions rarely needs more than a few words. You don't need to use full sentences on 1-mark questions. For example, in Figure 1:

- an answer to Question 1 'Moving from the countryside to cities' would be enough to earn 1 mark
- an answer to Question 2 could just be 'GDP' for 1 mark.

AO1 Knowledge

1 Define the term 'rural-urban migration'.

*2 Name **one** economic indicator used to measure the level of development in a country.*

*3 Identify which **one** of the following is an igneous rock:*

Limestone ☐
Sandstone ☐
Granite ☐
Slate ☐

AO4 Skills

*4 Identify **one** country with under US$1000 per capita in the table of figures.*

5 Calculate the increase in population between 2000 and 2020 from the graph.

6 Identify from the graph the three-month period in which most rain fell from the following:

April–June ☐
May–July ☐
June–August ☐
July–September ☐

▲ ***Figure 1*** *Examples of AO1 Knowledge and AO4 Skills 1-mark questions*

Tackling 2-mark questions

2-mark questions are of two types:

- those which do not require development of points
- those which do require development of points.

2-mark questions always have four lines of writing space.

Questions which do not require development

Some 2-mark questions are like two 1-mark questions bolted together. Question 1 in Figure 2 is one example.

- For the question '*Explain **two** reasons why a river's velocity increases with distance downstream*' a correct answer would give two separate reasons worth 1 mark each. No further development is needed.

Question 1

*Explain **two** reasons why a river's velocity increases with distance downstream. (2 marks)*

Comment: This is asking for two reasons for 2 marks, so no development is needed – see Example Answer A. As an example, if you **did** develop a point with only one reason, like Example Answer B below, you would only get 1 mark.

Example Answer A: worth 2 marks

i) the volume of water increases (1 mark)

ii) the channel gets smoother (1 mark)

Example Answer B: worth only 1 mark

i) the channel gets smoother (1 mark) so there are fewer boulders to provide friction which would slow the river. (No further marks – the question does not ask for development of a point)

▲ ***Figure 2*** *A 2-mark question which does not require further development of points*

Questions which do require development

Question 2 in Figure 3 is another example. Compare it with Figure 2 so you understand the difference in question style.

Question 2

*Explain **one** reason why a river's velocity increases with distance downstream. (2 marks)*

Comment: This is asking for a single reason for 2 marks, so a point **must** be developed. Two separate points will not earn a second mark.

Example Answer A: worth 1 mark

i) the volume of water increases

ii) the channel gets smoother (Neither point is developed – so it's just 1 mark)

Example Answer B: worth 2 marks

i) the channel gets smoother (1 mark) which means there are fewer boulders to provide friction. (1 mark)

▲ ***Figure 3*** *A 2-mark question which requires development of points*

How to develop your answers

Most questions carrying 2 marks need to show evidence of **developed** points. Take the example of the following question:

*Explain **one** possible economic impact of climate change. (2 marks)*

In this question, naming one economic impact earns you 1 mark. The key to success is knowing how to develop it, and turn it into 2 marks. Developing a point means that you should:

- either **explain a consequence**
  - e.g. Temperatures would increase (1 mark) **which could lead to** drought, killing crops. (1 mark)
  - e.g. Sea levels are rising because of melting ice (1 mark) **which increases** the likelihood of floods which would damage crops. (1 mark)
- or **give an example**
  - e.g. Increased droughts would kill farmers' crops (1 mark) **such as** rice which depends on water. (1 mark)

Further examples of 2-mark questions which require developed points are shown in Figure 4.

1 *Explain **one** way in which economic development can be measured. (2 marks)*

2 *Describe **one** feature of the quality of life in a megacity in an emerging or developing country. (2 marks)*

3 *Describe **one** feature of the structure of a tropical storm. (2 marks)*

▲ ***Figure 4*** *Examples of 2-mark questions which require developed points*

Developing points

To develop a point in an answer, use one of the following phrases:

- *'so that …'*
- *'which leads to …' (or 'which causes …')*
- *'which means that …'*
- *'therefore …'*
- *'for example …' (or 'e.g. …')*

11.4 Tackling 3- and 4-mark questions in Papers 1–3

In this section, you'll understand how to answer questions carrying 3–4 marks in Papers 1, 2 and 3. These questions are similar across all three exam papers.

Tackling 3-mark questions

Sometimes, examiners want more than one developed point, especially in questions which ask you to explain a detailed process or sequence of events. Explanations like this are called **chains of reasoning**, where one point follows logically from the other – i.e. *'A happens, which leads to B, and then to C'*. Questions like this carry 3 marks, where two developments are needed, or 4 marks, where three developments are required.

3-mark questions always have six lines of writing space. It's usual to have to write answers to 3-mark questions in full sentences.

Look at this question:

*Explain **one** way in which rainforests can be protected. (3 marks)*

With 3-mark questions, it is not enough to just name one way in which ecosystems can be protected. To earn 3 marks, you must:

- **Extend the point:** i.e. describing in more detail e.g. Make it illegal to carry out logging (1 mark) so that forest habitats are protected for animals. (1 mark)

And either:

- **Extend the point again:** e.g. Make it illegal to carry out logging (1 mark) so that forest habitats are protected for animals (1 mark) which helps to maintain biodiversity. (1 mark)

or:

- **Give an example:** of what you are describing e.g. for example like the conserved rainforests in Costa Rica. (1 mark)

Question 1 in Figure 1 shows another example.

Question 1

Explain how increased CO_2 emissions could lead to increased flood risk. (3 marks)

Comment: To earn 3 marks, you need to develop the point **and** extend it with further points (i.e. develop a chain of reasoning).

Example Answer A: worth 1 mark

Increased CO_2 emissions could cause increased temperatures. (1 mark)

Example Answer B: worth 2 marks

Increased CO_2 emissions could cause increased temperatures, (1 mark) **which would lead to** melting of ice caps and glaciers. (1 mark)

Example Answer C: worth 3 marks

Increased CO_2 emissions could cause increased temperatures, (1 mark) **which would lead to** melting of ice caps and glaciers (1 mark) **so that** sea level rises, increasing flood threats. (1 mark)

▲ ***Figure 1*** *A 3-mark question which requires a chain of reasoning*

Different versions of 'compare'

A question might use a different command word, but still require a comparison. For example:

*Explain **one** difference between the type of volcano found at constructive plate margins with those found at destructive margins. (3 marks)*

Just as with 'compare', you should state how volcanoes:

- at constructive margins are shield-type (1 mark), because they are basaltic and flow further (1 mark)
- at destructive margins are cone-type (1 mark).

This example in Figure 2 shows where marks are given.

Example Answer A: worth 3 marks

Volcanoes at destructive margins have cool lavas which don't run far (1 mark) so the volcano is a steep cone shape, (1 mark) whereas those at destructive margins have shallow sides. (1 mark)

▲ ***Figure 2*** *A 3-mark answer which gives a comparison*

Tackling 4-mark questions

4-mark questions are different from those carrying 1–2 marks because they can involve short paragraphs of writing. Whereas 1–2 mark questions can often be answered with short phrases, 4-mark questions must be written in full sentences. 4-mark questions always have eight lines of writing space.

4-mark questions are of two types:

- those which ask for two developed points, e.g. '*Explain **two** methods of hard engineering which are used to protect areas at risk from flooding.' (4 marks)* An example is given in Figure 2. Here, you must treat the question almost as if it were two 2-mark questions, because the mark scheme will credit 2 marks for each point.
- those which ask for a single chain of reasoning, e.g. '*Explain how a change of land use can increase the risk of flooding.' (4 marks)* An example is given in Figure 3. Here, you must treat the question as one stream of writing, explaining a process by which (for example) the removal of forest could lead to less interception (1 mark), faster saturation of the soil (1 mark) leading to quicker surface run-off (1 mark) which increases the likelihood of flooding (1 mark).

Question 2

Explain how an urban regeneration scheme you have studied has both benefited and caused a problem for people. (4 marks)

Comment: This is asking for two things for 4 marks (a benefit and a problem for people), so both points **must** be developed. To earn all 4 marks, you need to explain a benefit for people (developed for 2 marks) and a problem (developed for 2 marks). Two benefits, each developed, would earn only 2 marks – you must mention both a benefit and a problem.

Example Answer A: worth 2 marks

One benefit of regenerating Birmingham's Jewellery Quarter was lots of bars and restaurants. (1 mark) One problem was that many local people could not afford the flats built there. (1 mark)

Example Answer B: worth 4 marks

One benefit of regenerating Birmingham's Jewellery Quarter was lots of bars and restaurants, (1 mark) which created jobs for people. (1 mark) One problem was that the flats built there were expensive (1 mark) which meant that many local people could not afford them. (1 mark)

▲ ***Figure 2*** *An example of a 4-mark question requiring two developed points*

Question 3

*Explain **one** way in which volcanoes can form along constructive plate margins. (4 marks)*

Comment: To earn 4 marks, you need to go one stage further than the 3-mark question – i.e. develop the point and extend it with three further points to form a chain of reasoning.

Example Answer A: worth 2 marks

As the plates pull apart, (1 mark) a plume of magma rises to the surface to fill the gap. (1 mark)

Example Answer B: worth 4 marks

As the plates pull apart, (1 mark) a plume of magma rises to the surface to fill the gap. (1 mark) On constructive plate margins this is often basaltic lava which flows easily away from the boundary before it solidifies. (1 mark) More lava flows from the vent creating new layers on top of the first (1 mark) building up the volcano.

▲ ***Figure 3*** *An example of a 4-mark question requiring a chain of reasoning*

Further examples of 4-mark questions

As part of your revision, try tackling these 4-mark questions.

Question 4

Explain one strength and one weakness of using GDP to measure economic development. (4 marks)

Question 5

*Explain **two** ways in which geology can influence coastal landforms. (4 marks)*

Question 6

Explain how a change of land use could lead to an increased risk of flooding. (4 marks)

11.5 Tackling 8-mark questions in Paper 1

In this section, you'll learn how to answer 8-mark questions in Paper 1. Note that these are structured and assessed differently from those in Paper 2 (see sections 11.6, 11.7 and 11.8) and Paper 3 (see section 11.9).

What you need to do

Two command words – 'Assess' and 'Evaluate' – are used for 8-mark questions in Paper 1. They are very similar, and even have the same mark scheme, but in general:

- 'Evaluate' is used more to weigh up arguments or viewpoints
- 'Assess' is used to rank more serious or less serious impacts, like the example in Figure 3.

Figure 1 shows what you need to do in order to reach top marks.

Command word	No. of marks	What you need to do	AOs tested in Paper 1
Assess	8	**Think of relevant examples you've learned** • Use case studies and examples • Use detail of events (e.g. a tropical cyclone) or processes (e.g. rural-urban migration) **Analyse, make judgments and conclude** • Make about three points to create your argument using evidence • Analyse or rank these points in order of importance or severity • Pick out evidence to support each point • Write a brief conclusion	8 marks total These are split: a) **AO2 4 marks** (your understanding of key features, places and processes) b) **AO3 4 marks** (your interpretation and judgments you make based on evidence)
Evaluate	8	**Select relevant points** • Use case studies and examples • Use detail of events (e.g. a tropical cyclone) or viewpoints (e.g. the arguments in favour of climate change) **Analyse the evidence** • Why are there different views? • Why do people's opinions differ? **Make judgments and reach a conclusion** • Develop an argument using evidence by making about three points • Weigh up the evidence – does it support or reject the arguments? • Pick out evidence to support each point • Write a brief conclusion at the end	

▲ ***Figure 1*** *Command words and their meanings for 8-mark questions in Paper 1*

How to tackle questions using the command word 'Evaluate'

Consider the following 8-mark exam-style question:

Evaluate the judgment that the economic impacts of tectonic hazards upon developing or emerging countries are more damaging than those upon a developed country. (8 marks)

The 8 marks are split – 4 for AO2, 4 for AO3. To succeed, do three things:

- Think of three examples of economic impacts for either a volcanic eruption or an earthquake in each of a developing or emerging country and for a developed country. That's AO2 (understanding).
- Compare them – which was more serious: the event in the developing/emerging country or the one in the developed country? Why? That's AO3 (analysing the evidence).
- Think about the statement and whether you think it might be true or not. Reach a conclusion – are you agreeing with the viewpoint or not? That's also AO3.

A worked example of the command word 'Assess'

An example of an 8-mark exam-style question assessing AO2 and AO3 is:

Assess the economic impacts of a tropical storm in a developing or emerging country. (8 marks)

8-mark questions generally require three well-developed points, just like the chains of reasoning in 3- or 4-mark questions (see section 11.4). To reach Level 3, you need to:

- explain **three** economic impacts about a named tropical storm (AO2), **and**
- assess these – how serious were they? Were some more serious than others, and why? (AO3)

Now try planning the question shown in Figure 2.

Assess means make a judgment. Which economic impacts were more serious? Impacts on people and jobs, on costs of renewal, or on the economy? You need to judge how serious each impact was, and rank them.

Economic impacts: those affecting people's jobs or earnings (e.g. farm crops), power supplies or services, or renewal of buildings (e.g. schools or hospitals damaged in the storm). 8 marks means that you need to write about three of these economic impacts, developed in some detail.

Assess the **economic** impacts of a **tropical storm** in a **developing or emerging** country.' (8 marks)

You must **name** a tropical storm and the country that you wish to write about. If you don't name one, and just write generally, you would restrict your mark to the middle of Level 2.

Make sure you write about a **developing or emerging** country – NOT a developed one!

▲ ***Figure 2*** *Planning an exam question which uses the command word 'Assess'*

Figure 3 shows an answer that is worth the full 8 marks.

The storm is named – essential for Level 3. Note also how the answer makes a judgment straight away – 'were most serious among the poor'.

The economic effects of 'stormy winds and floods' are described. There is detailed explanation of their economic effects – 'damaged buildings', 'cost insurance companies and governments millions'. It assesses the storm as the 'worst known'.

In Cyclone Aila which affected Bangladesh in 2009 the economic effects were enormous, and were most serious among the poor. The cyclone brought some of the worst known stormy winds and floods. This damaged buildings, which cost insurance companies and governments millions. More storms also caused erosion of flood defences, which flooded villages and farmland, costing huge amounts to replace and repair, and destroyed crops. For many farmers and families, this meant loss of homes and crops, making their poverty worse and forcing some to leave the land and move to Dhaka, the capital, for work.

'Erosion of flood defences' is an economic impact which 'flooded villages and farmland', 'costing huge amounts to replace and repair'. The phrase 'costing huge amounts' is evidence of assessing how big an impact this is.

The 'loss of homes and crops' is evidence of an economic impact. The answer assesses the loss of crops, making poverty worse, forcing people to leave the land.

▲ ***Figure 3*** *A Level 3 answer to the question 'Assess the economic impacts of a tropical storm in a developing or emerging country.' (8 marks)*

11.6 Tackling 8-mark questions in Paper 2

In this section, you'll learn how to answer 8-mark Questions 4 (Section A) and 7 (Section B) in Paper 2. Note that these are assessed differently from 8-mark questions in Paper 1 (see section 11.5).

What you need to do

The most demanding command words, carrying the highest marks, are shown in Figure 1. Like Paper 1, only these two command words are used in 8-mark questions.

Command word	No. of marks	What you need to do	AOs tested in Paper 2
Assess	8	**Study the resources provided** • Look at what they are telling you • Pick out three points that are important **Make judgments and reach a conclusion** • Judge or rank these three points in order of importance or severity • Develop an argument explaining why these three are important • Write a brief conclusion at the end	8 marks total These are split: **a) AO4 4 marks** (your skills and ability to interpret resources) **b) AO3 4 marks** (your interpretation and judgments you make based on evidence)
Evaluate	8	**Study the resources provided** • Look at what they are telling you • Pick out three points that are important **Weigh up the evidence** • Which evidence looks strongest? • Why do people's opinions or views vary? **Make judgments and reach a conclusion** • Weigh up these three points in order of importance or severity • Explain why these three are important • Do they support or reject the arguments? • Write a brief conclusion at the end	

▲ ***Figure 1*** *Command words and their meanings for 8-mark questions in Paper 2*

What's different about 8-mark questions in Paper 2?

Questions 4 and 7 in Paper 2 differ from 8-mark questions in Paper 1.

- In Paper 1, AO2 (your understanding) is combined with your ability to develop an argument (AO3). You go to the exam having revised, and use what you know to argue a case.
- In Paper 2, you're given resources to interpret for Question 4 (UK Physical) and Question 7 (UK Human). Examiners assess your skills in interpreting these (AO4) and in applying what you see to develop an argument (AO3).

Tackling Questions 4 and 7

Questions 4 (UK Physical) and 7 (UK Human) begin by asking you to 'Analyse Figure X and Y'. This worked example of Question 4 uses two maps of the UK – one of its geology (Figure 2, page 111) and the other of its relief (Figure 1, page 110):

Analyse Figure 1 and Figure 2 on pages 110 and 111.

Assess the strength of the relationship between relief and geology in the UK. (8 marks)

To succeed, do two things:

- Analyse the two maps on pages 110 and 111 carefully, looking for distribution of different rock types, comparing this with distributions of highland and lowland areas. Are areas of most resistant rocks where relief is highest? That's AO4 (skills) for 4 marks.
- Consider explanations for any links between the maps, and why these exist. Then consider other factors that might be important and how strong the relationship is between geology and relief. That's AO3 (applying your knowledge) for another 4 marks.

Planning your answer

Using the maps on pages 110 and 111, plan your answer. Figure 2 below shows you how to plan. To get the highest marks you need to:

- identify **2–3** links between the maps (AO4), **and** explain these
- think about the relationship between geology and relief. Do other factors besides geology explain highland and lowland landscapes? (AO3)

Try planning the question as shown in Figure 2.

Assess means: spotting links between geology and relief. Are the links **strong**? Are other factors involved? You must assess how strong the relationship is to get top marks.

Key words: relief and **geology**. 8 marks means that you need to write about three developed points between relief and geology. Are the UK highland areas linked to rock type? Why is this?

***Assess** the **strength** of the **relationship** between **relief and geology** in **the UK**. (8 marks)*

Spot the **relationship** between geology and a) areas of high ground, and b) areas of low ground, and explain it. Then think about other things which affect the landscape, e.g. past glacial periods.

This question is about **the UK** – so name places you can see on the map as examples (e.g. the Lake District).

▲ ***Figure 2*** *Planning an exam question based on the maps on pages 110 and 111*

The answer in Figure 3 is worth full marks because:

- it makes three points, and at the end has a conclusion (this is AO3)
- it uses evidence from a range of places on the maps (AO4)
- the answer links relief to geology (AO3)
- finally, the student spots that geology is not the only explanation for the landscapes, but has referred to tectonic uplift, climate and weathering processes.

Note how the student makes a judgment straight away – 'Geology has a key role'. That's AO3. You don't have to give the answer at the end as a conclusion – this student has done so at the start!

Point 3
'However' brings a change of direction. There is detailed explanation of the reasons why the Pennines are an exception (AO3) – using sedimentary rocks as evidence (AO4). The cause of this is given (uplift – the student has learned this).

Geology has a key role in UK landscapes with a clear link between upland areas and rock types. In Scotland most geology is metamorphic and igneous, which explains why two thirds of Scotland is above 600m – these rocks are more resistant to erosion and weathering which leaves them standing as highland areas. This is also true in Northern Ireland where half the rocks are igneous and coincide with land over 200m. Most of the rest of England (except the south-west) has sedimentary rocks which are less resistant, so most landscapes are below 200m.

However, geology is not the only factor affecting landscapes. The Pennines are over 200m but are sedimentary. Here, tectonic activity has uplifted these to form the Pennines. Biological and chemical weathering occurs in lowland areas of England, whereas upland areas of Scotland and northern England are affected by freeze-thaw, creating scree slopes. So climate is a factor in landscape development. To conclude, geology is a major factor in forming landscapes but not the only one.

Point 1
'In Scotland most geology is metamorphic …' is interpreting the map correctly. (AO4)
The student then links this to the relief map 'two thirds of Scotland is above 600m'. (AO3)

Point 2
The student then uses the map (AO4) to make a second point about Northern Ireland (AO3) and then makes links to southern England (AO3) and lowland landscapes.

Conclusion
The student concludes in a single sentence – which is perfectly acceptable.

▲ ***Figure 3*** *A Level 3 answer to the question 'Assess the strength of the relationship between relief and geology in the UK.' (8 marks)*

11.7 Tackling familiar fieldwork questions in Paper 2

In this section, you'll understand how to answer exam questions about your own fieldwork – called *familiar fieldwork* – on Paper 2. *Unfamiliar fieldwork* (done by others) can be found in section 11.8.

Fieldwork in the GCSE course

As part of your GCSE course, you'll experience two days of fieldwork.

- One day **physical fieldwork** – either Coastal change and conflict or River processes and pressures.
- One day **human fieldwork** – either Dynamic urban areas or Changing rural areas.

Fieldwork is assessed in the exam in Paper 2.

- **Section C1** is on physical fieldwork – either Coastal change and conflict or River processes and pressures.
- **Section C2** is on human fieldwork – either Dynamic urban areas or Changing rural areas.

Make sure you do the right question in each section!

Four fieldwork terms you need to know!

- **Familiar fieldwork** – questions about your own fieldwork
- **Unfamiliar fieldwork** – questions about fieldwork done by other people
- **Quantitative data** – data involving hard numbers, tables and graphs
- **Qualitative data** – other non-numeric information (e.g. photos, sketches, interviews or reports)

The enquiry process and familiar fieldwork

Exam questions will be based on the enquiry process (see Figure 1). This shows six stages of a fieldwork enquiry, in the order that you would do them. Exam questions could be about any of these stages.

These stages, with examples, are:

Planning. The question you asked, and the location you visited. Was it a sound geographical question, and was the location suitable? *Example: Why you went to Walton or the Queen Elizabeth Olympic Park to carry out fieldwork.*

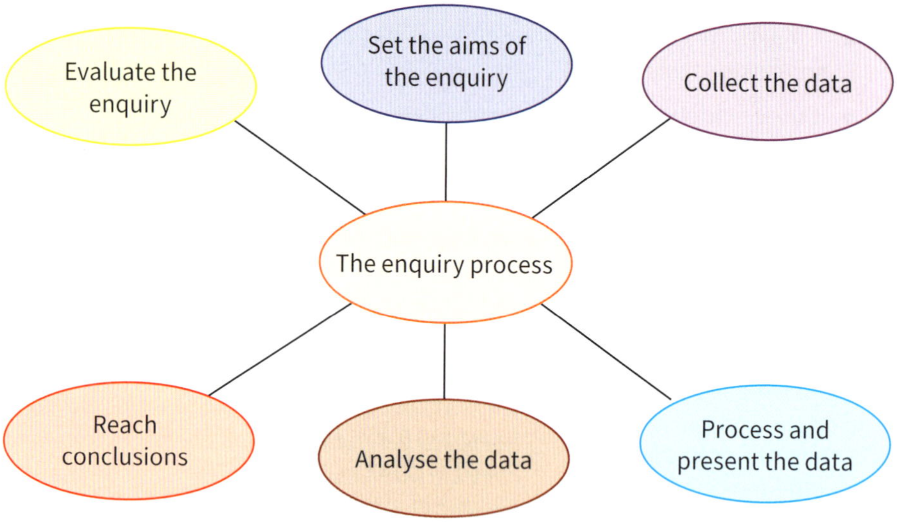

▲ ***Figure 1*** *The six stages in the enquiry process*

Methods. How suitable – and accurate – were the methods you used to measure and record data. Methods used to **sample** data collection. *Example: Why the fieldwork methods you used were chosen. Were they suitable? What sampling methods did you use?*

Presenting results. Can you justify the ways you presented your data? *Example: Why you chose particular type of graphs or maps.*

Analysing results. What trends or patterns did you find in your data? *Example: How you analysed your results and how strong or reliable the trends or patterns were.*

Conclusions reached. How certain can you be that your conclusions were sound? *Example: How far were your conclusions accurate or reliable?*

Evaluation. What went well in your fieldwork, and why? What went less well, and why? *Example: Explain how far your enquiry was reliable – i.e. would you get similar results if you went back on a different day?*

Five things you can do to get 'exam ready'

- Keep your fieldwork write-up safe.
- If you've lost or missed work, photocopy someone else's.
- Learn what your enquiry was about, and why the area you went to was suitable.
- Sketch out two results (e.g. a graph and map) which summarise your results and which you could draw under exam conditions **quickly**.
- Practice past exam questions.

Know the command words

It's important to remember command words that will be used:

- There are **no marks** for recall (AO1) or understanding (AO2).
- **All marks** are for AO3 (application) and AO4 (skills).

Therefore, focus on understanding these command words:

- **Explain** – for example, the reasons why the place you went to was suitable.
- **Suggest** – making a reasoned explanation.
- **Assess** – rank particular points or factors in terms of their importance.
- **Evaluate** – weigh up strengths and limitations.

You can see how these command words are used in exam questions about familiar fieldwork and the stages of enquiry in Figure 2.

For your physical fieldwork investigation:	For your human fieldwork investigation:
(a) Explain **one** reason why the method you used to measure sediment was suitable. (2 marks)	**(a)** Explain **one** reason why the location chosen for your human fieldwork was suitable. (2 marks)
(b) Suggest **one** source of error when you measured the gradient. (2 marks)	**(b)** Suggest **one** source of error when you measured environmental quality. (2 marks)
(c) You used a geology map in your fieldwork. Explain **two** ways in which this helped your investigation. (4 marks)	**(c)** You used census data in your fieldwork. Explain **one** way in which this helped your investigation. (2 marks)
(d) Evaluate the accuracy and reliability of your conclusions. (8 marks)	**(d)** Assess the accuracy of the methods you used in your data collection. (8 marks)

▲ ***Figure 2*** *A typical exam question sequence on the stages of enquiry*

Answering short exam questions – worked example

Question

*You have collected data as part of your physical fieldwork. Explain **one** limitation of one **quantitative** fieldwork method you used to collect data. (2 marks)*

Example Answer

When we measured river velocity, we found that the float used got stuck behind rocks. This increased the time taken to flow downstream, so we didn't always get an accurate result.

Examiner Comment

This gets 2 marks – 'the float used got stuck behind rocks' (1 mark) and a developed point 'this increased the time taken to flow downstream'. (1 mark)

Answering longer 8-mark questions – worked example

Consider the following question:

Evaluate the relative usefulness of primary and secondary data in your fieldwork investigation. (8 marks)

How to tackle this question

- Sketch out a table showing 'primary data' in one column, and 'secondary data' in the other.
- Under each heading, think of ways in which **a)** primary data and **b)** secondary data were useful.

Your answer could use the following phrases:

- *'Some primary data were more useful than others, for example …'*
- *'This was because …'*
- *'Some secondary data were more useful than primary data, for example …'*

Then think whether primary or secondary data helped you to **a)** get more reliable data, **b)** analyse your data, and **c)** draw conclusions.

11.8 Tackling unfamiliar fieldwork questions in Paper 2

In this section, you'll understand how to answer exam questions about fieldwork carried out by others – called *unfamiliar fieldwork* – in Paper 2. See section 11.7 for *familiar fieldwork*.

Unfamiliar fieldwork in Paper 2

Questions on unfamiliar fieldwork occur in:

- **Section C1** (physical fieldwork) – on either Coastal change and conflict or River processes and pressures.
- **Section C2** (human fieldwork) – on either Dynamic urban areas or Changing rural areas.

Examiners will give you situations where students carry out fieldwork on topics like yours. Examiners can ask you to apply what you've learnt by:

- thinking about methods used to collect data
- using skills in interpreting data
- judging the accuracy of data collection.

You'll be able to answer questions because these enquiries will be similar to yours, and you can judge based on experience. Questions will contain:

- tables, graphs and maps that you can interpret
- details about how students collected data.

For unfamiliar physical fieldwork:
(a) Explain **one** reason why the method the students chose to measure slopes may have been inaccurate. (2 marks)
(b) Suggest **one** source of error the students might have made in measuring sediment size. (2 marks)
(c) Explain how a named secondary data source might have improved the reliability of the students' data. (4 marks)
(d) Evaluate the accuracy and reliability of conclusions reached by the students. (8 marks)

▲ ***Figure 1*** *Examples of exam questions about unfamiliar fieldwork in physical geography*

Handy hints for physical fieldwork questions

- Was the place suitable for fieldwork? Was it safe?
- Were there any recent conditions that might have affected fieldwork e.g. prolonged rain or coastal storms?
- Was the group's equipment suitable for accurate data collection?
- Did the group sample enough data to make their study reliable?
- Were the results presented using suitable graphs?

The enquiry process

Exam questions will be about the six stages of enquiry (see section 11.7). Examples of exam questions are shown in Figure 1 (physical fieldwork) and Figure 2 (human fieldwork) below.

The six stages – and possible exam questions – are:

Planning. The question the students asked, and the location visited. Was the location suitable? *Expect maps or photos so you can judge.*

Methods. How suitable were methods used to measure and record data? How did they sample data collection? *Example: What were the strengths and weaknesses of sampling used? Time of day? Time of year?*

Presenting results. Did the group use suitable ways of presenting data? *Example: Did they use appropriate graphs or maps?*

Analysis. What trends or patterns did you see in their data? *Example: How strong or reliable were the trends or patterns?*

Conclusions. Were conclusions sound? *Example: How far were their conclusions accurate or reliable?*

Evaluation. What went well in their fieldwork, and why? What went less well? Has it affected the accuracy or reliability of their results? *Example: Assess how far their enquiry was accurate.*

For unfamiliar human fieldwork:
(a) Suggest **one** way in which the location chosen for their human fieldwork might have been unsafe. (2 marks)
(b) Suggest **one** source of error that the students may have made when measuring environmental quality. (2 marks)
(c) Explain how census data might have strengthened the students' conclusions. (4 marks)
(d) Assess the accuracy of sampling methods used by the students. (8 marks)

▲ ***Figure 2*** *Examples of exam questions about unfamiliar fieldwork in human geography*

Handy hints for human fieldwork questions

- How were data collected? Were enough data collected?
- Was there anything about the time of day or year that made data unreliable?

Answering 8-mark questions – worked example

Consider this exam question. It consists of information and a graph (Figure 3).

> *Ten students carried out questionnaires in an inner city and outer city suburb. They asked ten residents between them in each suburb on a Monday afternoon about quality of life in the suburb they lived in.*
>
> *They gave residents three statements about their suburb with which they could agree or disagree.*
>
> *1 – 'Streets are generally quiet in your suburb'*
>
> *2 – 'Open spaces are easily accessible from your suburb'*
>
> *3 – 'Your suburb is well served by public transport'*
>
> *Their results are shown in Figure 3.*
>
> *The students concluded that the outer city suburb had a better quality of life than the inner city suburb.*
>
> *Assess the evidence that this conclusion is correct. (8 marks)*

How to tackle this question

- Spend half your answer interpreting Figure 3 to see if the students' conclusion is correct (AO4), and the other half studying the text to decide **how far** their conclusion is correct (AO3).

1 Interpreting the graph

- Study Figure 3 to see whether the outer city suburb had a better quality of life. Start writing about this.
- You'll see that in two cases – Street noise and Open spaces – the conclusion is correct. Write about this, using data to support this.

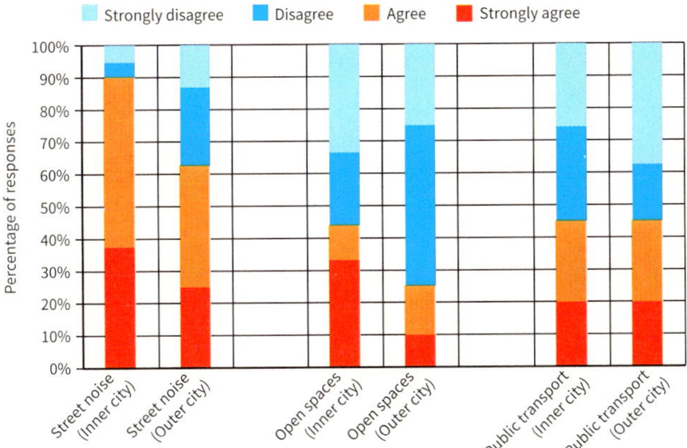

▲ ***Figure 3*** *Questionnaire survey results on quality of life in two suburbs*

- Now check for exceptions – there will always be one! In this case, for Public transport, the distinction between inner and outer city is less clear. Write about this, using data to illustrate.

2 Assessing whether their conclusion is correct

Consider when the survey was carried out.

- Is Monday afternoon a good time for fieldwork?
- Would students get a sample of the whole population?
- Is ten residents in each location a good sample?

When you have looked at the data, you may reach the view that the students got their conclusion wrong!

Example answer

Figure 4 shows a student answer to the question above. Read through and identify:

- where you think the student showed skills in interpreting the graphs of results
- where you think the student is judging the evidence showing whether the conclusion is correct.

> Overall, the conclusion looks wrong when you look at the graphs. The graphs about open space show that only 10% of people in the Outer city strongly agree, compared to 30% in the Inner city, which you would not expect and which goes against their conclusion. The Outer city also scored as highly as the Inner city (20% strongly agreeing) on public transport. The Inner city even had more people agreeing about open space (40% agreeing or agreeing strongly), which would be unexpected. The main problem is that the students were not measuring quality of life, but which area had the best environment, which is not the same thing. They only asked three questions about noise, open space and transport, and quality of life can include things like gardens or air quality. Perhaps the Outer city was a poorer area, and the Inner city a wealthier area like some of London's suburbs. We also don't know whether people who were asked were a fair sample – the survey was not large, and many would have been at work, so the students might have asked a higher proportion of elderly people, for example. That would distort the data.

▲ ***Figure 4*** *Student answer to the 8-mark question: 'Assess the evidence that this conclusion is correct.'*

This student scored 7 marks. The answer questions the accuracy of the conclusion and gives reasons why, and just needs a little more use of data to get it to 8 marks.

11.9 Tackling 8-mark questions in Paper 3

In this section, you'll learn how to answer 8-mark questions in Paper 3. Note that these are assessed differently from 8-mark questions in Paper 1 (section 11.5), and fieldwork (sections 11.7 and 11.8).

What you need to do

Two command words – 'Assess' and 'Evaluate' – are used for 8-mark questions in Paper 3. These are just the same as in Figure 1 on page 316.

What's different about 8-mark questions in Paper 3?

8-mark questions in Paper 3 are similar to those in Paper 2.

- They are based on resources in the booklet that accompanies Paper 3.
- Examiners will assess your skills in interpreting the resources (AO4) and in applying them to develop an argument (AO3).
- There is no command to 'Analyse Figure X and Y' as you would expect in Paper 2. The question you'll be given is something like 'Assess the impacts of X upon Y'.

A worked example

Consider the following exam-style question, based on the annotated map in Figure 1.

Study Figure 1. Assess the impacts of economic activity on Indonesia's environment. (8 marks)

Figure 1 tells you the environmental impacts of economic activity. All the information you need is here. To plan your answer, do the following:

- Pick out three environmental impacts in Figure 1 that you think are most serious. You need three impacts for an 8-mark answer.
- Take these three and for each one explain why it's serious. For example, smoke haze is not just serious for the rainforest, but also for human health.
- Now 'assess' these. Rank the ones that you selected from most to least serious. Work out your argument about how serious these impacts are.
- To guide you, think 'is the damage permanent?' or 'could the rainforest ever be restored?' This will help you assess the seriousness of each impact.

▼ ***Figure 1*** *The impacts of economic activity on Indonesia's environment*

1
- Much of Indonesia's rainforest has been cleared for palm oil plantations.
- By 2020, Indonesia had doubled palm oil production since 2008.
- Palm oil plantations have less biodiversity than rainforest.

2
- Greenpeace found that coal mining in Kalimantan had led to toxic river pollution.
- Half of Kalimantan's rivers were affected by pollution, with local environmental laws failing to prevent it.

3
- Since 1990, large areas of forest in Sumatra and Kalimantan have been cleared by burning.
- The smoke haze has spread across much of south-east Asia.
- In 2016 there were over 100 000 deaths from lung disease caused by smoke pollution.

4
- Drilling for oil offshore from the Seribu Islands has led to oil spills.
- Increased shipping is also a source of pollution, causing damage to fisheries, corals reef and tourism.

5
- Since 1990, rainforest in Indonesia has halved, clearing forest habitats from large areas.
- The number of orangutans has fallen from 230 000 in the 1920s to under 50 000 in 2020.

Judging a student answer

Read the answer in Figure 2. It earned all 8 marks. This is a strong answer because:

- the student identifies three problems – energy, deforestation, and toxic haze
- each problem is explained using detail in Figure 2
- the student assesses the severity of the problems, using words such as 'dramatic', 'irreversible', 'endangered'
- the student assesses how problems can be recovered from (e.g. through re-growing forest) but notes that some are near-impossible (toxic soil).

Exploiting Indonesia's energy resources has dramatic and some irreversible impacts on the environment. In Kalimantan, coal mines have scarred the landscape, polluting rivers and habitats. Deforestation for palm oil has meant orangutans have declined by over 80% due to habitat loss, and they are probably endangered now. Furthermore, deforestation in Indonesia is not only due to mining, they also burn forest to grow crops in the fertile soil, causing a toxic haze that spreads across the country and its neighbours. Most of these problems can be recovered from. They can regrow the forest and breeding programmes can increase orangutan numbers. However, the cleared landscape will take decades to recover, and re-growing the rainforest on toxic soil would be near impossible.

▲ ***Figure 2*** *Student answer to the exam-style question: 'Study Figure 1. Assess the impacts of economic activity on Indonesia's environment. (8 marks)'*

Evaluating viewpoints

Look at Figure 3. It shows three viewpoints about palm oil development in Indonesia. Consider an exam-style question about these views:

Study Figure 3. Evaluate the views about who is to blame for rainforest deforestation in Indonesia. (8 marks)

Organisation	View
1 WWF is a NGO and an environmental pressure group	"Large areas of rainforest have been cleared for palm oil plantations – destroying habitats for endangered species, including rhinos, elephants and tigers. In some cases, plantations have led to the eviction of forest-dwelling people."
2 World Growth is a pressure group promoting economic globalization	"Palm oil provides developing nations with a path out of poverty. Expanding sustainable agriculture such as palm oil plantations provides plantation owners and their workers with a means to improve their standard of living."
3 Cargill in a TNC based in the USA that grows, processes, and sells palm oil	"Millions of people depend on palm oil. We believe that palm oil should be produced sustainably. We have made a commitment that the palm oil products we supply will be certified as coming from sustainable forests."

▲ ***Figure 3*** *Three views about palm oil development in Indonesia*

To answer this question, think how Views 1 to 3 in Figure 3 differ:

- View 1 is from an environmental perspective.
- View 2 promotes global economic growth.
- View 3 is from a US-based TNC. What might its attitude be towards the future of rainforests?
- In this way you are evaluating – or weighing up – the views.

Figure 4 is an extract from an answer worth 8 marks.

One way that people's opinions differ is because of the economic gain of palm oil. WWF promotes the environment and blames plantations on economic interests. Their values are different from large TNCs clearing forest for palm oil. The US palm oil company's priority is economic gain. It promotes US interests in wealthy countries. It says it wants palm oil to be produced sustainably but how can that be if rainforest is destroyed?

View 2 also wants economic growth, but in the interests of countries like Indonesia as a way out of poverty. It has similar views to the US TNC (Cargill) but in the interests of developing countries. But it will probably never agree with WWF because it sees palm oil as sustainable and WWF blames palm oil for habitat loss.

▶ ***Figure 4*** *Student answer to the exam-style question: 'Evaluate the views about who is to blame for rainforest deforestation in Indonesia. (8 marks)'*

11.10 Tackling the 12-mark question in Paper 3

In this section, you'll learn how to answer the 12-mark question in Paper 3.

What you need to do

This is the question that many students get nervous about – the 12-mark essay! This spread should help to show you what you need to do in order to obtain top marks. You can read more about the format of Paper 3 in section 10.1.

Just a single command phrase – 'Select …. and justify' – is used for the 12-mark question in Paper 3. Figure 1 shows what you need to do to answer questions with this phrase.

Command phrase	No. of marks	What you need to do	AOs tested in Paper 3
Select … and justify	12	**Pick from a list of Options for the Future** • 'Select' the option you think is best **Develop an argument** • Decide which evidence from the Resource Booklet supports your case most strongly (i.e. 'justify') • Write a brief conclusion at the end **Make links to ideas and content you've learnt** • Use ideas and examples from the rest of your course to support your decision	12 marks total, split: **a) AO2 4 marks** (your knowledge and understanding) **b) AO3 4 marks** (your analysis, argument, and judgments you make based on evidence) **c) AO4 4 marks** (your use of evidence from the Resource Booklet)

▲ ***Figure 1*** *The single command phrase and its meaning for the 12-mark question in Paper 3*

What's different about the 12-mark question in Paper 3?

The 12-mark question in Paper 3 has some similarities to 8-mark questions, but for the most part it is unique (see Figure 1).

- It is focused on an issue about which resources are provided in the booklet that accompanies the exam. It's an unseen booklet – you won't have seen it before.
- Study the booklet carefully. Like the 8-mark questions in Papers 2 and 3, you'll be assessed in your skill in interpreting these (AO4) and in applying them to develop an argument (AO3).
- Unlike the other 8-mark questions, there are marks for AO2, in which you can use anything that you've been taught that helps you.

Getting your head around the question

Although the focus of the Resource Booklet will vary between one year and another, the 12-mark question will be very similar each year. It is likely to contain the same basic elements:

- You'll be asked to select an option (probably from a choice of three).
- You'll be asked to explore the impact of your choice on people, the environment and/or the economy.
- To reach your decision, you'll have to weigh up a range of advantages and disadvantages in a balanced way.
- You'll reach a conclusion, or decision, about which option is best.
- There is no preferred option from those given to you. You could make a case for any of them, and all will be justifiable. You'll be marked on how you justify your choice, not on the choice itself.

Making a choice

To make a choice, remember a few straightforward points. Often, the best place to start is **economic**, **social** and **environmental** or **political** factors (see Figure 2). None of these is straightforward – economic positives might lead to environmental negatives – or vice-versa. But all can be positives or negatives in helping you reach a decision.

Economic	£	• Jobs, employment, higher incomes • Higher skilled, higher paid jobs • Increase in GDP • Making an area more attractive to investment
Social		• Improving people's lives and quality of life • Better housing, reducing crime • Improved health and education • Greater cultural understanding
Environmental		• Improving air and water quality • Reducing pollution levels • Protecting/conserving wildlife/biodiversity • Improving the built environment

▲ ***Figure 2*** *Points to help you reach a decision*

What should your answer do?

Here are six points to guide you towards a top-rate 12-mark answer.

1 Give a range of points. Refer to at least two advantages and two disadvantages of your choice. These might include costs and benefits, or positive or negative impacts of your chosen option.

2 Give detailed evidence using detail from the Resource Booklet to support your choice.

3 Refer to a balance of social (people), environmental and economic factors to support your choice.

4 Be synoptic. This means using some knowledge and understanding from your course, or examples from Topics 7–9.

5 Give the counter-argument. Explain why you rejected the other options.

6 Reach a conclusion and overall judgment that is consistent with the evidence you have used.

Judging a student answer

Now read and judge the answer in Figure 3. Pick out what makes this a good answer – it obtained 10 marks out of 12.

These two extracts are part of a strong answer:

- The student identifies benefits and problems of the chosen option, for both the economy and the environment, and develops the argument to support their choice of option (AO3).
- The highlighted text in Extract 2 is material from the Resource Booklet (AO4).
- The underlined phrases show where the student is using what has been learned over the GCSE course – there was nothing in the Resource Booklet about this, so the student has written about this as part of their own understanding. (AO2).

Extract 1

Strategies such as eco-tourism and sustainable farming will help both the environment and the economy. By stopping palm oil production, deforestation decreases so less CO_2 is emitted. This slows down climate change as less greenhouse gases contribute to the greenhouse effect maintaining high biodiversity and creating stable ecosystems. This means plants can grow and soils will be rich and thick in nutrients as roots hold soils in place reducing soil erosion and surface run-off. This means more crops can grow and there's no extinction.

Extract 2

However, regenerating the rainforest may be hard as the land is already scarred and it costs money to clean up toxic waste and chemicals from soil or water. This means in the short-term the government loses money. But it has long-term benefits, because tourism is likely to increase. The closure of palm oil factories may cause a loss of jobs, leading to unemployment and a lower quality of life. However, this could be prevented by setting up government grants for businesses such as hotels. This would reduce the economic negative multiplier of this option.

▲ ***Figure 3*** *Two extracts from a student's answer to the exam-style question: 'Select and justify your choice of option for the economy and environment of Indonesia. (12 marks)' The student had chosen an option which halted all further deforestation of Indonesia's rainforest and would therefore prevent further palm oil plantations*

OS maps skills – what do I need to do?

This page lists Ordnance Survey (OS) map features and skills that you may need to use in the examination.

Map symbols

You don't need to learn these, as there will normally be a key (see page 327). But you do need to recognise physical and human features, such as:

- landforms along a coast (e.g. a cliff or beach) or river valley (e.g. steep or gentle slopes)
- human features such as towns and cities (e.g. streets and houses).

Map scale and distances

Scale is normally written as a ratio (e.g. 1:25000) – it means that the map is 1/25000 of real size on the ground. Maps used in GCSE use one of two scales:

- 1:50000 – where 1 cm = 50000 cm on the ground. Put more easily, 2 cm on the map equals 1 km.
- 1:25000 – where 1 cm = 25000 cm on the ground. Put more easily, 4 cm on the map equals 1 km.

The smaller the number, the larger the size of each feature on the map.

To measure distances:

a) Measure the distance between two points – e.g. 14 cm.

b) Divide by the number of cm to 1 km. So, on a 1: 25000 map, where 4 cm = 1 km, 14 cm is 3.5 km.

Grid references

The lines drawn across OS maps are called grid lines (see Figure 1). Each line is numbered; lines running left to right (west to east) are called eastings, because numbers increase to the east. Similarly, lines running across the map are called northings. The squares formed by grid lines are 1 km^2.

Grid lines are used to locate places – either a whole square (4-figures), or a point within a square (6-figures). Eastings are always given before northings.

- To give a **4-figure reference**, give the numbers of grid lines that cross at the bottom left (or south-west) corner of the square. Point B in Figure 1 is 2164.
- To give a **6-figure reference**, imagine each grid square is divided, like decimals except that you omit the decimal point. Easting 20 becomes 200, one tenth of the way across becomes 201, then 202 and so on. Point C in Figure 1 is at easting 230 and northing 640 – written as six figures (i.e. 230640). Look at Figure 1 and work out how to locate points D (245635) and E (232627).

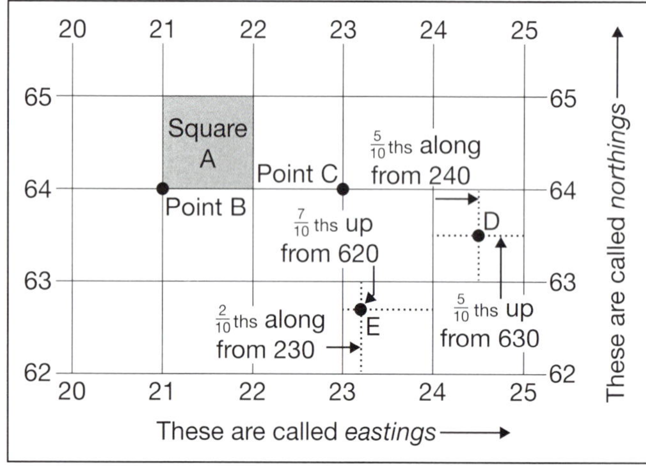

▲ ***Figure 1*** *Grid references*

Directions

You should know and be able to use 16 points of a compass (see Figure 2) in describing direction to or from somewhere.

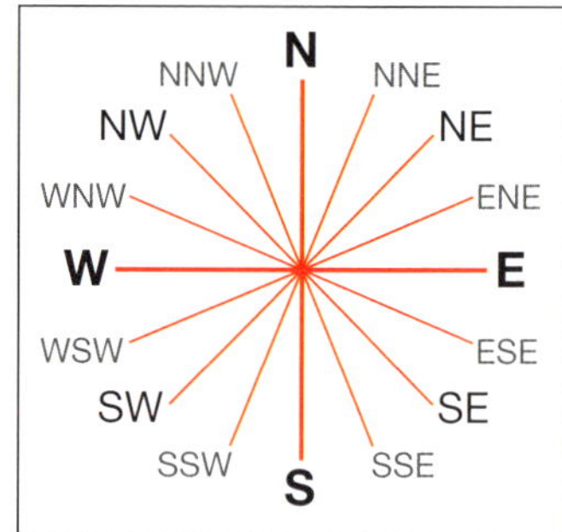

▲ ***Figure 2*** *Sixteen compass points to describe direction*

Height and slope

Contour lines are used to show two features of the landscape:

- its height (each contour line is at a particular height)
- its slopes (remember that contour lines which are close together show steep land, and widely spaced contours mean gently sloping land).

Symbols on Ordnance Survey maps (1:50 000 and 1:25 000)

ROADS AND PATHS

M I or A 6(M) Motorway
A 35 Dual carriageway
A 31(T) or A 35 Trunk or main road
B 3074 Secondary road
Narrow road with passing places
Road under construction
Road generally more than 4 m wide
Road generally less than 4 m wide
Other road, drive or track, fenced and unfenced
Gradient: steeper than 1 in 5; 1 in 7 to 1 in 5
Ferry Ferry; Ferry P – passenger only
Path

PUBLIC RIGHTS OF WAY

(Not applicable to Scotland)

1:25 000 1:50 000

Footpath
Road used as a public footpath
Bridleway
Byway open to all traffic

RAILWAYS

Multiple track
Single track
Narrow gauge/Light rapid transit system
Road over; road under; level crossing
Cutting; tunnel; embankment
Station, open to passengers; siding

BOUNDARIES

National
District
County, Unitary Authority, Metropolitan District or London Borough
National Park

HEIGHTS/ROCK FEATURES

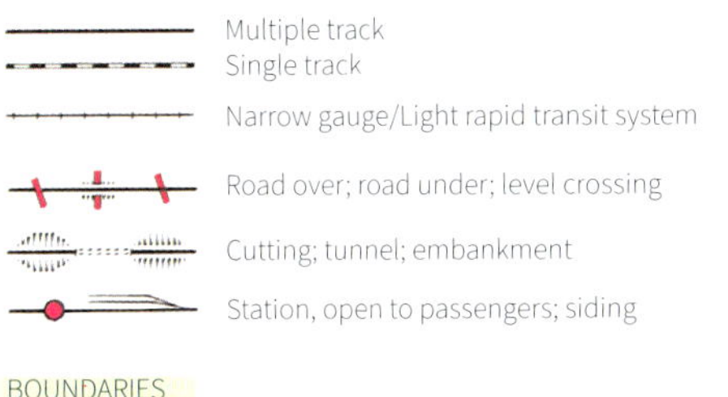

Contour lines
·144 Spot height to the nearest metre above sea level

ABBREVIATIONS

P	Post office	PC	Public convenience (rural areas)
PH	Public house	TH	Town Hall, Guildhall or equivalent
MS	Milestone	Sch	School
MP	Milepost	Coll	College
CH	Clubhouse	Mus	Museum
CG	Coastguard	Cemy	Cemetery
Fm	Farm		

ANTIQUITIES

VILLA Roman
Castle Non-Roman
Battlefield (with date)
Tumulus/Tumuli (mound over burial place)

LAND FEATURES

ruin Buildings
Public building
Bus or coach station
Place of Worship with tower
Place of Worship with spire, minaret or dome
Place of Worship without such additions
Chimney or tower
Glass structure
Heliport
Triangulation pillar
Mast
Wind pump / wind generator
Windmill
Graticule intersection
Cutting, embankment
Quarry
Spoil heap, refuse tip or dump
Coniferous wood
Non-coniferous wood
Mixed wood
Orchard
Park or ornamental ground
Forestry Commission access land
National Trust – always open
National Trust, limited access, observe local signs
National Trust for Scotland

WATER FEATURES

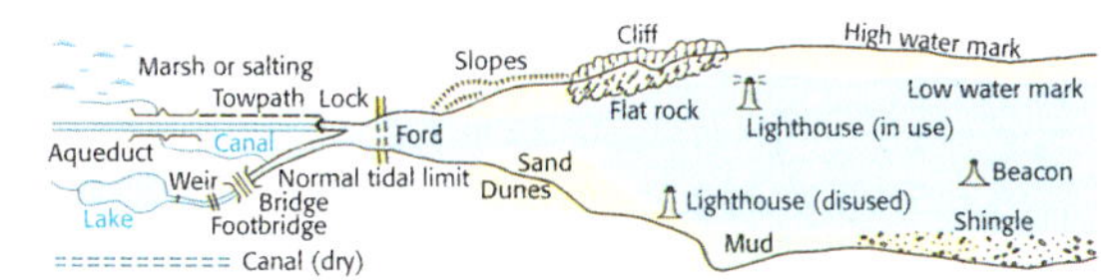

TOURIST INFORMATION

Parking
Park & Ride
Visitor centre
Information centre
Telephone
Camp site/ Caravan site
Golf course or links
Viewpoint
PC Public convenience
Picnic site
Pub/s
Museum

Castle/fort
Building of historic interest
Steam railway
English Heritage
Garden
Nature reserve
Water activities
Fishing
Other tourist feature
Moorings (free)
Electric boat charging point
Recreation/leisure/sports centre

Glossary

***cross reference**

A

abiotic non-living part of a *biome, includes the *atmosphere, water, rock and soil

abrasion the scratching and scraping of a river bed and banks by the stones and sand in the river

aftershocks follow an earthquake as the fault 'settles' into its new position

alluvium all deposits laid down by rivers, especially in times of flood

altitudinal zonation is the change in *ecosystems at different altitudes, caused by alterations in temperature, precipitation, sunlight and soil type

antecedent rainfall the amount of moisture already in the ground before a rainstorm

asthenosphere part of the Earth's *mantle. It is a hot, semi-molten layer that lies beneath the *tectonic plates

atmosphere the layer of gases above the Earth's surface

attrition the wearing away of particles of debris by the action of other particles, such as river or beach pebbles

B

bankful the *discharge or contents of the river which is just contained within its banks. This is when the speed, or *velocity, of the river is at its greatest

bar an accumulation of *sediment that grows across the mouth of a bay, caused by longshore drift

basalt a dark-coloured volcanic rock. Molten basalt spreads rapidly and is widespread. About 70% of the Earth's surface is covered in basalt *lava flows

biodiversity means the number of different plant and animal species in an area

biofuels any kind of fuel made from living things, or from the waste they produce

biogas a gas produced by the breakdown of organic matter, such as manure or sewage, in the absence of oxygen. It can be used as a *biofuel

biome a large-scale *ecosystem, e.g. tropical rainforest

biosphere the living layer of Earth between the *lithosphere and *atmosphere

biotic living part of a *biome, made up of plant (flora) and animal (fauna) life

black gold a term used for oil, as it is regarded as such a valuable commodity

bottom-up development experts work with communities to identify their needs, offer assistance and let people have more control over their lives, often run by *non-governmental organisations

brownfield sites former industrial areas that have been developed before

C

carbon dating uses radioactive testing to find the age of rocks which contained living material

carbon footprint a calculation of the total *greenhouse gas emissions caused by a person, a country, an organisation, event or product

carbon sequestration removing carbon dioxide from the atmosphere and locking it up in biotic material

carbon sinks natural stores for carbon-containing chemical compounds, like carbon dioxide or methane

Central Business District (CBD) the heart of an urban area, often containing a high percentage of shops and offices

channel refers to the bed and banks of the river

climatologist a scientist who is an expert in climate and climate change

collision zone where two *tectonic plates collide – forming mountains like the Himalayas

communism system of government, based on Karl Marx's theories; it believes in sharing wealth between all people

concordant coasts follow the ridges and valleys of the land, so the rock *strata is parallel to the coastline

connectivity how easy it is to travel or connect with other places

conservation means protecting threatened *biomes, e.g. setting up national parks or banning trade in endangered species

conservative boundary where two *tectonic plates slide past each other

constructive waves build beaches by pushing sand and pebbles further up the beach

continental crust the part of the Earth's crust that makes up land, on average 30-50 km thick

conurbation a continuous urban or built-up area, formed by merging towns or cities

convection currents transfer heat from one part of a liquid or gas to another. In the Earth's *mantle, the currents which rise from the Earth's core are strong enough to move the *tectonic plates on the Earth's surface

convergence a) the meeting of *tectonic plates; b) when air streams flow to meet each other

Coriolis force a strong force created by the Earth's rotation. It can cause storms, including hurricanes

cost-benefit analysis looking at all the costs of a project, social and environmental as well as financial, and deciding whether it is worth going ahead

counter-urbanisation when people leave towns and cities to live in the countryside

Covid-19 a highly contagious respiratory disease, first detected in 2019, that can lead to fatalities; in 2020 it spread rapidly around the world (as a *pandemic)

D

decentralisation shift of shopping activity and employment away from the *Central Business District (CBD)

deforestation the deliberate cutting down of forests to exploit forest resources (timber, land or minerals)

deindustrialisation decreased activity in manufacturing and closure of industries, leading to unemployment

delta a low-lying area at the mouth of a river where a river deposits so much *sediment it extends beyond the coastline

depopulation decline of total population of an area

deprivation lack of wealth and services. It usually means low standards of living caused by low income, poor health, and low educational qualifications

dip slope a gentle slope following the angle of rock *strata, found behind *escarpments

discharge the volume of water flowing in a river, measured in cubic metres per second

discordant coast alternates between bands of hard rocks and soft rocks, so the rock *strata is at right angles to the coast

dissipate means to reduce wave energy, which is absorbed as waves pass through, or over, sea defences

divergent plate boundary where two *tectonic plates are moving away from each other

diversification when a business (e.g. a farm) decides to sell other products or services in order to survive or grow

E

ecological debt when Earth's resources are being used up faster than Earth can replace them

ecological footprint is a calculation measured in global hectares (gha). It's the amount of land and water required to produce resources and deal with waste from each country

economic liberalisation when a country's economy is given the freedom of a 'market economy', consumers and companies decide what people buy based on demand

ecosystem a localized *biome made up of living things and their non-living environment. For example a pond, a forest, a desert

ecosystem services a collective term for all of the ways humans benefit from ecosystems

emerging economies countries that have recently industrialised and are progressing towards an increased role in the world economy

energy diversification getting energy from a variety of different sources to increase *energy security

energy security having access to reliable and affordable sources of energy

enquiry the process of investigation to find an answer to a question

epicentre the point on the ground directly above the focus (centre) of an earthquake

erosion means wearing away the landscape

escarpment a continuous line of steep slopes above a gentle *dip slope, caused by the erosion of alternate *strata

evacuate when people move from a place of danger to a safer place

evaporation the changing of a liquid into vapour or gas. Some rainfall is evaporated into water vapour by the heat of the sun

F

fault large cracks caused by past tectonic movements

fetch the length of water over which the wind has blown, affecting the size and strength of waves

fieldwork means work carried out in the outdoors

flood plain flat land around a river that gets flooded when the river overflows

focus the point of origin of an earthquake

food miles the distance food travels from the producer to the consumer. The greater the distance, the more carbon dioxide is produced by the journey

food web a complex network of overlapping food chains that connect plants and animals in *biomes

formal economy means one which is official, meets legal standards for accounts, taxes, and workers' pay and conditions

fossil fuels a natural fuel found underground, buried within sedimentary rock in the form of coal, oil or natural gas

free trade the free flow of *goods and *services, without the restriction of tariffs

friction the force which resists the movement of one surface over another

G

gentrification high-income earners move into run-down areas to be closer to their workplace, often resulting in the rehabilitation and *regeneration of the area to conform with middle class lifestyles

geographical conflict means disagreement and differences of opinion linked to the use of places and resources

geographical information systems (GIS) a form of electronic mapping that builds up maps layer by layer

geothermal heat from inside the Earth

glacial a cold period of time during which the Earth's glaciers expanded widely

global circulation model a theory that explains how the *atmosphere operates in a series of three cells each side of the Equator

global shift change in location of where manufactured goods are made, often from developed to developing countries

globalisation increased connections between countries

goods physical materials or products that are of value to us

green belt undeveloped areas of land around the edge of cities with strict planning controls

greenhouse effect the way that gases in the atmosphere trap heat from the sun. Like the glass in a greenhouse – they let heat in, but prevent most of it from escaping

greenhouse gases gases like carbon dioxide and methane that trap heat around the Earth, leading to global warming

gross domestic product (GDP) the total value of *goods and *services produced by a country in one year

groundwater flow movement of water through rocks in the ground

H

hard engineering building physical structures to deal with natural hazards, such as sea walls to stop waves

helicoidal flow a continuous corkscrew motion of water as it flows along a river channel

holistic management takes into account all social, economic and environmental costs and benefits. In coastal management this means looking at the coastline as a whole instead of an individual bay or beach

hot spot columns of heat in Earth's *mantle found in the middle of a tectonic plate

Human Development Index (HDI) a standard means of measuring human development

hydraulic action the force of water along the coast, or within a stream or river

hydrological cycle the movement of water between its different forms; gas (water vapour), liquid and solid (ice) forms. It is also known as the water cycle

hyper-urbanisation rapid growth of urban areas

I

Index of Multiple Deprivation (IMD) means of showing how deprived some areas are

indigenous peoples are the original people of a region. Some indigenous groups still lead traditional lifestyles, e.g. a tribal system, hunting for food

industrialisation where a mainly agricultural society changes and begins to depend on manufacturing industries instead

infiltration the soaking of rainwater into the ground

informal economy means an unofficial economy, where no records are kept. People in the informal economy have no contracts or employment rights

infrastructure the basic services needed for an industrial country to operate e.g. roads, railways, power and water supplies, waste disposal, schools, hospitals, telephones and communication services

interception zone the capture of rainwater by leaves and branches. Some *evaporates again and the rest drips from the leaves to the soil

interglacial a long period of warmer conditions between *glacials

interlocking spurs hills that stick out on alternate sides of a V-shaped valley, like the teeth of a zip

intermediate technology uses low-tech solutions using local materials, labour and expertise to solve problems

Inter-Tropical Convergence Zone (ITCZ) a narrow zone of low pressure near the Equator where northern and southern air masses converge

invasive species (or alien species) is a plant, animal or disease introduced from one area to another which causes ecosystem damage

irrigation is the artificial watering of land that allows farming to take place

J

jet streams high level winds at around 6-10km that blow across the Atlantic towards the UK

joints small and usually vertical cracks found in many rocks

L

lagoon a bay totally or partially enclosed by a *spit, *bar or reef running across its entrance

landslide a rapid *mass movement of rock fragments and soil under the influence of gravity

latitude how far north or south a location on the Earth's surface is from the Equator, measured in degrees

lava melted rock that erupts from a volcano

lava flows *lava flows at different speeds, depending on what it is made of. Lava flows are normally very slow and not hazardous but, when mixed with water, can flow very fast and be dangerous

level of development means a country's wealth (measured by its GDP), and its social and political progress (e.g. its education, health care or democratic process in which everyone can vote freely)

lithosphere the uppermost layer of the Earth. It is cool and brittle. It includes the very top of the *mantle and, above this, the crust

M

magma melted rock below the Earth's surface. When it reaches the surface it is called *lava

magnitude of an earthquake (how much the ground shakes), an expression of the total energy released

mantle the middle layer of the Earth. It lies between the crust and the core and is about 2900 km thick. Its outer layer is the *asthenosphere. Below the asthenosphere it consists mainly of solid rock

mass movement the movement of material downslope, such as rock falls, *landslides or cliff collapse

megacity a many centered, multi-city urban area of more than 10 million people. A megacity is sometimes formed from several cities merging together

middle course the journey of a river from its source in hills or mountains to mouth is sometimes called the course of the river. The course of a river can be divided into three main sections a) upper course b) middle course and c) lower course

migration movement of people from one place to another

Milankovitch cycles the three long-term cycles in the Earth's orbit around the sun. Milankovitch's theory is that *glacials happen when the three cycles match up in a certain way

mudflats flat coastal areas formed when mud is deposited by rivers and coasts

multicultural a variety of different cultures or ethnic groups within a society

multiplier effect when people or businesses move to an area and invest money on housing and services, which in turn creates more jobs and attracts more people

N

natural increase the birth rate minus the death rate for a place. It is normally given as a % of the total population

natural resources are materials found in the environment that are used by humans, including land, water, fossil fuels, rocks and minerals and biological resources like timber and fish

net primary productivity (NPP) a measure of how much new plant and animal growth is added to a biome each year

non-governmental organisation (NGO) NGOs work to make life better, especially for the poor. Oxfam, the Red Cross and Greenpeace are all NGOs

non-renewable energy sources that are finite and will eventually run out, such as oil and gas

northern powerhouse a major core region of cities (with a similar population to London) that has the potential to drive the economy of northern England

nutrient cycle nutrients move between the biomass, litter and soil as part of a continuous cycle which keeps both plants and soil healthy

O

ocean currents permanent or semi-permanent large-scale horizontal movements of the ocean waters

oceanic crust the part of the Earth's crust which is under the oceans, usually 6-8 km thick

Organisation of Petroleum Exporting Countries (OPEC) established to regulate the global oil market, stabilize prices and ensure a fair return for its 12 member states who supply 40% of the world's oil

outsourcing using people in other countries to provide services if they can do so more cheaply e.g. call centres

ox-bow lake a lake formed when a loop in a river is cut off by floods

P

pandemic disease which spreads to (almost) every country

Pangea a supercontinent consisting of the whole land area of the globe before it was split up by continental drift

peak oil the theoretical point at which half of the known reserves of oil in the world have been used

plate boundaries where *tectonic plates meet. There are three kinds of boundary a) *divergent – when two plates move apart b) *convergent – when two plates collide c) conservative – when two plates slide past one another

plumes upwelling of molten rock through the *asthenosphere to the *lithosphere

plunging waves typically tall and close together, created by strong winds

population density the average number of people in a given area, expressed as people per km^2

population structure the number of each sex in each age group usually displayed in a population pyramid diagram

poverty line the minimum level of income required to meet a person's basic needs (US$1.90)

predict saying that something will happen in the future. A scientific prediction is based on statistical evidence

prevailing winds the most frequent direction the wind blows in a certain area

primary effects the direct impacts of an event, usually occurring instantly

primary products raw materials

Purchasing Power Parity (PPP) shows what you can buy in each country, now used to measure *GDP

pyroclasts fragments of volcanic material that is thrown out during explosive eruptions

Q

quality of life a measure of how 'wealthy' people are, but measured using criteria such as housing, employment and environmental factors, rather than income

Quaternary the last 2.6 million years, during which there have been many *glacials

R

radioactive decay atoms of unstable elements release particles from their nuclei and give off heat

rebranded a change of image

recurved hooked

regeneration means re-developing former industrial areas or housing to improve them

renewable a resource that does not run out and can be restored, such as wind or solar

re-urbanisation when people who used to live in the city and then moved out to the country or to a suburb, move back to live in the city

Richter scale a scale for measuring the magnitude of earthquakes

rock outcrop a large mass of rock that stands above the surface of the ground

rockfalls a form of *mass movement where fragments of rock fall freely from a cliff face

rural-urban fringe the area where a town or city meets the countryside

rural-urban migration the movement of people from the countryside to the cities, normally to escape from poverty and to search for work

S

Saffir-Simpson Hurricane Scale a scale that classifies hurricanes into five different categories according to their wind strength

salt marsh salt-tolerant vegetation growing on mud flats in bays or estuaries. These plants trap sediments which gradually raise the height of the marsh. Eventually it becomes part of the coast land

saltation the bouncing of material from and along a river bed or a land surface

sand dune onshore winds blow sand inland, forming a hill or ridge of sand parallel to the shoreline

saturated soil is saturated when the water table has come to the surface. The water then flows overland

scree angular rock pieces created by freeze-thaw weathering

secondary effects the indirect impacts of an event, usually occurring in the hours, weeks, months or years after the event

secondary products manufactured goods

sediment material such as sand or clay that is transported by rivers

seismometer a machine for recording and measuring an earthquake using the *Richter scale

services functions that satisfy our needs

shore-line platform the flat rocky area left behind when waves erode a cliff away

Shoreline Management Plan (SMP) this is an approach which builds on knowledge of the coastal environment and takes account of the wide range of public interest to avoid piecemeal attempts to protect one area at the expense of another

slope processes cause *mass movement or *soil creep

soft engineering involves adapting to natural hazards and working with nature to limit damage

soil creep the slow gradual movement downslope of soil, *scree or glacier ice

solution chemicals dissolved in water, invisible to the eye

spatial means 'relating to space' e.g. the spatial growth of a city means how much extra space it takes up as it grows

spit a ridge of sand running away from the coast, usually with a curved seaward end

stakeholders a person with an interest or concern in something, such as those who are likely to be affected by natural hazards

storm hydrograph a graph which shows the change in both rainfall and discharge from a river following a storm

storm surge a rapid rise in the level of the sea caused by low pressure and strong winds

strata distinctive layers of rock

stratosphere the layer of air 10-50km above the Earth's surface. It is above the cloudy layer we live in, the troposphere

strip mining (or open-pit, opencast or surface mining) involves digging large holes in the ground to extract ores and minerals that are close to the surface

studentification communities benefit from local universities which provide employment opportunities and a large student population which can regenerate pubs, shops and buy-to-let properties

sub-aerial processes occurring on land, at the Earth's surface, as opposed to underwater or underground

subduction describes *oceanic crust sinking into the *mantle at a convergent *plate boundary. As the crust subducts, it melts back into the mantle

subsistence farming where farmers grow food to feed their families, rather than to sell

suburbanisation the movement of people from the inner suburbs to the outer suburbs

surface run-off rainwater that runs across the surface of the ground and drains into the river

suspension tiny particles of *sediment dispersed in water

Sustainable Development defined by the Brundtland Commission as that which 'meets the needs of the present without compromising the ability of future generations to meet their own needs'

sustainable management meeting the needs of people now and in the future, and limiting harm to the environment

T

tar sands sediment that is mixed with oil, can be mined to extract oil to be used as fuel

tectonic hazards natural events caused by movement of the Earth's plates that affect people and property

tectonic plate the Earth's surface is broken into large pieces, like a cracked eggshell. The pieces are called tectonic plates, or just plates

terms of trade means the value of a country's exports relative to that of its imports

thalweg the line of the fastest flow along the course of a river

thermal expansion as a result of heating, expansion occurs. When sea water warms up, it expands

throughflow the flow of rainwater sideways through the soil, towards the river

till sediment deposited by melting glaciers or ice sheets

top-down development when decision-making about the development of a place is done by governments or large companies

topography the shape and physical features of an area

traction force that rolls or drags large stones along a river bed

transnational companies (TNCs) those which operate across more than one country

transpire when plants lose water vapour, mainly through pores in their leaves

tsunami earthquakes beneath the sea bed generate huge waves that travel up to 900km/h

U

urbanisation means a rise in the percentage of people living in urban areas, compared to rural areas

V

velocity the speed of a river, measured in metres per second

Volcanic Explosivity Index (VEI) measures the explosiveness of volcanic eruptions on a scale of 1 to 8

W

water table the upper limit of saturated rock below the ground

weathering the physical, chemical or biological breakdown of solid rock by the action of weather (e.g. frost, rain) or plants

wildfire uncontrolled burning though forest, grassland or scrub. Such fires can 'jump' roads and rivers and travel at high speed

world cities trade and invest globally e.g. London and New York

Index

Page numbers in **bold** refer to glossary terms.